# Electrochemical Process Engineering

## A Guide to the Design of Electrolytic Plant

# Electrochemical Process Engineering

## A Guide to the Design of Electrolytic Plant

F. Goodridge and
K. Scott
University of Newcastle upon Tyne
Newcastle upon Tyne, England

Springer Science+Business Media, LLC

**Library of Congress Cataloging-in-Publication Data**

Goodridge, F., 1924-
Electrochemical process engineering : a guide to the design of electrolytic plant / F. Goodridge and K. Scott.
p. cm.
Includes bibliographical references and index.

DOI 10.1007/978-1-4899-0224-5
1. Electrochemistry, Industrial. I. Scott, K. (Kenneth), 1951-
. II. Title.
TP255.G66 1994
660'.297--dc20 94-42922
CIP


Originally published by Plenum Press, New York in 1995
MyCopy version of the original edition 1995

10 9 8 7 6 5 4 3 2 1

Our sincere thanks are due to our colleague Dr. R. E. Plimley for his considerable help during the preparation of this book.

# Preface

As the subtitle indicates, the overriding intention of the authors has been to provide a practical guide to the design of electrolytic plant. We wanted to show that the procedures for the design and optimization of such a plant are essentially simple and can be performed by readers comparatively new to the electrochemical field. It was important to realize that electrochemical engineering should not be confused with applied electrochemistry but had to be based on the principles of chemical engineering. For this reason, reference is often made to standard chemical engineering texts.

Since this is a practical guide rather than a textbook, we have included a large number of worked examples on the principle that a good worked example is worth many paragraphs of text. In some examples we have quoted costs, e.g., of chemicals, plant or services. These costs are merely illustrative; current values will have to be obtained from manufacturers or journals. If this is not possible, approximate methods are available for updating costs to present-day values (see Refs. 1 and 3, Chapter 6).

At the end of the book you will realize that there has been no recommendation or discussion of particular software packages for costing or design. There are two reasons for this omission. First, we wanted to emphasize the basic principles of design so that readers would be in a position to question in an informed way the basic assumptions of any design package. Second, changes and updates in design software are so frequent that any recommendations would soon be out of date. Although design software packages including electrochemical stages are still not numerous, any final design today will be computer-based. This has been no handicap to the present book, which is concerned mainly with preliminary cost estimation and design.

We have emphasized the use of reaction, reactor, and process models. We have tried to show—and this is a message close to our hearts—that even simple models can lead to significant savings. For example, minimizing pilot plant experimentation can reduce development cost considerably. A reaction model can tell the development scientist or engineer which parameters under investigation are particularly important, in this way shortening experimentation time.

One might ask, "If design methods are based on chemical engineering, is it worthwhile to present a text specializing in electrochemical plant?" The answer is that, in our experience of running short courses in electrochemical engineering, scientists and engineers have found it useful to be presented with detailed examples of the application of chemical engineering principles to an electrochemical situation. The reason can be found in university training, where electrochemistry is still the Cinderella of reaction techniques, particularly in chemical engineering courses.

The treatment to be found below is loosely based on the short courses we have been running for many years in the Department of Chemical and Process Engineering at the University of Newcastle upon Tyne, England. Enthusiastic response from participants all over the world has led us to hope that this book will be considered as useful as, apparently, the courses have been.

The book has been written for undergraduate and graduate students of science and engineering, as well as for the practicing chemist and chemical engineer who have had little previous experience of designing and scaling up electrochemical processes.

Our sincere thanks are due to our colleague Dr. R. E. Plimley for his considerable help during the preparation of this book.

F. Goodridge
K. Scott

*University of Newcastle upon Tyne*

# Contents

# Chapter 1

# Introduction to Electrochemical Engineering

## 1.1. WHAT IS ELECTROCHEMICAL ENGINEERING?

Electrochemical engineering is the knowledge required to either design and run an industrial plant which includes an electrolytic stage for the production of chemicals or to produce an electrolytic device for the generation of power. The former involves the use of electric power for the production of chemicals and the latter the use of chemicals for the production of electric power. There are also a number of processes based on electrolysis, for example, electrochemical machining and electrophoretic painting, which clearly belong to the domain of electrochemical engineering. Nonelectrolytic processes such as gas discharge phenomena can justifiably be taken as outside the realm of the definition. It would be impossible to cover all these topics in a single volume, so this book confines itself to the production of chemicals. We have therefore called it a guide to electrochemical process engineering (E.P.E).

The overlap between E.P.E. and chemical process engineering is considerable. For example, fluid dynamics, heat and mass transfer, thermodynamics, separation processes, process modeling and optimization, and control are based on the same principles whether an electrolytic or a catalytic reactor is used. We cannot provide extensive treatment of these subjects in a single volume. We are therefore providing sufficient references to standard textbooks and original papers to enable the reader to deepen his knowledge of these topics. Since this is a practical guide, we have for convenience included introductory matter on a number of topics such as mass transfer and kinetics, in order to avoid the irritating necessity of having to consult reference books too often in order to follow this text.

One problem facing the student of E.P.E. literature is that the field has often been mistaken for applied electrochemistry. Many supposedly engineering papers have reported results of synthetic processes using electrolytic cells with electrodes a few square centimeters in surface area. Such results cannot be safely applied to process design since (see Chapter 5) apart from current distribution, the mass transfer regime in the small cell would be radically different from any in the industrially sized one. Again, many "optimization studies" lose considerable value if you look at a reactor in isolation from the later separation stage. Let us illustrate this by considering briefly a scheme for the design of electrochemical reactors (Fig. 1.1). The first step is a reaction model (Chapter 3) which gives us information on current density as a function of the kinetic and mass transfer elements. To determine the dependence of current density on electrode potential we must obtain

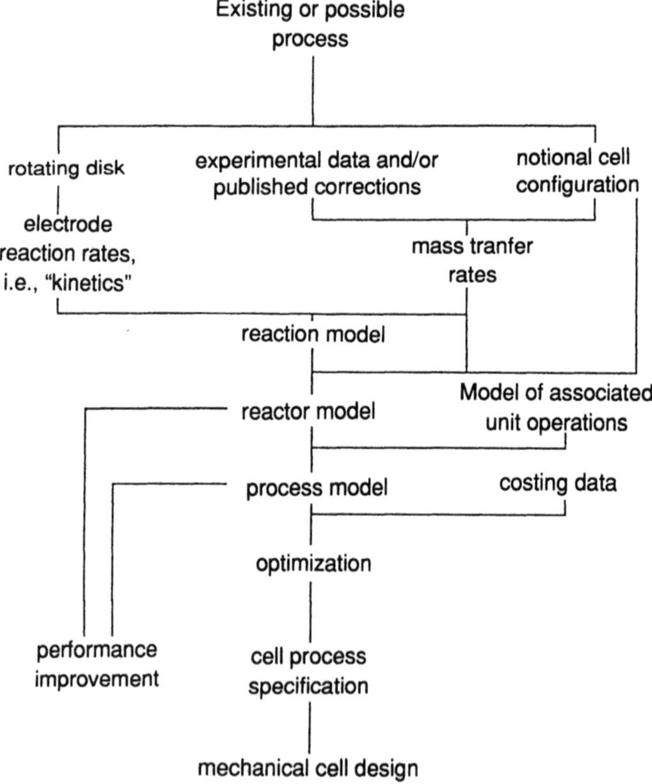


**FIGURE 1.1.** Scheme for the design of electrolytic reactors.

kinetic constants from polarization and preparative runs using preferably rotating disk electrodes for both or, alternatively, a small glass cell for the latter (Chapter 3). Also, mass transfer coefficients for the laboratory cell used in these experiments must be available (Chapter 2). A reactor model is developed (Chapter 4) which gives us such parameters as the required electrode area as a function of process variables, e.g., current density and conversion. As Fig. 1.1 indicates, to do this we must select a notional industrial cell configuration (see Section 5.2). Again, mass transfer data for the selected reactor must be available. Combining this reactor model with models for associated unit processes such as distillation results in a process model. A detailed examination of models for actual processes is beyond the scope of this book—indeed, this is the province of chemical engineering texts—but general considerations are presented in Chapter 6. Using costing data along with the process model enables us to optimize the process (discussed in Chapter 6) and obtain process specifications for the cell. Should these turn out to be satisfactory, a mechanical design can be undertaken; if not, a different cell geometry must be selected and a new reactor model developed.

The reader may ask why all this emphasis on modeling during the development and scale-up of an electrolytic process? The answer is that the alternative procedure, namely a factorial type of approach where the affects of changing process variables are investigated, is not only time-consuming but very expensive, particularly when operating pilot plant equipment.

## 1.2. AIMS AND INTENTIONS OF THIS BOOK

It follows that there will be little descriptive matter about industrial practice in this book. Instead emphasis is placed on a quantitative analysis of processes occuring in various industries. The aim is not only to provide either directly or by means of references a systematic treatment of design procedures for electrolytic reactors but also to emphasize their interaction with other units of the flow sheet, as indicated in Section 1.1. It must be stressed, however, that our intention is to provide a guide to process development, not a textbook.

## 1.3. THE INDUSTRIAL IMPORTANCE OF ELECTROLYTIC PROCESSES

In spite of our remarks in Section 1.2 it will be useful, before considering some of the basic concepts used in the rest of the book, to touch briefly on some industries using electrolytic techniques. This will help bring the importance of the latter into perspective.

Of the industrial activities under the heading of process engineering the chloralkali industry is probably the most important. Chlorine world capacity in 1985 was estimated as 43 million metric tons[1] and total world production in 1983 as 31 million tons. Over 80% of the chlorine produced is used to produce other chemicals.[1] Large-tonnage products associated with chlorine include caustic soda and soda ash (anhydrous sodium carbonate).

Another important area is metal winning including aluminum production at 20 million metric tons per year. Copper and zinc are other metals won in major amounts electrolytically, using an aqueous electrolyte instead of a molten salt. Amounts of copper and zinc produced in this way approach 1 and 10 million tons per annum respectively.[2] Other metals won from aqueous electrolytes include manganese, cadmium, nickel, cobalt, chromium, gallium, and antimony. In contrast to the sophisticated design of cells for chlorine production, where minimizing ohmic losses is of overriding importance, cells for aqueous metal winning are relatively undeveloped and in consequence mass transfer is poor, and energy efficiency and space-time yields low (see Section 1.5). More advanced cells, incorporating, e.g., fluidized-bed electrodes (see Chapter 5), are, however, under industrial development. Many metals, whether won electrolytically or not, need further purification. One way to purify is by electrolytic refining using aqueous electrolytes. Metals refined in this way include copper, nickel, cobalt, lead, indium, tin, gold, and silver. Of these, copper accounts for the largest tonnage by far, about 10 million tons per year. Cells used are similar in design to those employed for metal winning; in consequence, performance indicators such as space-time yield and energy efficiency are relatively low.

Although the chloralkali and metal industries are the most important ones using electrolytic processes, the production of organic chemicals has occupied the attention of academics and industrialists for some time. In principle the advantages of the technique are the specificity of reaction products, the relative ease of preforming reactions not possible by chemical means, and the minimizing of effluent problems by the use of electricity as a "reagent." So far the only large-tonnage electrochemical process of importance is Monsanto's hydrodimerization of acrylonitrile to adiponitrile[3] since the production of tetra-alkyl lead[4] as a petrol additive will presumably be phased out. This is not surprising: when you cost a number of electrochemical processes it becomes clear that electrochemical techniques are more likely to score in the production of more expensive fine chemical and pharmaceutical products. Indeed, as an unpublished survey by one of the authors (F.G.) showed, a number of companies are producing so-called specialty chemicals electrolytically.

## 1.4. SOME BASIC CONCEPTS AND DEFINITIONS

In this section some of the concepts used in the book to define the performance of electrochemical processes are introduced.[5-10] A number of these will be further discussed later.

### 1.4.1. Electrolytic Cells and Faraday's Law

#### 1.4.1.1. Basic Terminology

Consider the simple circuit shown in Fig. 1.2 in which a direct current $I$ is supplied to two electrolytic cells in series, one comprising two electrodes of silver and the other two of copper. The electrodes are immersed in aqueous silver nitrate and aqueous cupric sulfate respectively.

The electrode at which current enters each cell is the anode, and the one at which it leaves is the cathode. The liquid phase which completes the electric circuit between each pair of electrodes is called the electrolyte.

The passage of current through cell (1) causes silver to dissolve from the anode and deposit on the cathode. In cell (2) a similar process occurs with copper. These observations are described in chemical terms as follows. Cathode:

$$Ag^+ + e \rightarrow Ag \tag{1.1}$$

$$Cu^{2+} + 2e \rightarrow Cu \tag{1.2}$$

Anode:

$$Ag - e \rightarrow Ag^+ \tag{1.3}$$

$$Cu - 2e \rightarrow Cu^{2+} \tag{1.4}$$

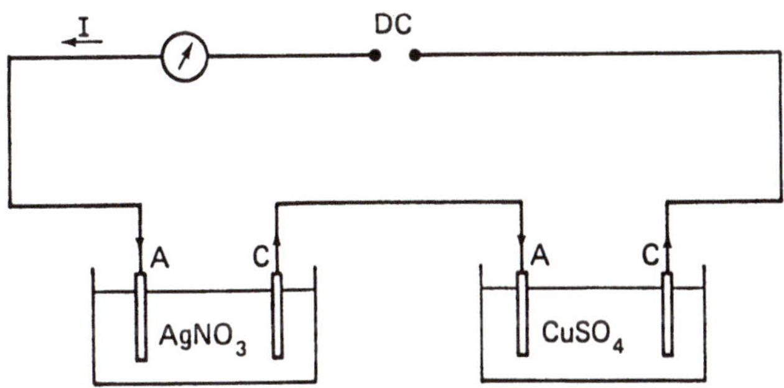


**FIGURE 1.2.** Diagram depicting two electrolytic cells in series. $A$ = anode; $C$ = cathode; $I$ = current.

where $e$ is an electron. Equations (1.1) and (1.2) are cathodic reactions and Eqs. (1.3) and (1.4), anodic reactions. Evidently a cathode supplies electrons to the electrode whereas an anode withdraws them. This is why cathodic reactions are reductions and anodic reactions are oxidations.

#### 1.4.1.2. Faraday's Law

By weighing the cathodes in Fig. 1.2 before and after an interval of time $t$ during which the current $I$ is passed we can determine the amount of deposition which has accompanied the passage of the amount of charge, $It$ coulombs (ampere seconds), through each cell. The results show that:

$$\text{mass of silver deposited in gmols} = It/F \tag{1.5}$$

$$\text{mass of copper deposited in gmols} = It/2F \tag{1.6}$$

where $F$, the Faraday, equals 96,487 coulombs. More generally, for the reduction

$$M^{n+} + ne \rightarrow M \tag{1.7}$$

$m$, the deposition in gmols, will be given by:

$$m = It/nF \tag{1.8}$$

With more experimentation we would find that the above equation (Faraday's law) is universal, not confined to metal deposition. For instance if a third cell is included in the circuit in which the following reaction is carried out:

$$C_2H_5OH + 6OH^- - 12e \rightarrow 2CO_2 + 3H_2O + 6H^+ \tag{1.9}$$

we will find that the amount of ethanol converted, $m_E$ in gmol is given by:

$$m_E = It/12F \tag{1.10}$$

For any cathodic reaction:

$$aA + bB + \cdots + ne \rightarrow xX + yY + \cdots \tag{1.11}$$

or anodic reaction:

$$aA + bB + \cdots - ne \rightarrow xX + yY + \cdots \tag{1.12}$$

the amount of $A$ converted is:

$$m_A = aIt/nF \tag{1.13}$$

Strictly speaking, $F$ has the units of coulombs/g-equivalent where 1 g-equiv. = $a/n$ gmol. Where reasonable we shall, however, use $10^3F$ in calculations and equations and denote it by the symbol $\mathfrak{F}$.

### 1.4.2. Current Density, Electrode Potential, and Current Efficiency

Figure 1.3 shows the region at the interface between an electrode and an electrolyte. The potentials in the electrode and adjacent to the interface (see also Section 1.4.4) are denoted $E_m$ and $E_s$. Their difference is referred to as the electrode potential, $E'$, i.e.:

$$E' = E_m - E_s \tag{1.14}$$

At a cathode, positive current flows from the electrolyte to the electrode. Therefore, by the above definition $E'$ for a cathode tends to be negative, and it is positive for an anode.

If the electrode potential $E'$ is kept constant and the area of the electrode is doubled, the current passing between the electrode and electrolyte is also doubled. In other words, the current supported by an electrode is proportional to its area, other parameters remaining unchanged. This is why the current supported by unit area of electrode is a more fundamental property than the total current; it is referred to as the current density ($i$).

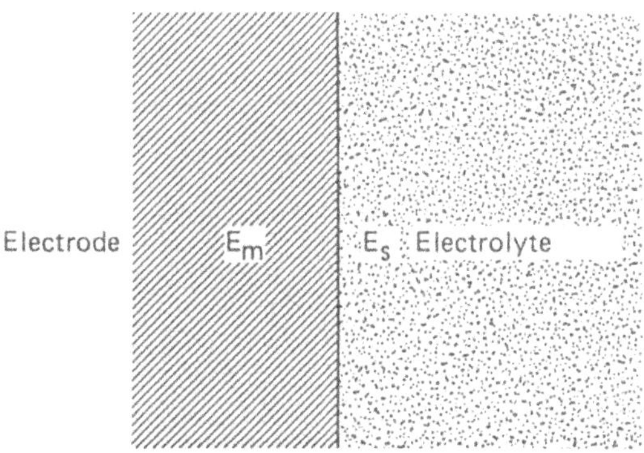


**FIGURE 1.3.** Region near an electrode interface. $E_m$ = metal potential; $E_s$ = solution potential.

Information about electrode reactions can often be obtained from plotting $i$ as a function of potential. In the experiment in Fig. 1.2 the electrode potential $E'$ when no current is flowing is $E$, the equilibrium potential. Consider the cathode where silver is being deposited. Once a net current is flowing, the cathode, or working electrode, is said to be polarized. The degree of polarization for such a system is measured by the overpotential $\eta$, which is defined as $\eta = E' - E$. A plot of $i$ or ln $[i]$ against $\eta$ is called a polarization curve: an idealized example is shown in Fig. 1.4. Defining the rate of reaction $r_A$ as the rate of conversion of $A$ per unit time per unit area of electrode, it follows from Eq. (1.13) that:

$$r_A = ai/nF \tag{1.15}$$

or

$$i = (nFr_A)/a \tag{1.16}$$

The current density for an electrochemical reaction is therefore equivalent to the rate of reaction. As we shall see in Chapter 3 when deriving

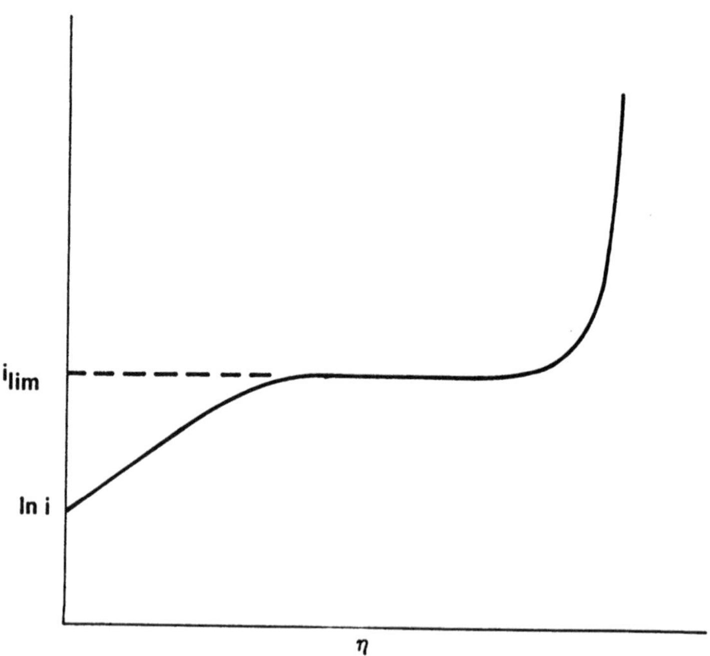


**FIGURE 1.4.** Idealized polarization curve. $i$ = Current density; $i_{lim}$ = limiting current density; $\eta$ = overpotential.

expressions for the rate of reaction, the initial linear rise in current density in Fig. 1.4 corresponds to conditions when the limiting step is the rate of charge transfer. The horizontal portion, where the current density becomes independent of the overvoltage, represents the situation when mass transport is rate-controlling; $i$ is termed $i_{lim}$, the limiting current density. The further increase in current density beyond the plateau is due to a second reaction beginning.

Consider now a process for the formation of product $B$ where two simultaneous cathodic reactions are occurring:

$$A + ne \rightarrow B \tag{1.17}$$

and

$$2H^+ + 2e \rightarrow H_2 \tag{1.18}$$

The number of kmols of $B$ formed per unit electrode area ($nm_B$) will be according to Eq. (1.13):

$$nm_B = i_B t/\mathfrak{F} \tag{1.19}$$

Similarly, $2m_H$, the number of kmols of hydrogen, will be:

$$2m_H = i_H t/\mathfrak{F} \tag{1.20}$$

where $i_B$ and $i_H$ are the partial current densities for Eqs. (1.17) and (1.18).

We can also write:

$$i_B + i_H = i \tag{1.21}$$

where $i$ is the current density for the whole process.

Current efficiency C.E., sometimes called Faradaic yield of the process, can be defined by use of Eqs. (1.19) and (1.20):

$$\text{C.E.} = nm_B/(nm_B + 2m_H) = i_B/i \tag{1.22}$$

### 1.4.3. Exchange Current Density

If the electrode potential is changed from an appropriate negative value to a positive one, the direction of current flow across the interface will be reversed (Fig. 1.5). For an elementary reaction (see Section 3.1.1), it seems clear that by reversing the current flow we have reversed the direction of reaction. It seems reasonable to suppose therefore that the transfer of an

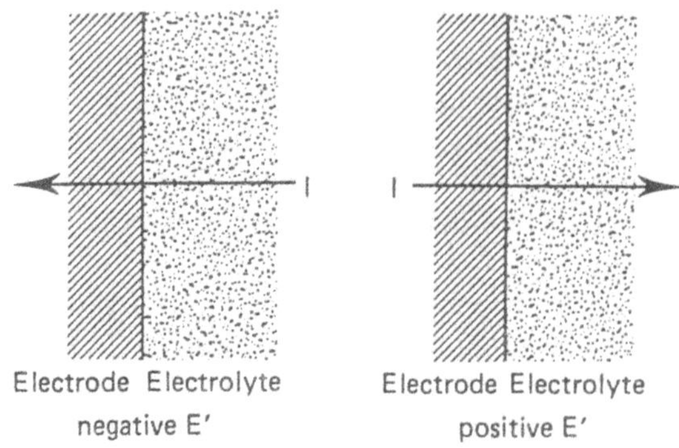


**FIGURE 1.5.** Direction of current for a negative and positive electrode potential $E'$.

electron is reversible. Denoting the oxidized and reduced species by $O$ and $R$ respectively, we can write:

$$O + e \rightleftharpoons R \tag{1.23}$$

and consider that the current $I_c$, which is observed in the external circuit, is a net effect given by:

$$I_c = I_- - I_+ \tag{1.24}$$

where $I_-$ is the current associated with the forward (cathodic) reaction, e.g., $O + e \rightarrow R$, and $I_+$ is that associated with the back (anodic) reaction, e.g., $O + e \leftarrow R$.

Considering unit area of electrode, we should have:

$$i_c = i_- - i_+ \tag{1.25}$$

Between the more negative and the more positive values of potential there will be some value, referred to as the equilibrium potential $E$, at which the current density is zero, i.e., the forward and backward reactions occur at an equal rate. We can write:

$$i_0 = i_- = i_+ \tag{1.26}$$

where $i_0$ is the exchange current density. By moving toward potentials more negative than $E$ we enhance $i_-$ and depress $i_+$, and vice versa. It follows from Eq. (1.19), therefore, that the value of $i$ and its direction are potential-dependent. We shall return to a quantitative treatment of Eq. (1.25) in Chapter 3.

### 1.4.4. The Double Layer

Before discussing the effect of electrode potential on reaction rate in Chapter 3, we will look briefly at the structure of the region adjacent to the electrode surface.[7,11,12] We will consider only the elementary reaction:

$$Ag^+ + e \rightleftharpoons Ag \tag{1.27}$$

by which silver is deposited or dissolved electrolytically.

If a silver electrode is immersed in a aqueous solution of a silver salt and rendered cathodic it will contain an excess of electrons (Fig. 1.6). These electrons and the silver cations existing in the surrounding electrolyte will be mutually attracted, and the anions in the electrolyte will be repelled. As a result, an excess of positive charge will occur in the region adjacent to the electrode. This region consists of two entities, a compact layer (the Helmholtz layer) and a diffuse layer (Fig. 1.6). The compact layer is some 10 Å thick, and the diffuse layer up to several hundred angstroms, depending on the potential of the electrode and the concentration of ionic species at the electrode. At the concentrations often encountered in commercial operations the diffuse layer may be only a few angstroms thick.

The region between the ions located in the compact layer and the electrode surface is occupied by solvent molecules. If the solvent is water the layer of solvent molecules is thought to be two molecules deep, with the layer nearer the electrode tending to be oriented so that the positive end of the dipoles points toward the electrode. This tendency increases as the electrode potential is made more negative. Intermingled with the solvent molecules may be other species, particularly anions, which may be specifically adsorbed on the electrode surface. Such adsorption requires displacement of the solvent molecules. The adsorption of anions creates a layer of charged species which is closer to the electrode surface than the compact layer just described. The planes at which the respective charges are sited are the outer Helmholtz plane ($x_H$ in Fig. 1.6), and the inner Helmholtz plane for the adsorbed species. Figure 1.6 also shows how $E_s$, the potential in the electrolyte, varies as a function of distance from the electrode. Remembering Eq. (1.14), this is also a measure of the electrode potential $E'$, if $E_m$ is constant.

### 1.4.5. Cell Voltage

Although the cell voltage $V^c$ has no fundamental theoretical significance affecting the electrode reaction, it is of great practical importance since electricity costs are directly proportional to it.

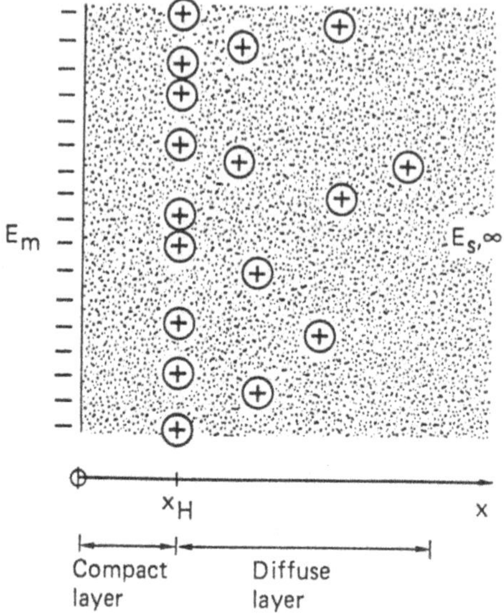


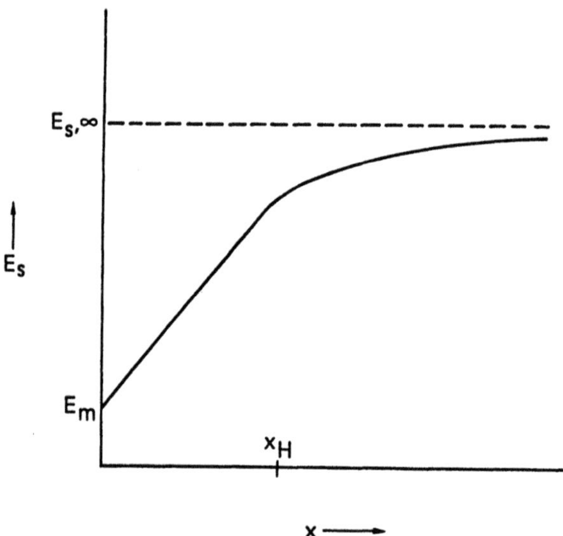


**FIGURE 1.6.** Simple model of the electrical double layer and the resultant potential profile. $E_m$ = metal potential; $E_s$ = solution potential; $E_{s,\infty}$ = bulk solution potential.

Consider a cell where anolyte and catholyte are separated by a diaphragm. From Fig. 1.7 we can write:

$$V^c = E'_A + V_{RA} + V_D + V_{RC} + |E'_C| \tag{1.28}$$

where $E'_A$ and $|E'_C|$ are the anodic and the modulus of the cathodic electrode potential respectively and $V_{RA}$, $V_D$, and $V_{RC}$ are the voltage drops caused by ohmic resistance in the anolyte, diaphragm, and catholyte. Equation (1.28) defines the cell voltage as positive.

To minimize the cell voltage for a given current density, the following steps can be taken:

1. Electrode materials are selected that minimize the electrode potential required for the desired reaction.
2. Distance between anode and cathode is kept to a mininum.
3. Conductivities of the anolyte and catholyte are maximized.
4. A diaphragm is selected that produces the minimum voltage drop. If possible, the diaphragm is dispensed with.

Some of these aspects are discussed further in Chapters 2, 5, and 6.

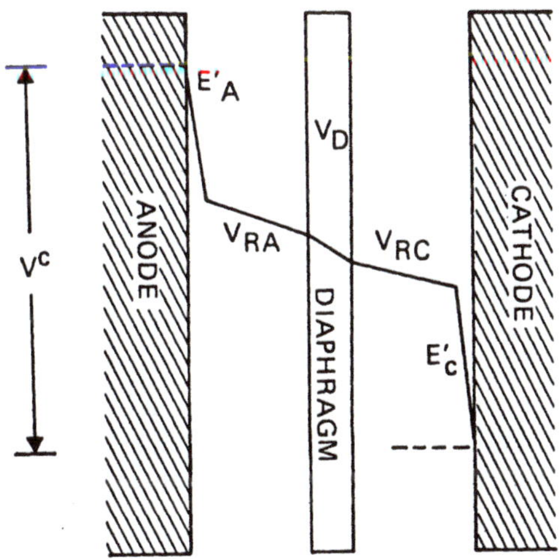


**FIGURE 1.7.** Voltage distribution in an electrochemical reactor $V^C$ = cell voltage; $E'_A$ = anodic electrode potential; $V_{RA}$ = voltage drop in anolyte; $V_D$ = voltage drop in diaphragm; $V_{RC}$ = voltage drop in catholyte; $E'_C$ = cathodic overpotential.

## 1.5. CRITERIA FOR REACTOR PERFORMANCE

Reactor performance can be measured by a number of parameters. For electrolytic cells two important parameters are current efficiency C.E. and space-time yield $Y_{ST}$. (Current efficiency or current yield has been defined in Section 1.4.2.)

The space-time yield of a reactor is defined as the mass of product produced by unit reactor volume in unit time. For an electrochemical reactor, it is:

$$Y_{ST} = (\text{C.E. } iaM)/(1000nF)\ \text{kg/m}^3\text{s} \tag{1.29}$$

where $a$ is the specific electrode area (electrode area per unit volume of reactor) and $M$ the relative molar mass. Some authors define $a$ as the electrode area per unit volume of electrode system. This not surprisingly leads to unrealistically high values of space-time yield.

Other important performance indicators are energy consumption $E_C$ (see Section 5.2.1.1), in KWh/kg, and $Y_C$, the chemical yield, which can be defined as:

$$Y_C = m_{ACT}/m_{MAX} \tag{1.30}$$

where $m_{ACT}$ is the actual amount of product and $m_{MAX}$ is the maximum amount of product that can be obtained for 100% conversion of the reactant.

As a simple example to clarify the difference between C.E. and $Y_C$, consider an aqueous acidic electrolyte containing nitrobenzene ($A$) which is being reduced according to Eq. (1.31) to phenylhydroxylamine ($B$), the latter undergoing chemical

$$\begin{array}{ccccc} \phi NO_2 & \xrightarrow[4\mathfrak{F}]{4e} & \phi NHOH & \xrightarrow[2\mathfrak{F}]{2e} & \phi NH_2 \\ (A) & & (B) & & (D) \\ & & \downarrow & & \\ & & HO\phi NH_2 & & \\ & & (C) & & \end{array} \tag{1.31}$$

rearrangement to the desired product $p$-aminophenol ($C$). Also a by-product is formed due to the further reduction of phenylhydroxylamine to aniline ($D$). A second by-product, hydrogen, is formed according to the equation:

$$2H_2O \xrightarrow[2\mathfrak{F}]{2e} H_2 + 2OH^- \tag{1.32}$$

Chemical analysis shows that for every kmol of $A$ reacted 1 kmol of hydrogen is evolved, and that the mol ratio of $C$ to $D$ is 3 to 1. It follows from Eq. (1.31) that to produce 3 kmol of $C$ will require 12$\mathfrak{F}$. To this we have to add 6$\mathfrak{F}$ due to the kmol of by-product $D$. Finally the 4 kmol of hydrogen will according to Eq. (1.32) account for another 8$\mathfrak{F}$, making a total of 26$\mathfrak{F}$. Thus one requires $26/3 = 8.7\mathfrak{F}$ per kmol of $C$. Therefore:

$$\text{C.E.} = 4/8.7 = 0.46 \qquad \text{and} \qquad Y_C = 3/4 = 0.75$$

Generally $E_C$ is important for high-tonnage chemicals whereas $Y_C$ gains in relative importance with increasing value of product. Some authors consider chemical yield to be synonymous with selectivity. We prefer to define selectivity as the ratio of wanted to unwanted product.

## 1.6. ELECTROCHEMICAL AND CATALYTIC REACTIONS, AND REACTORS

We end the chapter by considering some similarities between electrochemical and catalytic processes. We want to emphasize that electrolytic cells are just one type of a heterogeneous reactor and can be appropriately analyzed by the techniques of reactor engineering as demonstrated in Chapter 4. Indeed, as we shall see in Chapter 5, there are a number of cell types which employ aspects of conventional chemical reactors, e.g., fluidized beds. Throughout the book, we shall consider the terms cell and reactor to be synonymous.

Since we are dealing with heterogeneous electrochemical and catalytic processes, the various steps of the reactions will be similar. To begin with, the reactant has to approach a surface where reaction takes place. For this reason mass transport is very important and a knowledge of values of mass transfer rates, essential. This is treated in Chapter 2.

Apart from the reaction steps involved in electrochemical and catalytic processes, there are other similarities. As we shall see in Chapter 3, electrochemical reaction rates can be increased by applying an electrical potential to the reacting surface. Indeed, a relatively modest electrode potential will be equivalent to raising the reaction temperature by several hundred degrees Celsius. The reason is that the applied potential reduces the height of the potential energy barrier that has to be overcome to make the reaction proceed. The analogous phenomenon in catalysis is the formulation of the catalyst to achieve the same objective. Further study of electrode processes indicates that intermediates adsorbed on an electrode surface will be able to modify a particular reaction route.[13] We shall return to this phenomenon, called electrocatalysis, in Chapter 3.

## REFERENCES


1. Brennan, C. K., 1986, "The world chlor-alkali outlook 1975–1990," in *Modern Chlor-Alkali Technology*, (K. Wall, ed.), Ellis Horwood, vol. 3, pp. 20–35.
2. Pletcher, D., and Walsh, F. C., 1990, *Industrial Electrochemistry*, Chapman & Hall, p. 220.
3. Danly, D. E., and Campbell, C. R., 1982, "Experience in the scale-up of the Monsanto adiponitrile process," in *Techniques of Electroorganic Synthesis, Part III: Scale-up and engineering*, (N. L. Weinberg and B. V. Tilak, eds.), John Wiley & Sons, New York, pp. 283–384.
4. Bott, L. L., 1965, "How nalco makes lead alkyls," *Hydrocarbon Process. Pet. Refiner*, **44**: 115–118.
5. Moore, W. J., 1972, *Physical Chemistry*, Longman.
6. Bockris, J. O'M., and Reddy, A. K. N., 1970, *Modern Electrochemistry*, Plenum Press, New York.
7. Bard, A. J., and Faulkner, L. R., 1980, *Electrochemical Methods*, John Wiley & Sons, New York.
8. Albery, J., 1975, *Electrode Kinetics*, Oxford University Press.
9. Vetter, K. J., 1967, *Electrochemical Kinetics*, Academic Press, San Diego.
10. Koryta, J., Dvorak, J., and Bohackova, V., 1970, *Electrochemistry*, Methuen & Co., New York.
11. Delahay, P., 1965, *Double Layer and Electrode Kinetics*, Interscience Publishers, John Wiley & Sons, New York.
12. Parson, R., 1961, "The structure of the electrical double layer and its influence on the rates of electrode reactions," in *Advances in Electrochemistry and Electrochemical Engineering, Vol 1*, (P. Delahay and C. W. Tobias, eds.), pp. 1–64.
13. Pletcher and Walsh, pp. 32–42.

Chapter 2

# Aspects of Mass and Heat Transfer and the Energetics of Electrolytic Cell Systems

## 2.1. INTRODUCTION

As will be shown in Chapter 3, formation of a reaction model depends on knowing the rate at which reactants arrive at the electrode surface from the bulk of the electrolyte and the rate at which the products formed by reactions disappear back into the bulk of the electrolyte, i.e., mass transfer or the synonymous term mass transport. Since mass transfer is conditioned by the hydrodynamic behavior of the electrolyte, we begin by considering basic aspects of fluid dynamics; for further study the reader is referred to a standard text on the subject.[1] Our treatment of mass transfer is simplistic, but will nevertheless allow the reader to understand and use relevant mass transfer relationships.

In Section 1.6, similarities between electrochemical and catalytic reactors were briefly considered. One major difference would be the considerable amount of energy that can be required by the electrochemical reactor. This is exemplified by the production of chlorine in the chloralkali industry where minimizing the cell voltage is most important. In contrast, energy costs in many organic electrosynthetic processes amount to less than 10% of the total costs. An energy balance using simple thermodynamic relationships is essential when trying to arrive at the cost of a process, as will be shown in Chapter 6. Energy use and heat balances are considered in the final section of this chapter under the heading of energetics.

## 2.2. SOME BASIC ASPECTS OF FLUID DYNAMICS

Fluid dynamics, as the name implies, deals with the dynamic behavior of fluids, i.e., gases and liquids. We shall confine ourselves to the behavior of liquids in general and electrolytes in particular. When a liquid flows through a channel or along a surface, you can distinguish between two extremes of types of flow; laminar flow and turbulent flow. In the former, the flow is streamline, i.e., there is no bulk motion perpendicular to the direction of flow and any transfer is due to the motion of single molecules. In the latter, fluid in the form of eddies moves rapidly in random directions across the direction of flow.

Let us look at an electrolyte flowing through a rectangular channel composed of two parallel solid electrodes. One of these electrodes is shown in Fig. 2.1. The layer of the electrolyte next to the solid surface is at rest. When a liquid flows past a stationary plate with a velocity $V$, it follows there will be electrolyte in motion next to the electrolyte at rest. Viscous forces will retard the moving electrolyte; the region over which retardation occurs is called the boundary layer. If the forward velocity of the electrolyte $u$ is measured at a number of points along a normal to the electrodes, a plot will give a velocity profile (Fig. 2.1). Within the region $0 < y < \delta$ the velocity of flow is less than that of the free stream $V$, and the boundary layer thickness at the place where we made our observation is $\delta$. Further measurements would show that $\delta = 0$ at the leading edge of the electrode (Fig. 2.2a) and increases as we proceed along the electrode in the direction of flow. The character of flow in the boundary layer is initially laminar, but at some point down the electrode it changes to a transitional character (the fluid is sometimes laminar sometimes turbulent), although a thin layer adjacent to the plate remains laminar (Fig. 2.2b). Still further along the electrode the

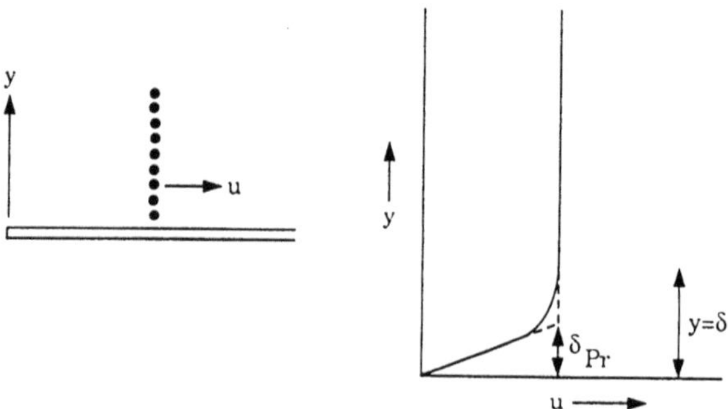


**FIGURE 2.1.** Velocity profile for fully developed turbulent flow. $u$ = Fluid velocity; $\delta$ = boundary layer thickness; $\delta_{Pr}$ = thickness of Prandtl boundary layer.

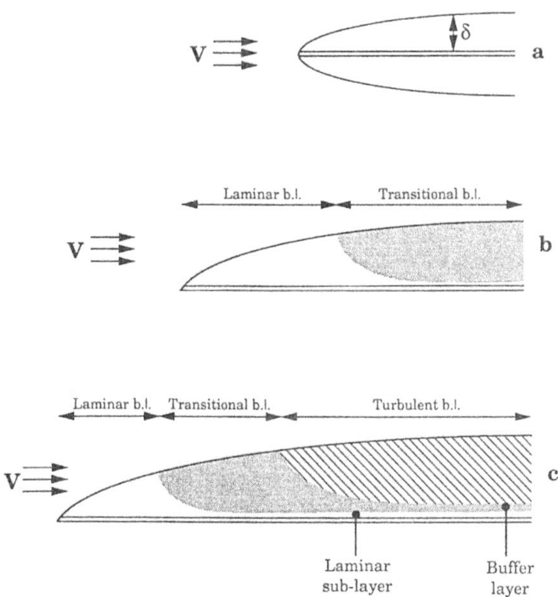


**FIGURE 2.2.** Formation of boundary layers. (a) Initial laminar layer ($\delta$ = boundary layer thickness); (b) formation of buffer layer; (c) formation of turbulent layer.

boundary layer becomes turbulent, with transitional and laminar flow persisting in the buffer layer and laminar sublayer respectively (Fig. 2.2c).

The behavior of the boundary layer explains the character of flow in the channel formed by the two electrodes. When the liquid enters the channel it immediately begins to form a boundary layer at the inside of the two electrodes and as it thickens it closes in on itself to meet in a cusp (Fig. 2.3). Before the cusp the flow pattern changes progressively as the boundary layer builds up; this region is called the entrance region. Beyond the cusp, the flow pattern suffers no further change and is called fully developed.

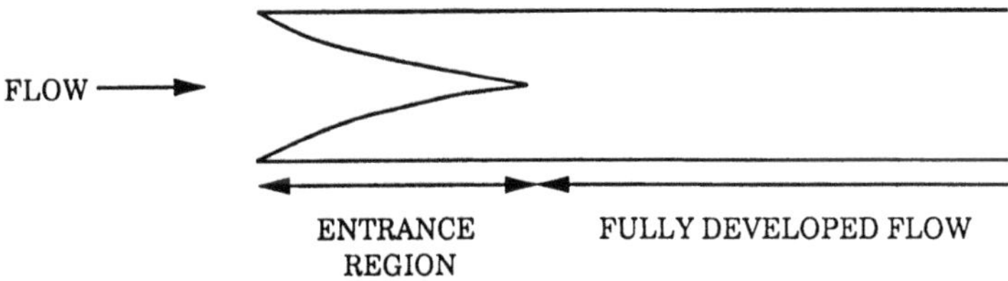


**FIGURE 2.3.** Formation of fully developed flow.

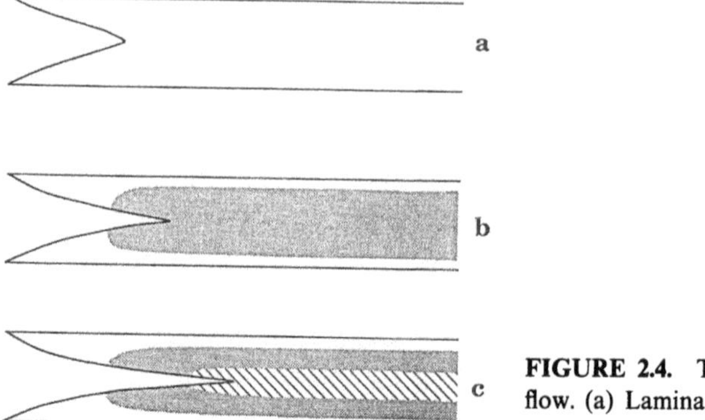


FIGURE 2.4. Types of fully developed flow. (a) Laminar; (b) transitional; (c) turbulent.

The character of the fully developed region is determined by the character of the flow in the boundary layer at the cusp (Fig. 2.4). If it is laminar, the fully developed flow will be laminar (Fig. 2.4a). Or the flow may be transitional (Fig. 2.4b) or completely turbulent (Fig. 2.4c). The latter, in a pipe, channel, or annulus, has three zones of flow behavior which reflect those of the turbulent boundary layer. The velocity distribution depicted in Fig. 2.1 corresponds to the latter regime.

## 2.3. MASS TRANSFER

### 2.3.1. Mass Flux in a Fully Developed Turbulent Regime

The transfer of a solute $O$ through a liquid can be effected by three mechanisms:

- Molecular diffusion under the influence of a concentration gradient
- Migration of a charged species under the influence of a potential or electric field
- Eddy diffusion due to a turbulent regime

The behavior of the boundary layer discussed in Section 2.2 can be used to obtain a quantitative expression for $N_D$, the diffusional flux of solute $O$ to the electrode surface. Let us use again the example from Section 2.2, namely that of a channel formed by two parallel solid electrodes. Figure 2.5 replaces the velocity profile of Fig. 2.1 by a concentration profile at the same location of the electrode for a fully developed turbulent flow. There are three regions: a near horizontal portion corresponding to the fully mixed turbulent bulk;

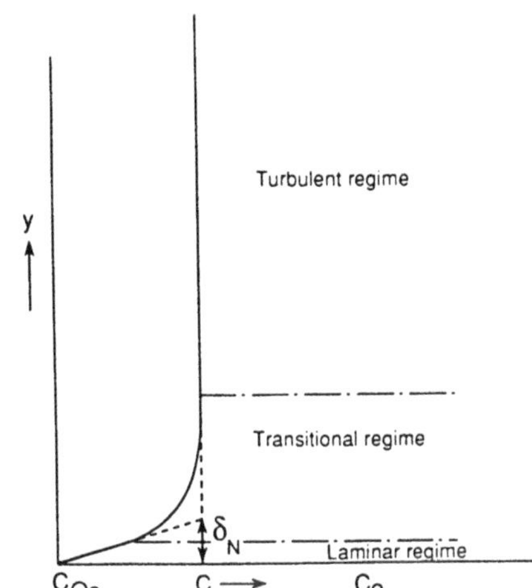


**FIGURE 2.5.** Concentration profile for a fully developed turbulent flow. $C_{Os}$ = Concentration of species $O$ at the electrode surface; $c_O$ = bulk concentration of species $O$; $\delta_N$ = thickness of diffusion layer.

a linear decrease occurring in the laminar sublayer; and a curved portion connecting the two.

Disregarding for the moment any migration phenomena, a rigorous calculation of the mass flux would have to find quantitative expressions for mass transport in the turbulent bulk. Unfortunately, the treatment of turbulent flow of liquids is much more complex than that for laminar flow. Turbulence is characterized by random and rapid fluctuations in pressure, velocity, and concentration about some mean values. Attempts have been made to characterize the flow in terms of Reynolds stresses, which describe the turbulent momentum flux in a way analogous to that for shear stress in laminar flow. Some success has been achieved, but a tractable analysis of mass transport is not yet available to the practicing engineer. Instead, the following simple model, termed the film model or theory,[2] is used.

The first assumption is that mass transport due to eddy diffusion in the turbulent bulk is so fast when compared to molecular diffusion that the latter can be ignored and a uniform concentration of the bulk can be assumed. The second assumption is that all the resistance to the mass transfer process can be expressed in terms of molecular diffusion in the laminar sublayer. Third, this resistance can be expressed quantitatively by assuming a completely linear concentration gradient (Fig. 2.5), including in this way the effect of the buffer layer. The point where the linear concentration gradient meets the horizontal bulk concentration defines the thickness $\delta_N$ of a layer called the diffusion boundary layer. The flux $N_D$ can be calculated by deriving an expression for the molecular diffusion of $O$ through a stagnant layer of thickness $\delta_N$.

The general differential equation describing one-dimensional molecular diffusion can be written as:[3]

$$\frac{\partial c}{\partial t} = D \frac{\partial^2 c}{\partial y^2} \tag{2.1}$$

where $c$ is the concentration at time $t$ and coordinate $y$ of species $O$, and $D$ its diffusivity.

For a steady state, i.e., $\partial c/\partial t = 0$, Eq. (2.1) reduces to:

$$\frac{d^2 c}{dy^2} = 0 \tag{2.2}$$

The boundary conditions depicted in Fig. 2.5 are:

$$c = c_{Os} \qquad y = 0 \tag{2.3}$$

and:

$$c = c_O \qquad y = \delta_N \tag{2.4}$$

The diffusional flux of $O$ to the electrode surface $N_D$ is given by Fick's law:[4]

$$N_D = -D(dc/dy)_{y=0} \tag{2.5}$$

Integrating Eq. (2.2) twice:

$$c = Ay + B \tag{2.6}$$

Substituting Eqs. (2.3) and (2.4) in Eq. (2.6) results in:

$$B = c_{Os} \qquad \text{and} \qquad A = (c_O - c_{Os})/\delta_N$$

giving Eq. (2.6) as:

$$c = \left[\frac{c_O - c_{Os}}{\delta_N}\right] y + c_{Os} \tag{2.7}$$

Differentiating Eq. (2.7) with respect to $y$ and substituting in Eq. (2.5) (remembering that the concentration gradient in the direction of $y$ is negative) gives:

$$N_D = \frac{D}{\delta_N}(c_O - c_{Os}) \tag{2.8}$$

Mechanism (2), namely migration, will add to the molar flux given by Eq. (2.8), as discussed in Section 3.1.5.3.

Looking again at Eq. (2.8), $\delta_N$ is identical with the Nernst diffusion layer thickness. Unfortunately, this quantity is often discussed and used as if it had physical reality when as we have seen from Fig. 2.5 it is a mathematical artifact arising from a particular mass transfer model. For this reason diffusion layer thickness is not normally found in design equations but Eq. (2.8) is written in the form:

$$N_D = k_L(c_O - c_{Os}) \tag{2.9}$$

where $D/\delta_N$ is replaced by $k_L$, a mass transfer coefficient. This is not a trivial change since whatever the mass transfer model used, you get a relationship of the form:

$$\text{mass flux} = \text{mass transfer coefficient} \times \text{driving force}$$

The concept of a mass transfer coefficient is generally applicable.

Before replacing the concept of diffusion layer with that of mass transfer coefficient we will consider the order of magnitude of the former relative to that of the other boundary layers. The normal range of thickness of the laminar sublayer is $10^{-5}$ to $10^{-3}$ m; that of the diffusion layer is an order of magnitude less, i.e., $10^{-6}$ to $10^{-4}$ m. This means that in a liquid the diffusion layer in fully developed flow is well inside the hydrodynamic boundary layer. Finally, the electrical double layer, which as stated in Chapter 1 is of the order of $10^{-8}$ m, is very small compared to the diffusion layer.

We have confined our argument to the flow of liquids. In many situations (usually in the laboratory rather than in industrial practice) turbulence is introduced into the electrolyte by stirring. The argument advanced above applies in principle to a stirred liquid. The problem is that it is difficult to characterize mass transfer coefficients in a small stirred cell, so it is not easy to obtain results which can be applied in practice to a flow system.

### 2.3.2. Entrance and Exit Effects

The concentration profile in Fig. 2.5 was of a fully developed turbulent flow. It follows from Section 2.2 that until a fully developed flow has been reached the situation remains radically different. At the leading edge of the electrode the velocity profile is horizontal and the mass transfer coefficient is infinite and will gradually decrease to a value corresponding to the fully developed flow; this is called an entrance effect. Similarly, the region approaching the point where the electrolyte leaves the electrode will not be representative of the hydrodynamic regime of the bulk of the electrode and will exhibit unrepresentatively high mass transfer coefficients due to an exit effect.

It is clear that experiments conducted with a small cell, where entrance and exit effects occupy an unduly high proportion of the electrode area, give results that in terms of mass transfer will not apply to a larger, scaled-up cell. We shall return to this important topic in the next section and in Chapter 5.

### 2.3.3. Obtaining Numerical Values of $k_L$ by Calculation

It is essential for modelling purposes to be able to obtain numerical values of a mass transfer coefficient that can be introduced in the equations derived in Chapter 3. We shall first discuss methods of obtaining values of $k_L$ by calculation and then in Section 2.3.4 give an account of the limiting current technique for obtaining $k_L$ experimentally.

Mass transfer coefficients for fully developed laminar and turbulent flow can be calculated from correlations involving dimensionless quantities. In an industrial cell one aims for a turbulent flow regime since, as we have just seen, the thickness of the laminar sublayer is much smaller than that of the boundary layer for laminar flow and therefore gives much higher values of $k_L$. Apart from exhibiting low values, mass transfer coefficients in a laminar regime also vary significantly with position along the electrode. It is not always practicable to obtain turbulent flow, say when dealing with viscous liquids or narrow channels, so for completeness' sake we will consider some correlations for laminar flow. Strictly speaking, these correlations (laminar and turbulent) are applicable to flow in a pipe only, but they have been applied to other geometries, including packed and fluidized beds. The correlations are based on Reynolds's analogy or other analogies for pipe flow; their derivations, often by analogy with heat transfer, are treated satisfactorily in standard chemical engineering texts.[4] We will only discuss correlations that apply to some of the most important geometries in electrolytic cell design.

#### 2.3.3.1. Rectangular Flow Channels

We begin with the rectangular flow channel depicted in Fig. 2.6 and used to describe the nature of boundary layers in Section 2.2. This represents

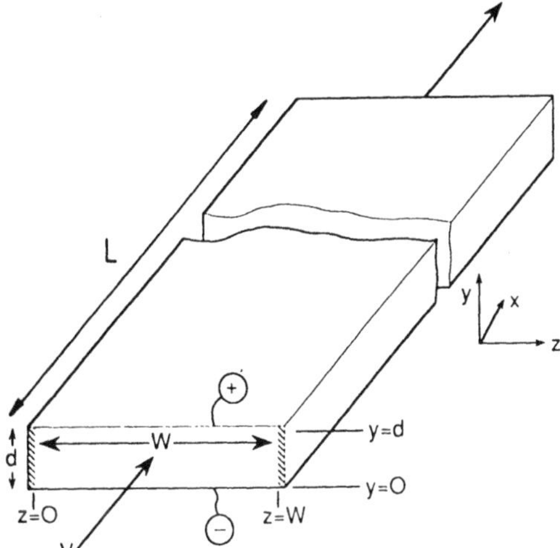


FIGURE 2.6. Rectangular flow channel.

the most common electrochemical reactor design, exemplified in the plate and frame cell. Here an electrode almost constitutes the complete wall or boundary of the reactor. As indicated in Fig. 2.6 the distance $d$ separating the electrodes is small compared with the electrode width $W$ and length $L$. It was mentioned in Section 2.2 that the flow characteristics in this geometry resemble closely those of a pipe.

(i) Laminar flow. Early attempts at the analysis of mass transport during laminar flow in a parallel plate system considered the flow to be fully developed and the electrode to be of infinite width. In this case the Navier–Stokes equation or momentum balance[5] is easily solved to give the one-dimensional variation of velocity in the direction $x$ at point $y$, as:

$$u_x = 6u_{av}\left[\frac{y}{d} - \frac{y^2}{d^2}\right] \tag{2.10}$$

where $u_{av}$ is the average velocity. Combining Eq. (2.10) with the convective diffusion equation[6] gives:

$$6u_{av}\left[\frac{y}{d} - \frac{y^2}{d^2}\right]\frac{\partial c}{\partial x} = D\left[\frac{\partial^2 c}{\partial x^2} + \frac{\partial^2 c}{\partial y^2}\right] \tag{2.11}$$

The solution of Eq. (2.11) with appropriate boundary conditions requires sophisticated numerical methods to obtain what is in principle a reactor design. Exact solutions of Eq. (2.11), or of similar forms, have been obtained by a number of workers. Picket in his very useful book *Electrochemical Reactor Design*[7] summarizes some of the work to be found in the literature.

For a simpler but more approximate approach the following assumptions can be made:

- That mass transport in the $x$ direction is dominated by convection and that molecular diffusion in that dimension can be ignored
- That the velocity $u_x$ varies linearly with $y$
- That suitable boundary conditions can be chosen for the concentration or concentration flux at the surface(s); typically this involves a constant surface concentration, other boundary conditions simulating electrorefining or redox processes.

On the basis of these assumptions Eq. (2.11) can be simplified to:

$$\frac{6u_{av}y}{d}\frac{\partial c}{\partial x} = D\frac{\partial^2 c}{\partial y^2} \qquad (2.12)$$

Solution of this equation is based on the combination of variables technique and can be written in the form of a dimensionless correlation:

$$Sh_x = k_{L,x}d_e/D = 1.23[ReSc(d_e/x)]^{1/3} \qquad (2.13)$$

where $k_{L,x}$ is the mass transfer coefficient at location $x$, $d_e = 2d$ and is the equivalent diameter of the channel, $Re$ is the Reynolds number $= ud_e/v$, and $Sc$ is the Schmidt number $= v/D$.

Since the mass transfer coefficient varies in the direction of flow, we integrate Eq. (2.13) in order to obtain an average Sherwood number, $Sh = k_L d_e/D$:

$$Sh = \frac{1}{L}\int_0^L Sh_x\,dx = 1.85[(Re)(Sc)(d_e/L)]^{1/3} \qquad (2.14)$$

For electrodes of finite width, correlations are obtained as a function of $\gamma$, the aspect ratio $d/W$, of the reactor:[8]

$$Sh = 1.47[(Re)(Sc)\,(d_e/L)]^{1.3}[2/(1+\gamma)]^{1/3} \qquad (2.15)$$

where for the finite width reactor $d_e = 2Wd/(W + d)$.

Experimental data substantiate correlations of the form of Eqs. (2.14) and (2.15), although the data of Picket *et al.*[8] agree better with the empirical equation:

$$Sh = 2.54[(Re)(Sc)(d_e/L)]^{0.3} \tag{2.16}$$

In general, Eq. (2.14) to (2.16) give comparable predictions of mass transfer coefficients and can be recommended subject to a maximum electrode length of about 35 equivalent diameters.

Mass transfer data in a region of developing laminar flow can be estimated from:

$$Sh'_x = 0.96 Re_x^{0.5} Sc^{1/3} (d_e/L)^{-0.05} \tag{2.17}$$

where $Sh'_x = k_{L,x}x/D$ and $Re_x = ux/\nu$, but for design purposes it is better to use a form of Eq. (2.16) to ensure a conservative estimate of $k_L$.

(ii) Turbulent flow. In the fully developed region the Chilton–Colburn correlation[9] originally obtained for heat transfer has been found to apply to mass transfer:

$$Sh = 0.023 Re^{0.8} Sc^{1/3} \tag{2.18}$$

For the entry region the following correlation can be used:

$$Sh = 0.276 Re^{0.58} Sc^{1/3} (d_e/L)^{1/3} \tag{2.19}$$

To design an electrochemical reactor of known performance, two questions have to be answered about the entry region: what is its length, and more important, what are its conditions of flow. We shall return to this when considering scale-up in Chapter 5.

---

EXAMPLE 2.1. Calculation of Mass Transfer Rates in a Parallel Plate Reactor

THE PROBLEM: Electrosynthesis is to be carried out in a parallel plate reactor with electrodes 10 cm wide, the electrolyte gap being 0.5 cm. The synthesis is to take place in a fully developed turbulent hydrodynamic regime so that mass transfer characteristics will be well defined. Calculate the minimum flow rate required and the value of the mass transfer coefficient if the electrolyte density is 1100 kg/m$^3$ and its viscosity is $3 \times 10^{-3}$ Ns/m$^2$. The diffusivity of the diffusing species is $0.92 \times 10^{-9}$ m$^2$/s.

THE SOLUTION: To achieve a turbulent regime the Reynolds number must be at least 2000, i.e., $Re = (ud_e\rho)/\mu = 2000$, and:

$$d_e = 2Wd/(W + d) = 2 \times 0.1 \times 0.005/(0.1 + 0.005) = 0.00952\,\text{m}$$

$$u = Re\mu/\rho d_e = 2000 \times 3 \times 10^{-3}/(1100 \times 0.00952) = 0.573\,\text{m/s}$$

The corresponding flow rate $Q$ is given by:

$$Q = uWd = 0.573 \times 0.1 \times 0.005 = 2.86 \times 10^{-4}\,\text{m}^3/\text{s}$$

The Schmidt number $Sc$ is:

$$\mu/(\rho D) = 3 \times 10^{-3}/(1100 \times 0.92 \times 10^{-9} = 2960$$

The mass transfer coefficient $k_L$ can be derived from a slightly rearranged Eq. (2.18), remembering that $Sh = (k_L d_e)/D$:

$$k_L = 0.023Re^{0.8}Sc^{1/3}D/d_e$$
$$= 0.023 \times 2000^{0.8} \times 2960^{1/3} \times 0.92 \times 10^{-9}/0.00952 = 1.40 \times 10^{-5}\,\text{m/s}$$

This value of $k_L$ is typical for an electrochemical reactor; it generally lies in the range $10^{-6}$ to $10^{-4}$ m/s.

---

#### 2.3.3.2. The Annulus

An electrochemical reactor containing two concentric cylindrical electrodes (Fig. 2.7), with or without a diaphragm, represents a practical and attractive geometry since it offers a uniform primary current distribution (see Chapter 5). Axial flow through the annular space between the two electrodes, or an electrode and a diaphragm, has characteristics between those of a pipe and a rectangular channel. A theoretical analysis for the condition of laminar or turbulent flow is approached in the same way as that for pipes and channels; hence, we merely identify appropriate experimental correlations.

(i) Laminar flow. For fully developed laminar flow Ross and Wragg[10] have obtained the following correlations for local and overall Sherwood numbers:

$$Sh_x = 1.076[\Phi ReSc(d_e/x)]^{1/3} \tag{2.20}$$

and

$$Sh = 1.614[\Phi ReSc(d_e/L)]^{1/3} \tag{2.21}$$

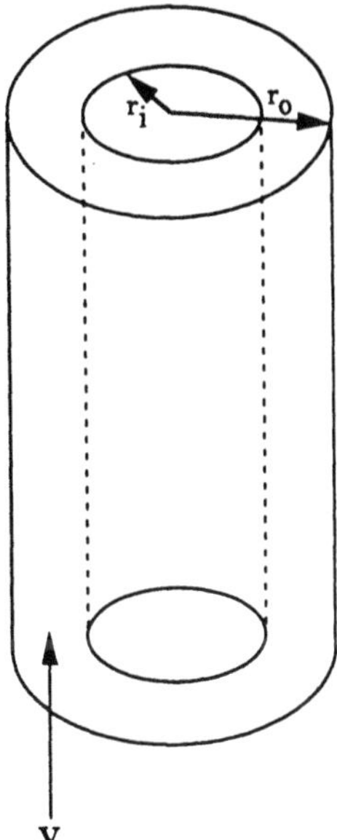


FIGURE 2.7. Annulus between two concentric cylinders with axial flow. $r_i$ = Radius of inner cylinder; $r_o$ = radius of outer cylinder.

where the equivalent diameter $d_e$ (also used in defining the Sherwood and Reynolds numbers) is given by $d_e = 2(r_o - r_i)$, and $\Phi$ is a geometric parameter defined as:

$$\Phi = \left[ \left( \frac{1 - r}{r} \right) \frac{\left( 0.5 - \frac{r^2}{1 - r^2} \ln [1/r] \right)}{\left( \frac{1 + r^2}{1 - r^2} \ln [1/r] - 1 \right)} \right] \tag{2.22}$$

where $r = r_i / r_o$.

The correlations provide good predictions and are suitable for electrode lengths of up to about $20d_e$. Equation 2.21 can also be used for developing laminar flow. For design purposes, entry effect can be ignored.

(ii) Turbulent flow. For the estimation of mass transfer coefficients for turbulent flow in an annulus the Chilton–Colburn Eq. (2.18) can be used for fully developed flow and Eq. 2.19 for developing flow.

### 2.3.3.3. Rotating Cylinder Electrodes

An important application of the concentric cylinder arrangement is when one of the cylinders rotates. In the more popular system the inner cylinder is rotating (Fig. 2.8b), the outer one being stationary (Fig. 2.8a). This is preferred to the situation where the outer cylinder is rotating, for two reasons: (1) lower rotational speeds are required to produce turbulence and hence good mass transfer characteristics; (2) the design is mechanically less complicated.

The hydrodynamics of the flow between the two cylinders can be characterized by a dimensionless quantity *Ta*, the Taylor number, named after the mathematician G. I. Taylor.[11] The Taylor number incorporates two dimensionless groups, the Reynolds number and a geometric ratio, and is defined as:

$$Ta = \left[\frac{u_i(r_o - r_i)}{v}\right]\left[\frac{r_o - r_i}{r_i}\right]^{1/2} = Re_w\left[\frac{r_o - r_i}{r_i}\right]^{1/2} \tag{2.23}$$

where $u_i$ is the peripheral velocity of the rotating cylinder. On the basis of the Taylor number one can distinguish three flow regimes:

1. At low flow rates a simple laminar, or Couette, flow prevails in which the velocity of the fluid is tangential. This region is character-

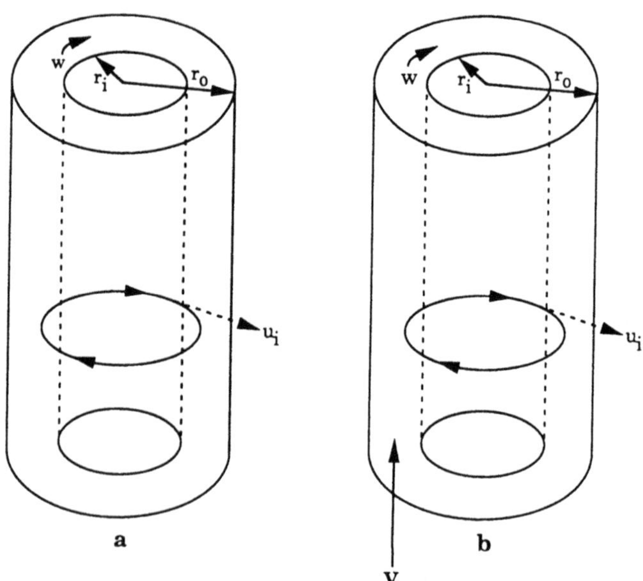


FIGURE 2.8. Rotating cylinder. (a) No axial flow; (b) with axial flow. $r_i$ = Radius of inner cylinder; $r_o$ = radius of outer cylinder; $u_i$ = peripheral velocity; $w$ = angular velocity.

ized by $Ta < 41.3$. This offers little enhancement of mass transport and is of no consequence to reactor design.

2. At higher flow rates the simple tangential flow becomes unstable and a cellular motion is superimposed upon the tangential flow. The so-called Taylor vortices carry material from one cylinder to the other, thereby increasing mass transfer rates. However, because of the cellular or vortex nature of the flow, the mass transfer characteristics are not uniform. The region is characterized by $41.3 < Ta < 400$.
3. At high flow rates ($Ta > 400$) the flow becomes turbulent, characterised by random and rapid fluctuations of velocity. Theoretical analyses of this regime[12] agree in general with experimentally obtained correlations, and in particular with the extensive data of Eisenberg and collaborators[13] for a range of $Re_w$ (defined as $2r_i u_i/\nu$) of 100–160,000. Their correlation for an annular space of >8 mm is:

$$St_w = k_L/u_i = 0.0791 Re^{-0.3} Sc^{-0.644} \tag{2.24}$$

where the Stanton number $St_w = Sh_w/Re_w Sc$.

Equation (2.24) shows no dependence on the dimensions of the outer cylinder.

---

EXAMPLE 2.2. Mass Transfer Requirements for a Rotating Electrolyzer

THE PROBLEM: An undivided rotating cylinder reactor is operated in a batch recycle loop to process a batch of $0.2\,\text{m}^3$ every 8 hours. The cylinder is 1 m long and 0.1 m in radius; the interelectrode gap is 0.05 m. The reaction takes place under mass transfer limiting conditions. If the maximum rotation speed of the cylinder is 150 rpm, determine whether the operation is feasible if the required conversion is to be 80%. If so, what is the required rotation speed and what is the maximum conversion possible? Assume that axial flow in the reactor can be ignored. The kinematic viscosity $\nu$ is $1.1 \times 10^{-6}\,\text{m}^2/\text{s}$ and the Schmidt number $Sc$ is 2500.

THE SOLUTION: First we see whether the mass transfer coefficient can be calculated from Eisenberg's Eq. (2.24), by checking the value of the Taylor number, Eq. (2.23), and the value of the Reynolds number. The rotation speed $w = 2\pi(150/60) = 15.7$ radians/s.

$$Ta = [r_i^{1/2} w (r_o - r_i)^{3/2}]/\nu$$
$$= [0.1^{1/2} \times 15.7(0.05)^{3/2}]/(1.1 \times 10^{-6}) = 5.05 \times 10^4$$

And remembering that $wr_i = u_i$:

$$Re_w = (2r_i^2 w/\nu) = (2 \times 0.1^2 \times 15.7)/(1.1 \times 10^{-6}) = 285{,}000$$

We are well into the third flow regime [$Ta \gg 400$], but $Re$ is somewhat larger than the 160,000 investigated by Eisenberg. Hoping for the best, we nevertheless use Eq. (2.24) (some optimism is essential in design when data are not available):

$$k_L = u_i \times 0.0791(2r_i u_i/\nu)^{-0.3} Sc^{-0.644}$$

A slight rearrangement of Eq. (2.24) gives:

$$k_L = 0.0791 w^{0.7} r_i^{0.4} (2/\nu)^{-0.3} Sc^{-0.644}$$

$$w = (2\pi \times 150)/60 = 15.7 \text{ radians/s}$$

$$k_L = 0.0791 \times 15.7^{0.7} \times 0.1^{0.4} [2/(1.1 \times 10^{-6})]^{-0.3} (2500)^{-0.644}$$

$$= 1.86 \times 10^{-5} \text{ m/s}$$

As will be shown in Chapter 4, for a process approximating to a batch operation under mass transfer control the fractional conversion $X_O$ of reactant $O$ is given by:

$$X_O = 1 - \exp(-Sk_L t/V)$$

where $S$ is the electrode area and $V$ the electrolyte volume to be treated in time $t$.

Hence:

$$k_L = -V/(St) \ln [1 - X_O]$$

$$= -0.2/(2\pi \times 0.1 \times 1 \times 8 \times 3600) \ln [1 - 0.8] = 1.78 \times 10^{-5} \text{ m/s}$$

Since this mass transfer coefficient is lower than the one obtained from Eq. (2.24), we conclude that the operation is feasible. The rotational speed corresponding to the lower mass transfer coefficient of $1.78 \times 10^{-5}$ is obtained by simple proportion:

$$w^{0.7} = (1.78/1.86) \times 15.7^{0.7} = 6.6$$

Hence $w = 14.7$ radians/s or 140 rpm.

The maximum possible conversion is calculated from:

$$X_o = 1 - \exp[(-2\pi \times 0.1 \times 1 \times 1.86 \times 10^{-5} \times 8 \times 3600)/0.2] = 0.81$$

---

Other studies of mass transfer for rotating cylinders have appeared in the literature.[14] These include outer rotating cylinders, as well as fanned, wiped, and rough cylinders. With rough surfaces in the turbulent regime, the Stanton number becomes independent of the Reynolds number.

For the rotating cylinder electrode to be adopted as a continuous reactor, some degree of axial flow has to be superimposed on the tangential and turbulent motion in the annulus (Fig. 2.8b). If the rate of mass transport due to axial flow exceeds that due to rotation then the reactor will exhibit approximate plug-flow characteristics. If the reverse is true the behavior will tend to approach that of a continuous stirred-tank reactor (see Section 5.1.1.1).

Estimation of mass transfer coefficients for rotating cylinders with axial flow can be made using the equations which have already been given for rotation or axial flow alone, taking the higher mass transfer coefficient of the two as the design value. Alternatively, the following can be used for the range of parameters indicated:

$$Sh = 0.38 Ta_m^{1/2} Sc^{1/3} \qquad 25 < Re_a < 300 \qquad (2.25)$$

$$200 < Ta_m < 1600$$

$$Sh = 0.12 Ta_m^{2/5} (Re_a Sc)^{1/3} \qquad 300 < Re_a < 800 \qquad (2.26)$$

where

$$Ta_m = \frac{w r_o}{\nu} \frac{(r_o - r_i)^{3/2}}{r_i^{1/2}}$$

and $w$ = the peripheral velocity in radians/s.

$Re_a$, the axial Reynolds number, is based on the axial velocity and on an equivalent diameter of $2(r_o - r_i)$.

---

EXAMPLE 2.3. Mass Transfer Characteristics of a Rotating Cylinder Reactor with Axial Flow

THE PROBLEM: An inner cylinder 1 m long and with a radius of 0.15 m rotates at 25 rpm. It is to be used to process an electrolyte with a Schmidt number of 3000 and a kinematic viscosity of $10^{-6}$ m$^2$/s. The I.D. of the outer cylinder is 0.32 m and the cell is undivided. If the reactor operates with an axial flow of electrolyte with velocities varying between 0.01 and 0.1 m/s, investigate the mass transfer characteristics of the system.

THE SOLUTION: Consider the correlations available for annular systems.
First, rotation alone: here we shall again use Eisenberg's Eq. (2.24):

$$k_L = 0.0791 u_i (2 r_i u_i / \nu)^{-0.3} Sc^{-0.644}$$

$$u_i = 2\pi \times 0.15 \times 25/60 = 0.393\,\text{m/s}$$

$$k_L = 0.0791 \times 0.393 (2 \times 0.15 \times 0.393/10^{-6})^{-0.3} (3000)^{-0.644} = 5.4 \times 10^{-6}\,\text{m/s}$$

Second, rotation and axial flow: First, we calculate the range of the axial Reynolds numbers $Re_a$:

$$Re_a = [2(r_o - r_i)/\nu] u_a = [2(0.16 - 0.15)/10^{-6}] u_a = 20000 u_a$$

$Re_a$ therefore ranges from 200 to 2000. We determine $Ta_m$ from Eq. (2.26), noting that strictly speaking it is only applicable for $Re_a < 800$:

$$Ta_m = \frac{w r_o (r_o - r_i)^{3/2}}{\nu \times r_i^{1/2}} = \frac{2\pi \times 25 \times 0.16 (0.16 - 0.15)^{3/2}}{60 \times 10^{-6} \times 0.15^{1/2}} = 1080$$

which is in the middle of the applicable range for Eqs. (2.26) and (2.25). The former is now rearranged:

$$\begin{aligned} k_L &= 0.12 Ta_m^{2/5} (d_e/\nu)^{-2/3} Sc^{-2/3} u_a^{1/3} \\ &= 0.12 \times 1080^{2/5} (2 \times 0.1/10^{-6})^{-2/3} (3000)^{-2/3} u_a^{1/3} = 1.28 \times 10^{-5} u_a^{1/3} \end{aligned}$$

The values for $k_L$ are calculated in (Table 2.1).

TABLE 2.1. Values of $k_L$

| $u_a$ (m/s) | $Re_a$ | $k_L$ (m/s) |
|---|---|---|
| 0.01 | 200 | $2.7 \times 10^{-6}$ |
| 0.015 | 300 | $3.2 \times 10^{-6}$ |
| 0.04 | 800 | $4.4 \times 10^{-6}$ |
| 0.1 | 2000 | $5.9 \times 10^{-6}$ |

Since for $Re_a < 300$, Eq. (2.25) is recommended, we rearrange it and use it for $Re_a = 200$. This gives:

$$k_L = 0.38 u_a Ta_m^{1/2} Re_a^{-1} Sc^{2/3}$$
$$= 0.38 \times 0.01 \times 1080^{1/2} \times 200^{-1} \times 3000^{-2/3} = 3.0 \times 10^{-6}\,\text{m/s}$$

Clearly the difference between the values obtained from Eqs. (2.25) and (2.26) is not significant.

Third, axial flow alone: Even though Eq. (2.21) is limited to an electrode length of $20d_e$, we shall have to use it to calculate $k_L$ under laminar conditions. Rearranging the equation:

$$k_L = 1.614(\Phi d_e/L)^{1/3}(d_e/\nu)^{-2/3} Sc^{-2/3} u_a^{1/3}$$

Note that $\Phi \to 1.5$ as $r_i/r_o \to 1$. In this example, $\Phi = 1.52$. Hence:

$$k_L = 1.614(1.52 \times 0.02)^{1/3}(0.02/10^{-6})^{-2/3}(3000)^{-2/3} u_a^{1/3} = 3.29 \times 10^{-6} u_a^{1/3}$$

For turbulent flow the Chilton–Colburn Eq. (2.18) can be rearranged:

$$k_L = 0.023(d_e/\nu)^{-0.2} Sc^{-2/3} u_a^{0.8}$$
$$= 0.023(0.02/10^{-6})^{-0.2}(3000)^{-2/3} u_a^{0.8} = 1.53 \times 10^{-5} u_a^{0.8}$$

We are now in a position to calculate (in Table 2.2) mass transfer coefficients from the two equations just given for laminar and turbulent flow.

TABLE 2.2. Mass Transfer Coefficients

| $u_a$ (m/s) | $Re_a$ | $k_L$ (m/s) (laminar) | $k_L$ (m/s) (turbulent) |
|---|---|---|---|
| 0.01 | 200 | $7.1 \times 10^{-7}$ | |
| 0.015 | 300 | $8.1 \times 10^{-7}$ | |
| 0.04 | 800 | $1.12 \times 10^{-6}$ | $1.16 \times 10^{-6}$ |
| 0.1 | 2000 | $1.5 \times 10^{-6}$ | $2.4 \times 10^{-6}$ |

It is instructive to compare the values of the mass transfer coefficients $k_L$(m/s) obtained from the various correlations:

| | |
|---|---|
| Rotation alone | $5.4 \times 10^{-6}$ |
| Axial flow + rotation | $2.7 \times 10^{-6}$–$5.9 \times 10^{-6}$ |
| Axial flow alone | $7.1 \times 10^{-7}$–$2.4 \times 10^{-6}$ |

Two conclusions can be drawn. First, rotation produces the dominant mass transfer effect. Second, the values of the mass transfer coefficients are on the low side. For a reaction where mass transport is important one would look for a mass transfer coefficient that is at least in the $10^{-5}$ m/s range.

---

### 2.3.3.4. Rotating Disk Electrodes

The rotating disk electrode (RDE), although best known to the electrochemist as an analytical tool, has been considered as the basis for an electrochemical reactor. Apart from this, as will be seen in Section 3.2.2.2, it is the preferred experimental equipment for determining kinetic constants when setting up a reaction model. To do this, however, the value of $k_L$ for the particular RDE cell arrangement must be known.

It is possible to treat the velocity profile like that leading to the concept of the Nernst diffusion layer (Fig. 2.1). The distance corresponding to the point when $u$ reaches the value $V$ is called the Prandtl boundary layer, $\delta_{Pr}$.[15] This is a hypothetical although useful quantity. The unique feature of the RDE is that $\delta_{Pr}$ is independent of the radial distance $r$, i.e., the velocity of the fluid normal to the disk is dependent only on the distance $z$ from it and not on the radial distance. Flow in the boundary layer remains laminar for a Reynolds number $Re_w$, (defined as $r^2 w/v$) of up to $2 \times 10^5$. For a diffusion limited flux, $\delta_N$ is independent of position and as a first approximation is given by:

$$\delta_N = 1.61 w^{-1/2} D^{1/3} v^{1/6} \tag{2.27}$$

The corresponding Sherwood number for laminar flow is:

$$Sh_w = 0.62 Re_w^{1/2} Sc^{1/3} \tag{2.28}$$

This well-defined and experimentally confirmed relationship is one reason why RDEs are preferred as electroanalytical devices and for determining kinetic constants (see Section 3.2.2.2). It is important to remember, however, that Eq. (2.27) and (2.28) assume an electrode rotating in an infinite amount of electrolyte, which is not true of a real RDE cell. It is better to obtain an actual value for $k_L$ for the system being modeled (Example 3.2.3.1), or if this is not possible, to determine $k_L$ by using a limiting current technique (see Section 2.3.4). This value of $k_L$ is corrected for the Schmidt number of the system to be modeled.

A second limitation of Eq. (2.27) and (2.28) is the diameter of the rotating disk. At high rotational speeds the flow characteristics of a larger disk become turbulent in the region near the periphery of the disk, while a laminar region still exists at the center. A concentric zone of transitional flow exists between the laminar and turbulent regions. If the disk is large enough the central laminar region can usually be neglected and turbulent eddies determine the diffusion layer thickness. Under these circumstances, mass transfer to the disk is:[16]

$$Sh_w = 0.0117 Re_w^{0.9} Sc^{1/4} \tag{2.29}$$

### 2.3.4. Obtaining Numerical Values of $k_L$ by Experiment

As indicated in Section 1.4.2, on increasing the electrode potential the current density will increase until the rate at which the reactant molecule reaches the electrode becomes rate-determining. The current density is said to have reached a limiting value $i_{lim}$ and for a reduction is given by Eq. (3.111). Hence:

$$i_{lim} = nFk_L c_O \tag{2.30}$$

where $n$ is the number of electrons involved in the reaction, $F$ the Faraday, and $c_O$ the concentration of the molecule to be reduced.

Measuring $i_{lim}$ for a given $n$ and $c_O$ enables you to determine $k_L$. This limiting current technique[17] is a simple and reliable way to obtain numerical values of $k_L$, if certain experimental precautions are taken. One great advantage of this is that the surface area over which the mass transfer rate is measured can be made as small or as large as you like. Thus it is easy to determine point values of the mass transfer coefficient $k_{L,x}$ or to obtain the distribution of mass transfer over larger sections of the electrode for scale-up (see Chapter 5). This limiting current technique is not confined to an electrochemical context, but is in general use whenever mass transfer rates in a liquid have to be determined.

#### 2.3.4.1. The Chemical System

There are a number of systems that can be used to determine a limiting current density but in the authors' opinion the ferri-ferrocyanide redox system dissolved in aqueous alkali (the latter stabilizes the complex ion and acts as a supporting electrolyte) is as good as any. The reduction of ferricyanide to ferrocyanide is used as the measured reaction and is represented by

$$Fe(CN)_6^{3-} + e \rightarrow Fe(CN)_6^{4-} \tag{2.31}$$

At the anode the reverse of Eq. (2.31) occurs, keeping the concentration of the redox system and hence that of the reactant constant. To ensure that limiting current conditons are not reached at the counterelectrode, i.e., the anode, the ferrocyanide concentration is well above that of the ferricyanide. It is advisable, to ensure that the total anode area is greater than that of the cathode. One disadvantage of the ferri-ferrocyanide system is its sensitivity to light. Any transparent part of the equipment has to be shielded and made-up solutions stored in the dark. Even then, reagents should be renewed regularly.

The most commonly used electrode materials are nickel, stainless steel, and platinum. Workers once used KOH most often as the supporting electrolyte. It has been found, however, that with stainless steel in particular,[18] but to some extent also with nickel, limiting currents decrease with time. The use of $K_2CO_3$ avoids this difficulty. A typical system is therefore:

0.01 M potassium ferricyanide
0.03 M potassium ferrocyanide
0.50 M potassium carbonate

Because of the high cost of platinum, nickel or stainless steel are the preferred cathodic materials. It is vital that the cathode be cleaned frequently. The anode with a small cathode, e.g., an RDE, is normally platinum. When larger cathode areas are under test, e.g., during scale-up studies of cells, a platinized titanium counterelectrode can be used.

#### 2.3.4.2. The Experimental Arrangement

Two examples will be given, one that determines $k_L$ in the context of evaluating kinetic constants of a reaction model using an RDE and one for determining mass transfer coefficients in an electrochemical reactor.

(i) Mass transfer to a rotating disk. Figure 2.9 shows a typical cell incorporating an RDE. The counterelectrode is a platinum wire. Nitrogen gas is blown through the electrolyte to minimize the contribution of the reduction of oxygen to the limiting current. Usually, limiting current measurements

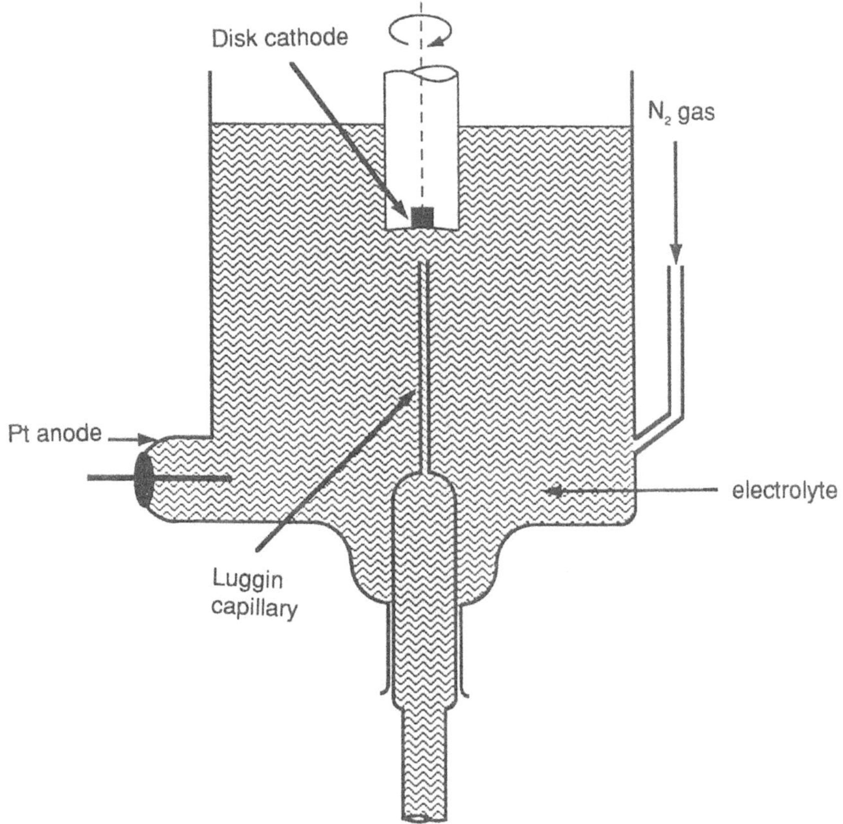


**FIGURE 2.9.** Electrolytic cell with RDE.

are done galvanostatically (at constant current). In Fig. 2.9 the electrode potential is measured with an adjustable Luggin capillary (made from the body and plunger of a glass syringe) and connected via a salt bridge to a reference electrode (Fig. 2.10). To determine mass transfer coefficients it is not necessary to know the electrode potential as long as you are certain that you are working on the horizontal part of the polarization curve. For modeling purposes it is essential, however, that an accurate value of the electrode potential be known, which is why methods of determining the "ohmic correction" are discussed in Section 3.2.2.3.

Figure 2.11 shows a cross section of a typical RDE. Electrical contact is made with a rod and a spring at the back of the electrode which is let into a polypropylene cylinder. Take care that there is no electrolyte leakage to the back of the electrode which could result in corrosion and a high

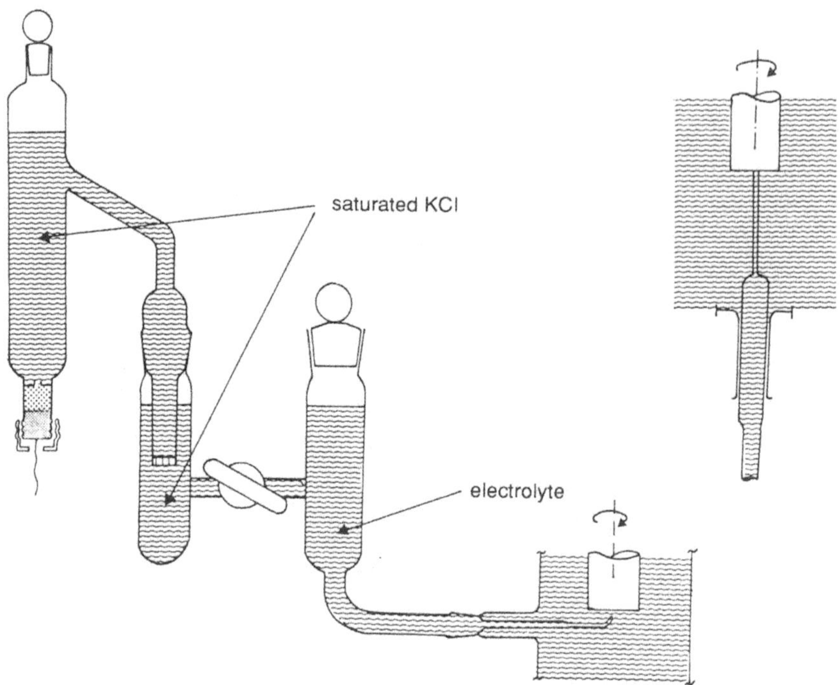


**FIGURE 2.10.** Saturated calomel reference electrode with salt bridge.

resistance which in turn could produce an undesirable voltage drop (see Section 3.2.2.3.ii) in the circuit.

A simple cell circuit for supplying current from a potentiostat to a cell is shown in Fig. 2.12. In the potentiostatic mode the required potential is set on the potentiometric control. The potentiostat compares the actual potential with the set point and adjusts the current until the values agree. In the present application the potentiostat is used as an ordinary current source and the current is manually set to correspond with a current density on the middle of the plateau of the polarization curve.

Results are fed into a computerized data logging system which can work out appropriate correlations. But periodic checks on the raw results are essential to ensure that meaningful values of mass transfer coefficients and hence correct correlations are obtained. Checks should include verification that there has been no significant shift of electrode potential in the current plateau.

(ii) Mass transfer in a synthesis cell. The RDE of the previous example is replaced by a segmented electrode (Fig. 2.13). The cell in this example[19] is

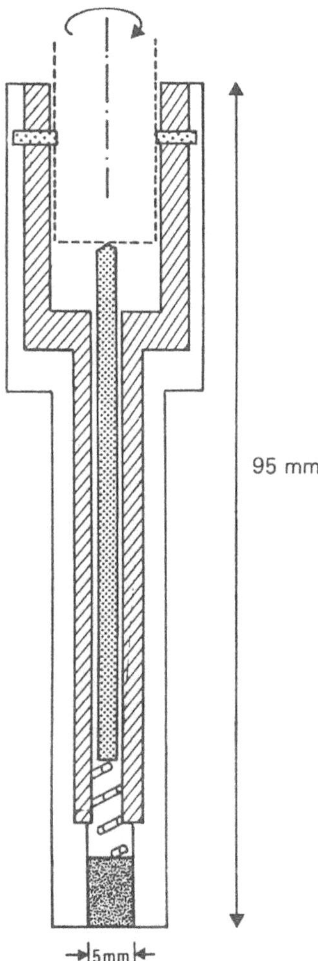


FIGURE 2.11. Cross-section through typical RDE.

undivided and incorporates a nickel or a stainless steel cathode, divided into ten segments (each 0.3 m × 0.05 m). Segments are machined from blocks of the respective metal and cemented together using epoxy resin or similar adhesive. A stainless steel Luggin capillary is inserted into the middle segment and linked to a reference electrode to ensure that all segments are operated within the limiting current plateau. When measuring mass transfer coefficients in such a flow cell it is advisable to place a Nafion® membrane between the Luggin and the reference electrode in order to prevent any flow of electrolyte, caused by variations of pressure in the cell, into the reference electrode. The electrolyte gap between the two electrodes can be varied by

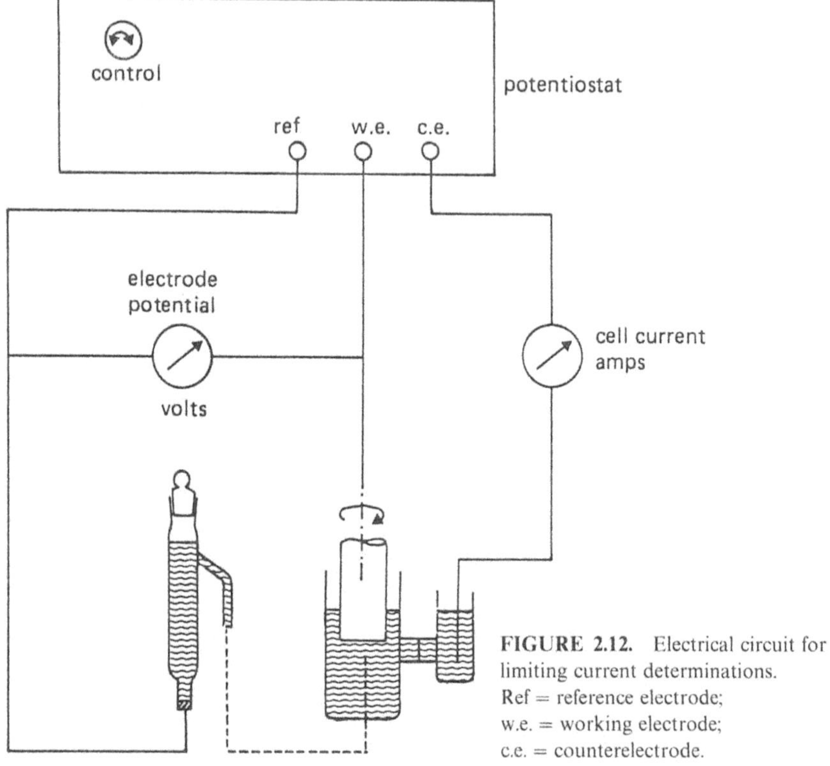


FIGURE 2.12. Electrical circuit for limiting current determinations. Ref = reference electrode; w.e. = working electrode; c.e. = counterelectrode.

inert gaskets of appropriate thickness. Turbulence promoters (discussed in the next section) can be placed in this narrow gap when flow would otherwise remain laminar.

Power to the cell can be provided by a potentiostat or a rectifier. All of the electrode must be electrochemically active at all times; otherwise, a lack of concentration profiles will cause unrealistically high mass transfer coefficients. Relatively large segments are used instead of small probes because a large number of probes would produce an unmanageable complexity of data when used with a pilot plant–size electrode. The current from each segment is converted into a voltage drop and fed into the computerized data logging system. In this case[19] it consisted of a Torch Z80 disk drive, a Cu-Dac 12 analog/digital converter, a 100 gain amplifier, a multiplexer, and a BBC microscomputer (model B), linked to a printer. Developments in the field of computerized data logging are so rapid that further detail would not be helpful.

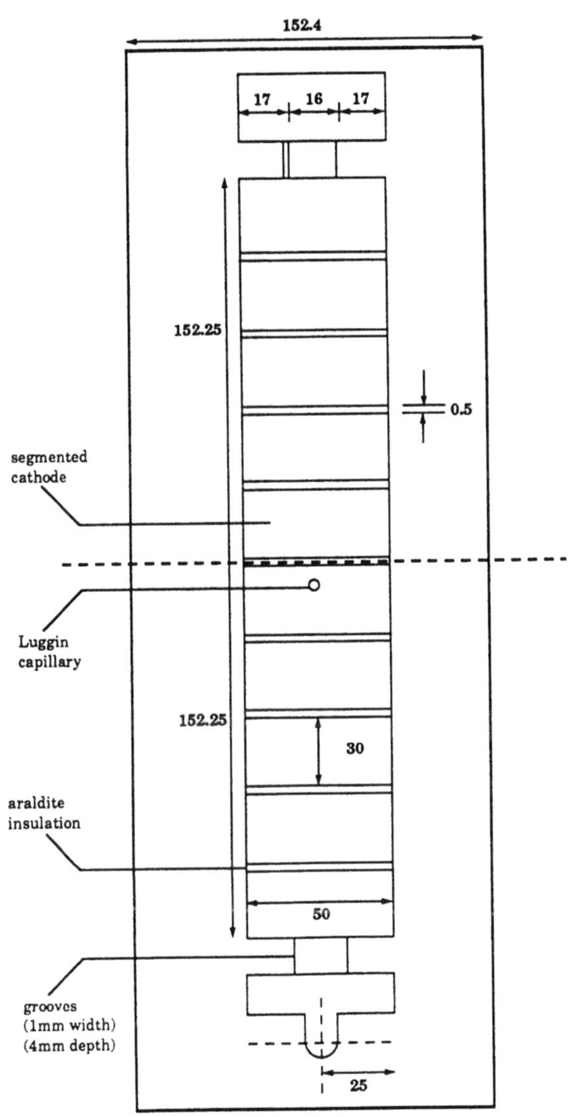


**FIGURE 2.13.** Segmented electrode for limiting current determinations (dimensions in mm).

### 2.3.5. Turbulent Flow Promoters

As discussed in Section 1.4.5, electricity costs are directly proportional to cell voltage. One way of minimizing voltage is to make the gap between anode and cathode as small as possible. This is especially important when an electrolyte with low electrical conductivity is used. Because of a very narrow electrolyte gap the hydrodynamic regime is laminar even at very high electrolyte flow rates. Industrial practice requires high mass transfer rates and this is done by inducing turbulent flow. One way to achieve turbulent flow is to introduce "turbulence promoters" which disturb the flow, causing local variations in magnitude and direction of fluid velocity. All turbulence promoters obstruct the fluid flow path, inducing increased eddy transport of momentum and mass. The exact mechanism for the increased mass transport is not known, although boundary layer separation, flow reversal, and vortex formation and shedding may play a part.

Sticking to flow in channels (the characteristics of reactors involving "three-dimensional" electrodes are considered in Chapter 5), turbulence promoters can be divided into two categories which involve electrochemically inert or active materials.

#### 2.3.5.1. Inert Promoters

(i) Mesh and Cloth promoters. Where possible commercially, available plate and frame cells use some kind of open mesh between the electrodes (or between the electrode and the diaphragm if a divided cell is involved) to increase mass transport. Presence of a mesh increases the mass transfer coefficient by a factor of 2 to 7, and coefficient values are no longer dependent on electrode length. Figure 2.14 presents some correlations for

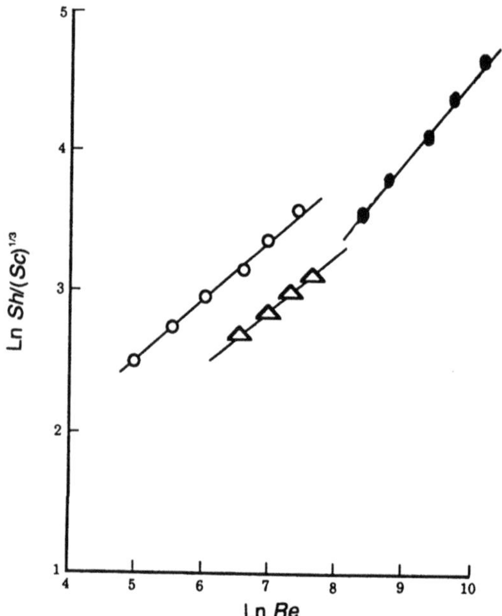


**FIGURE 2.14.** Mass transfer correlations for mesh type turbulence promoters. ○ = SU mesh; Δ = 0.93 mm mesh; ● = baffled cell.

plate and frame cells. $Sh$ values obtained by Coeuret and coworkers[20] using the mesh from an SU cell[21] and those determined for a somewhat different mesh, 0.93 mm thick, in an electrolyte gap of 1 mm,[19] give similar results (see also Example 2.4). The correlation for the SU mesh[20]:

$$Sh = 1.25Re^{0.46}Sc^{1/3} \qquad (150 < Re < 1500) \tag{2.32}$$

And that for the 0.93 mm mesh[19]:

$$Sh = 1.47Re^{0.43}Sc^{1/3} \qquad (500 < Re < 1470) \tag{2.33}$$

Included in Fig. 2.14 are the results for a baffled cell[22] which are not inconsistent with the correlations for the mesh-type turbulence promoters, the correlation for the baffled cell being:

$$Sh = 0.20Re^{0.63}Sc^{1/3} \qquad (3000 < Re < 15000) \tag{2.34}$$

Robertson and collaborators[23] investigated the performance of a cloth turbulence promoter in a narrow gap (0.8 mm) associated with the "Swiss Roll cell." Again mass transfer rates increased by a factor of 5 compared with those in the empty channel and were independent of electrode length. The mass transfer data were correlated by:

$$Sh = 0.258Re^{0.62}Sc^{1/3} \tag{2.35}$$

---

EXAMPLE 2.4. Mass Transfer Rates with and without a Mesh-Type Turbulence Promoter

THE PROBLEM: A synthesis is to be performed in an undivided narrow gap parallel plate cell. Calculate the mass transfer coefficients in the absence and presence of a mesh-type turbulence promoter. The following data are available: electrode height = 1 m; electrode width = 0.1 m; electrolyte gap = 0.001 m; $Sc = 1500$; $\nu = 10^{-6}\,\mathrm{m^2/s}$; superficial electrolyte velocity = 0.36 m/s.

THE SOLUTION: First we calculate the Reynolds number.

$$d_e = 2Wd/(W + d) = 2 \times 0.1 \times 0.001/(0.1 + 0.001) = 0.00198\,\mathrm{m}$$

Hence:

$$Re = ud_e/\nu = 0.36 \times 0.00198/10^{-6} = 713$$

The hydrodynamic regime will therefore be well into the laminar region. Although the length of the electrode is well beyond 35 equivalent diameters, the best we can do is use Eq. (2.16), rearranged:

$$k_L = 2.54uRe^{-0.7}Sc^{-0.7}(d_e/L)^{0.3}$$
$$= 2.54 \times 0.36(713)^{-0.7}(1500)^{-0.7}(1.98 \times 10^{-3})^{0.3} = 8.5 \times 10^{-6}\,\text{m/s}$$

For mass transfer with the mesh we shall use Eqs. (2.32) and (2.33) just to see how different their predictions are.

Rearranging Eq. (2.32):

$$k_L = 1.25uRe^{-0.54}Sc^{-2/3} = 1.25 \times 0.36(713)^{-0.54}(1500)^{-2/3}$$
$$= 9.9 \times 10^{-5}\,\text{m/s}$$

From Eq. (2.33):

$$k_L = 1.47uRe^{-0.57}Sc^{-2/3} = 1.47 \times 0.36(713)^{-0.57}(1500)^{-2/3}$$
$$= 9.5 \times 10^{-5}\,\text{m/s}$$

The two mass transfer coefficients are very close because the two meshes are very similar. The warning that different meshes will produce divergent mass transfer rates still stands.

An important point is that with turbulence promoters, $k_L$ values are an order of magnitude greater than those for an open channel, although significantly more pumping energy is expended in achieving this increase.

---

**(ii) Particulate bed promoters.** For a cell with wide electrolyte gaps one of the simplest turbulence promoters is a particulate bed of inert particles in the flow channel next to the working electrode. These beds, packed or fluidized, cause an enhancement of the mass transfer rate when compared with that for similar flow rates in the open channel. Packing materials include spheres, cylinders (e.g., Raschig rings), and Berls saddles. For spheres packed in a rectangular channel a typical correlation is:

$$j_D = StSc^{2/3} = 0.97[Re_p/(1 - \text{e})]^{-0.40} \tag{2.36}$$

where e, the voidage of the bed, plays a significant part in all correlations for this type of promoter and $Re_p = ud_p/v$. Caution should be exercised on the applicability of the dimensionless parameters and the shape and size of the particulate material and channel when using this equation for $k_L$.

A commercially available cell[24] uses this type of particulate turbulence promoter in the form of a fluidized bed of electrochemically inert particles, typically glass spheres. This Chemelec cell is aimed at the secondary recovery of metals and their removal from effluents. Little reliable information is available on mass transfer to spheres in a fluidized bed. Nassif,[25] using a limited current technique, investigated mass transfer to the particles of a fluidized bed as well as to the wall enclosing them. For the latter he derived:

$$j_D e = k_L Sc^{2/3} e/u = 0.71[ud_p/\nu(1-e)]^{-0.33} \tag{2.37}$$

where e is the voidage of the fluidized particles and $d_p$ their diameter.

The low index of the Reynolds number indicates that values of $k_L$ change little with increases in $u$ of as much as 200%. This reflects the unusual hydrodynamic characteristics of fluidization where particle motion within the bed is quite random and increasing $u$ causes an increase in the "circulation" of particulates. Relative velocities between solid and liquid, which in fact determine hydrodynamics, do not increase linearly with $u$. Although Eq. (2.37) appears to correlate Nassif's results,[25] it suffers from a basic disadvantage: $u$ occurs on both sides of Eq. (2.37); be careful that values of $u$ do not swamp the influence of other variables. Nevertheless one can use Eq. (2.37) to get an order-of-magnitude value for $k_L$.

---

EXAMPLE 2.5. Enhancement of Mass Transfer Using a Particulate Bed as a Turbulence Promoter

THE PROBLEM: An undivided parallel plate electrochemical reactor is used to carry out a mass transfer limited reaction. Mass transfer rates are enhanced with a packed or fluidized bed of electrochemically inert spheres. Compare the mass transfer rates and pressure drops encountered with those for the open channel.

Technical data: electrode width = 0.1 m; electrolyte gap = 0.03 m; density of electrolyte = $10^3$ kg/m$^3$; Schmidt number = 1300; kinematic viscosity = $10^{-6}$ m$^2$/s; sphere diameter = 0.002 m; sphere density = 1800 kg/m$^3$; voidage of loosely packed spheres = 0.42.

THE SOLUTION: We begin with the fluidized-bed turbulence promoter because this will give us a flow velocity to compare with the packed bed.

First, the fluidized bed: Mass transfer coefficients will be calculated from Eq. (2.37), in the form:

$$k_L = \frac{0.71}{\mathrm{e}} u^{0.67}[d_p/\nu(1-\mathrm{e})]^{-0.33} Sc^{-2/3}$$

The first step is to estimate the minimum velocity required for fluidization $u_{mf}$[26]:

$$u_{mf} = 0.0055 \frac{\mathrm{e}_{mf}^3}{1-\mathrm{e}_{mf}} \frac{d_p^2(\rho_s - \rho)g}{\mu}$$

where $e_{mf}$ is the bed voidage at minimum fluidization and $\rho_s$ and $\rho$ are the density of the solid particles and the fluidizing medium. Substituting values:

$$\begin{aligned} u_{mf} &= 0.005 \times 0.42^3 \times 0.002^2(1800-1000) \times 9.81/[(1-0.42) \times 10^{-3}] \\ &= 2.21 \times 10^{-2}\ \mathrm{m/s} \end{aligned}$$

When velocity is increased beyond that for minimum fluidization, the bed's expansion is given by $(\mathrm{e} - \mathrm{e}_{mf})/(1-\mathrm{e})$, where e is the bed voidage at the appropriate expansion. The fluid velocity at any expansion can be estimated from $u/u_{mf} = (\mathrm{e}/\mathrm{e}_{mf})^n$, where $n$ is a function of the Reynolds number and the ratio $d_p/D$. At low Reynolds numbers, $n$ is independent of this ratio and is approximately 4.6. The effect of bed expansion on the mass transfer coefficient can thus be determined. The Reynolds number, here is too high for a constant value of $n$; as a result, calculated values of the electrolyte velocity $u_c$ are too high compared with velocities determined by experiment[25] $u_e$ (Table 2.3). Values of $k_L$ for various expansions have been calculated using $u_e$ in the rearranged Eq. (2.37).

TABLE 2.3. Calculated Values of Electrolyte Velocity and Mass Transfer Coefficient

| % Expansion | e | $u_c$ (m/s) | $u_e$ (m/s) | $k_L$ (m/s) |
|---|---|---|---|---|
| 0 | 0.42 | $2.2 \times 10^{-2}$ | $2.2 \times 10^{-2}$ | $7.5 \times 10^{-5}$ |
| 10 | 0.47 | $3.7 \times 10^{-2}$ | $3.2 \times 10^{-2}$ | $8.3 \times 10^{-5}$ |
| 20 | 0.52 | $5.9 \times 10^{-2}$ | $3.9 \times 10^{-2}$ | $8.3 \times 10^{-5}$ |
| 40 | 0.59 | $1.1 \times 10^{-1}$ | $5.3 \times 10^{-2}$ | $8.9 \times 10^{-5}$ |
| 80 | 0.68 | $2.0 \times 10^{-1}$ | $8.1 \times 10^{-2}$ | $9.1 \times 10^{-5}$ |

Pressure loss per unit length of electrode $\Delta P/L$ equals the apparent weight of the fluidized bed:

$$\Delta P/L = (\rho_s - \rho)(1 - e)g$$

For the lowest mass transfer coefficient, i.e., $7.5 \times 10^{-5}$, is:

$$\Delta P/L = (1800 - 1000)(1 - 0.42) \times 9.81 = 4.5\,\mathrm{kN/m^3}$$

Second, the packed bed: Mass transfer coefficients can be estimated from Eq. 2.36 in the form:

$$k_L = 0.97[d_p/\nu(1 - e)]^{-0.4} Sc^{-2/3} u^{0.6}$$

With a velocity of $2.2 \times 10^{-2}$ m/s:

$$k_L = 0.97[0.002/10^{-6}(1 - 0.42)]^{-0.4}(1300)^{-2/3}(2.2 \times 10^{-2})^{0.6}$$
$$= 3.2 \times 10^{-5}\,\mathrm{m/s}$$

At a velocity of $8.1 \times 10^{-2}$ m/s the mass transfer coefficient is $6.9 \times 10^{-5}$ m/s. These values are lower than those for the fluidized bed but represent respectable mass transfer coefficients.

Third, the open channel: One can use the Chilton–Colburn Eq. (2.18) in the form:

$$k_L = 0.023(d_e/\nu)^{-0.2} Sc^{-2/3} u^{0.8}$$

To apply this equation the hydrodynamic regime must be turbulent, i.e., the minimum velocity must correspond to a Reynolds number of 2000:

$$2000 = d_e u/\nu$$
$$d_e = Wd/(W + d) = (2 \times 0.1 \times 0.03)/(0.1 + 0.03) = 4.62 \times 10^{-2}$$

Hence:

$$u = 2000 \times 10^{-6}/(4.62 \times 10^{-2}) = 4.33 \times 10^{-2}\,\mathrm{m/s}$$

This gives:

$$k_L = 0.023(4.62 \times 10^{-6}/10^{-6})^{-0.2}(1300)^{-2/3}(0.0433)^{0.8} = 1.8 \times 10^{-6}\,\mathrm{m/s}$$

To get the lowest mass transfer coefficient obtained by the fluidized bed promoter, i.e., $7.5 \times 10^{-5}$ m/s, we can calculate the required electrolyte velocity from the rearranged Chilton–Colburn equation:

$$k_L = 2.25 \times 10^{-5} u^{0.8}$$

Hence:

$$u^{0.8} = 7.5/2.25$$

The velocity $u$ is 4.5 m/s. Pressure loss can now be compared to the fluidized bed pressure loss. Since the regime is fully turbulent we can use a friction factor ($C_f$) correlation:

$$C_f = \frac{\Delta P}{L} \frac{d_e}{2\rho u^2} = \frac{0.184}{Re^{0.2}}$$

Rearranging:

$$\begin{aligned} \Delta P/L &= 0.184 \times 2 \times \rho u^{1.8} \times \nu^{0.2}/d_e^{1.2} \\ &= 0.184 \times 2 \times 10^3 \times 4.5^{1.8} \times (10^{-6})^{0.2}/0.0462^{1.2} = 9.8 \text{ kN/m}^3 \end{aligned}$$

The pressure loss in the open channel is significantly higher than in the fluidized-bed.

---

Inert turbulence promoters, whether fluidized-bed or packed-bed, improve mass transport. Before using them, you should check whether this is significantly better than what could be achieved by increasing the electrolyte flow rate. Watch out for a possible increase in pressure drop and decrease in effective electrolyte conductivity, and also for greater reactor costs and problems with electrode fouling unless the characteristics of the turbulence promoter are carefully chosen.

Many inert turbulence promoters are available and only a few have been mentioned. (See also Example 2.4.) Our correlations are specific to one material or system and should not be extrapolated to other types of parameters. This is particularly true for meshes and cloths: the only safe procedure is to determine mass transfer coefficients experimentally using the limiting current technique described in Section 2.3.4.

#### 2.3.5.2. Electroactive Promoters

There are many electrode geometries. Particulate electrodes are discussed in Chapter 5, but there is also the single electrode that has macroscopic contours built into, such as expanded metal, screens, mesh, or fins. There are no general guidelines in choosing among these shapes to enhance mass transfer. Choice depends on the specific requirements of a given reactor, electrode material, hydraulic sealing of the electrode, gas release, flow channel size, and so on. Even when a type has been selected, its detailed properties, e.g., mesh size, will affect the mass transfer coefficient considerably; an appropriate correlation will have to be determined experimentally.

### 2.3.6. Mass Transfer in Two-Phase Flow

Two-phase flow often occurs in electrochemical reactors, usually in a gas/liquid or liquid/liquid form.

Gas can be present either because of evolution from the electrode surface or use of gas sparging in order to enhance mass transfer rates. An example of the latter is agitation of the electrolyte with air during copper winning. For the case when gas evolution influences mass transport to the electrode, Stephan and Vogt[27] looked at experimental results from over thirty authors over a wide range of conditions and found that the following correlation could be recommended, although 15% of the data diverged from it by a considerable margin (> + 100% and − 50%):

$$Sh = 0.93 Re^{0.5} Sc^{0.487} \tag{2.38}$$

On applying phase transfer agents, electrochemical reactions in liquid/liquid systems can attain industrial importance. Overall mass transfer limiting currents have been found to be the linear sum of three effects:

- Mass transfer due to the continuous phase flow $k_L$
- Mass transfer due to the disruption of the boundary layer by the dispersed phase $k'_L$
- Mass transfer enhancement resulting from the extraction of the reacting species from the dispersed phase in intimate contact with the electrode surface $k_e$

This can all be written as:

$$\frac{i}{nF} = k_L c_c + k'_L c_c + k_e c_d \tag{2.39}$$

where $c_c$ and $c_d$ are concentrations of reactant in the continuous and dispersed phase. For further information see, e.g., Lu and Alkire.[28]

## 2.4. ENERGETICS

We shall see in Chapter 6 that energy consumption during industrial synthesis increases in importance with the decrease of the unit price of a particular product. In this section we shall address the fundamental concepts of thermodynamics to the calculation of "minimum" energy requirements and open-cell voltages. We shall consider energy balances and heat transfer requirements, since both can be an essential part of practical reactor design.

### 2.4.1. Voltage Requirements

The minimum electrical energy requirement for an electrochemical reactor occurs when the reaction proceeds at an infinitely slow rate. The cell voltage then corresponds to the sum of the anodic and cathodic equilibrium potentials if the reaction is reversible and to the rest potentials if not. Unfortunately, this situation requires an infinite capital investment, so reactions have to be driven at reasonable rates. The reactor or cell voltage $V^C$ is what will achieve the desired rate; it must therefore overcome all cell resistances, electrodic and ohmic. As we have seen in Chapter 1, $V^C$ is:

$$V^C = E'_A + V_{RA} + V_D + V_{RC} + E'_C \tag{2.40}$$

The electrode potentials are made up from a reversible or rest potential $E$ and an overpotential or overvoltage $\eta$. One can rewrite Eq. (1.28):

$$V^C = E_A + E_C + \eta_A + \eta_C + V_{\text{ohm}} \tag{2.41}$$

Here we shall consider two components of this voltage balance in detail, $(E_A + E_C)$ and $V_{\text{ohm}}$, and merely mention overpotentials (for a reversible charge transfer step) and overvoltages (nonreversible step) resulting from the polarization of electrodes. Overvoltages (overpotentials) are a function of electrode kinetics and electrode material and are discussed in Chapter 3.

### 2.4.2. Open-Circuit Voltage

Open-circuit voltage can be defined thermodynamically through the change in the Gibbs free energy of reaction. Consider again Eq. (1.11):

$$aA + bB + \cdots + ne \rightarrow xX + yY + \cdots$$

The Gibbs free energy change at constant temperature and pressure $(\Delta G)_{T,P}$ equals the sum of the chemical potentials $\mu_j$ of a species $j$:

$$(\Delta G)_{T,P} = \sum n_j \mu_j \tag{2.42}$$

where $n_j$ is the number of moles of the species $j$.

In an electrochemical reaction like Eq. (1.11) the Gibbs free energy change is:

$$(\Delta G)_{T,P} = -nFE \tag{2.43}$$

where $E$ is the open-circuit voltage, $F$ the Faraday, and $n$ the number of electrons involved in the reaction.

Traditionally a negative value of the free energy change is associated with a spontaneous reaction and is a measure of the maximum available work. However, in a typical electrosynthesis $\Delta G$ values are positive and need to be driven by a negative open-circuit voltage $E$. This highlights an advantage of electrochemical synthesis which allows us to drive thermodynamically unfavorable reactions. One use of Eq. (2.43) is that it allows us to calculate the minimum electrolyzing voltage from free energy data.

The chemical potential of a species of activity $a_j$ is related to its standard chemical potential $\mu_j^o$ by

$$\mu_j = \mu_j^o + RT \ln [a_j] \tag{2.44}$$

For simplicity's sake we leave out the indication in Eq. (2.42) for constant temperature and pressure, but the restriction still applies:

$$\Delta G = \sum n_j \mu_j^o + RT n_j \ln [a_j] \tag{2.45}$$

where

$$\sum n_j \mu_j^o = \Delta G^o$$

which is the standard Gibbs free energy at 1 atmosphere and 298 K.

Standard free energy can also be expressed in terms of the standard open-circuit voltage $E^o$:

$$\Delta G^o = -nFE^o \tag{2.46}$$

The open-circuit voltage $E$ can be related to the activities of the participating species by a combination of Eqs. (2.43), (2.45), and (2.46), remembering that $E$ is by definition negative:

$$E = E^o - \frac{RT}{nF} \sum n_j \ln [a_j] \tag{2.47}$$

In principle we are in a position to estimate $E$ either from standard Gibbs free energies or from standard electrode potentials, values of which are readily available in the literature for a range of common compounds. It however can be difficult to obtain activity data, particularly for ionic species. The only solution sometimes is to use concentrations instead of activities, but the result will only be a rough estimate which must be treated with caution. There is a second problem when using Eq. (2.47) to obtain a value for open-circuit voltages. The relationship has been derived on the basis of a reversible system which is in equilibrium. Many systems of industrial interest are not electrochemically reversible (hence the concept of a rest potential and of an overvoltage instead of a reversible potential and overpotential); also, a real electrolytic reactor with diaphragms separating anolyte from catholyte involves liquid junctions potentials which negate the concept of an equilibrium process. Nevertheless, the old engineering adage that a value however inaccurate is better than no value applies.

In engineering calculations one is rarely concerned with systems in their standard states and in consequence one has to use further thermodynamic calculations (see below) to determine the effect of say temperature on the open-cell voltage. The question is whether the required data are available in the literature.

#### 2.4.2.1. The Effect of Temperature on Open-Circuit Voltage

The effect of temperature on the open-circuit voltage can be determined by the application of the Gibbs–Helmholtz equation, written as:

$$\Delta S = -\left[\frac{\partial \Delta G}{\partial T}\right]_P \tag{2.48}$$

where $\Delta S$ is the change in entropy.

Introducing the thermodynamic definition of free energy:

$$\Delta G = \Delta H - T\Delta S \tag{2.49}$$

where $\Delta H$ is the associated heat of reaction of the electrochemical process, one obtains:

$$\Delta H = \Delta G - T\left[\frac{\partial \Delta G}{\partial T}\right]_P \tag{2.50}$$

Substituting Eq. (2.43) in Eq. (2.50):

$$\Delta H = -nF\left\{E - T\left[\frac{\partial E}{\partial T}\right]_P\right\} \tag{2.51}$$

By expressing conditions in the standard state, the temperature coefficient is obtained from Eq. (2.51):

$$\frac{\partial E^o}{\partial T} = \frac{(\Delta H^o/nF + E^o)}{T} \tag{2.52}$$

The coefficient is calculated knowing the identities

$$\Delta H^o = \sum n_j H_j^o \tag{2.53}$$

and

$$\Delta G^o = \sum n_j G_j^o \tag{2.54}$$

where $H_j^o$ and $G_j^o$ are standard heats of formation and standard free energies of formation which can be found in the literature. The standard enthalpy of formation $\Delta H^o$ of any compound is the $\Delta H$ of the reaction by which it is formed from its elements, the reactants and products all being in the standard state. A convenient standard state for a substance is the state in which it is stable at 298.15 K and 1 atm pressure, e.g., oxygen and chlorine as gases. By convention enthalpies of the chemical elements in their standard state are set to zero.

The standard potential $E^o$ at any required temperature is now found by expanding $E^o(T)$ as a Taylor series:

$$E^o = E_{298}^o + (T - 298)\frac{\partial E^o}{\partial T} + \frac{(T-298)^2}{2}\frac{\partial^2 E^o}{\partial T^2} \tag{2.55}$$

For moderate temperature changes the first two terms usually suffice, but if more accuracy is required the third term can be used by differentiating Eq. (2.51):

$$\frac{\partial \Delta H}{\partial T} = nFT\,\frac{\partial^2 E}{\partial T^2} \tag{2.56}$$

Substituting in Eq. (2.56) for $\Delta H$ in terms of the molar heat capacity change $\Delta C_p$[29]:

$$\frac{\partial^2 E}{\partial T^2} = \frac{\Delta C_p}{nFT} \tag{2.57}$$

where $\Delta C_p = \sum n_j C_{p(j)}$, the summation being applied as usual as products-reactants.

The temperature coefficient is also given by:

$$\frac{\partial E}{\partial T} = \frac{\Delta S}{nF} \tag{2.58}$$

The heat absorbed $Q'$ is equal to $T\Delta S = \Delta H - \Delta G$. For an endothermic process which is driven ($\Delta H$ and $\Delta G$ positive), with $\Delta H > \Delta G$, a heat input is required, i.e., $Q'$ is positive.

An alternative method of calculating $E^o$ is based on a rearranged Eq. (2.46):

$$E^o = -\Delta G^o/nF \tag{2.59}$$

$\Delta G^o$ can be found from Eq. (2.49). The variation of $\Delta H^o$ with temperature can be written as:

$$\Delta H^o = \Delta H_1^o + \int_{T_1}^{T} \Delta C_p\, dT \tag{2.60}$$

where subscript 1 denotes the standard state.

The variation of entropy with temperature is defined as:

$$\Delta S^o = \Delta S_1^o + \int_{T_1}^{T} \frac{\Delta C_p}{T}\, dT \tag{2.61}$$

$\Delta S_1^o$ is given by:

$$\Delta S_1^o = \frac{\Delta H_1^o}{T_1} - \frac{\Delta G_1^o}{T_1} \tag{2.62}$$

Combining Eqs. (2.49), (2.60), (2.61), and (2.62):

$$\Delta G^o = \left[1 - \frac{T}{T_1}\right]\Delta H_1^o + \frac{T}{T_1}\Delta G_1^o + \int_{T_1}^{T} \Delta C_p \, dT - T\int_{T_1}^{T} \frac{\Delta C_p}{T}\, dT \qquad (2.63)$$

Thus the standard free energy at any temperature $T$ can be evaluated from thermodynamic data of $H^o$, $G^o$, and $C_p$. Data for $C_p$ are often available as a power series function of temperature.

---

EXAMPLE 2.6. Calculation of Open-Circuit Voltage

THE PROBLEM: Calculate the standard open-circuit voltage of a chlorine electrolyzer when the cathodic reaction is either:

First, the evolution of hydrogen according to:

$$2H_2O + 2e \rightarrow H_2(g) + 2OH^-$$

or

Second, the reduction of oxygen according to:

$$2H_2O + O_2(g) + 4e \rightarrow 4OH^-$$

The chlorine evolution is:

$$Cl^- \rightarrow 0.5Cl_2(g) + e$$

For thermodynamic data, see Table 2.4.

TABLE 2.4. Thermodynamic Data for Examples 2.6, 2.9, and 2.14

| | Standard heat of formation $H^o$ (kJ/mol) | Standard free energy $G^o$ (kJ/mol) | $C_p$ (kJ/mol K) |
|---|---|---|---|
| $OH^-$ (aq) | −230 | −157 | −144 |
| $H_2O$ (l) | −285 | −237 | 75 |
| $Cl^-$ (aq) | −167 | −131 | −136 |
| $H_2$ (g) | 0 | 0 | 29 |
| $Cl_2$ (g) | 0 | 0 | 34 |

THE SOLUTION: Cathodic potentials are calculated as follows. For the evolution of $H_2$:

$$\Delta G^o_{298} = 2G^o_{OH^-} + G^o_{H_2} - 2G^o_{H_2O} = 2 \times -157 + 0 - 2 \times -237$$
$$= +160\,kJ/mol$$

Hence:

$$E^o_C = \Delta G^o/nF = -160 \times 10^3/(2 \times 96500) = -0.829\,V$$

For the reduction of O:

$$\Delta G^o_{289} = 4G^o_{OH^-} - 2G^o_{H_2O} - G^o_{O_2} = 4 \times -157 - 2 \times -237 - 0$$
$$= -154\,kJ/mol$$
$$E^o_C = 154 \times 10^3/(2 \times 96500) = +0.798\,V$$

The anodic standard free energy change is:

$$\Delta G^o_{289} = 0.5G^o_{CL_2} - G^o_{CL^-} = 0.5 \times 0 - (-131) = +131\,kJ/mol$$

Hence:

$$E^o_A = -131 \times 10^3/96500 = -1.358\ V$$

The standard open-circuit voltage is $E^o = E^o_C - (-E^o_A)$. For a cathodic reaction of evolution of H:

$$E^o = -0.829 - 1.358 = -2.187\,V$$

For a cathodic reaction of reduction of O:

$$E^o = +0.798 - 1.358 = -0.560\,V$$

From these values it is clear that the cathodic reduction of oxygen is much more energy-efficient. The search for a suitable cathode material that can perform under industrial conditions is receiving much attention. This is an example of how the adoption of a thermodynamically more favorable (spontaneous) reaction can benefit an industrial synthesis.

EXAMPLE 2.7. Temperature Dependence of the Standard Open-Circuit Voltage of a Chlorine Electrolyzer

THE PROBLEM: Calculate the standard open-circuit voltage at 338 K of the electrolyzer in Example 2.6, assuming that there is a cathodic reaction of the evolution of hydrogen.

THE SOLUTION: The temperature coefficient for the reaction is from Eqs. (2.46) and (2.52):

$$\frac{\partial E^o}{\partial T} = (\Delta H^o - \Delta G^o)/nFT$$

The free energy and enthalpy change for the overall process must equal the sum of the reactions occuring at the two electrodes. One can write, therefore:

$$\frac{\partial E_C^o}{\partial T} = (\Delta H_C^o - \Delta G_C^o)/nFT \qquad \text{and} \qquad \frac{\partial(-E_A^o)}{\partial T} = -(\Delta H_A^o - \Delta G_A^o)/nFT$$

Using the heat of formation data from Example 2.6 for the chlorine reaction:

$$\Delta H_A^o = 0.5H_{\mathrm{Cl_2}}^o - H_{\mathrm{Cl^-}}^o = +167\ \mathrm{kJ/mol}$$

Hence:

$$\frac{\partial(-E_A^o)}{\partial T} = -\frac{(167 - 131) \times 10^3}{1 \times 96500 \times 298} = +1.252 \times 10^{-3}\ \mathrm{V/K}$$

The second temperature coefficient in Eq. (2.55) is given by Eq. (2.57):

$$\frac{\partial^2(-E_A^o)}{\partial T^2} = \frac{-\Delta C_p}{nFT} = \frac{-[0.5 \times 34 - (-136)]}{1 \times 96500 \times 298} = -5.320 \times 10^{-6}\ \mathrm{V/K}$$

Applying Eq. (2.55), the variation of anode potential with temperature becomes:

$$(-E_A^o) = +1.357 - (T - 298)(1.252 \times 10^{-3}) - (T - 298)^2(2.660 \times 10^{-6})\ \mathrm{V}$$

For a 40 K increase in temperature this gives:

$$(-E_A^o) = +1.357 - 0.050 - 0.004 = +1.303\ V$$

For practical purposes the effect of the second coefficient is small.

For the hydrogen evolution reaction:

$$\Delta H^o_C = 2H^o_{OH^-} - 2H^o_{H_2O} = 2(-230 + 285) = +110 \text{ kJ/mol}$$

Hence:

$$\frac{\partial E^o_C}{\partial T} = \frac{(110 - 160) \times 10^3}{2 \times 96500 \times 298} = -8.69 \times 10^{-4} \text{ V/K}$$

Ignoring the second temperature coefficient, the temperature variation of the cathodic potential is given by:

$$E^o_C = -0.829 - (T - 298)(8.69 \times 10^{-4})$$

For an increase of 40 K this gives:

$$E^o_C = -0.829 - 0.035 = -0.864 \text{ V}$$

The minimum electrolyzing voltage at 298 + 40 = 338 K will be:

$$E^o = -0.864 - 1.303 = -2.167 \text{ V}$$

---

EXAMPLE 2.8. Calculation of the Standard Open-Circuit Voltage from the Standard Free Energy of the Reaction

THE PROBLEM: Calculate the standard open-circuit voltage for the electrolyzer in Example 2.7 from the free energy of the overall reaction.

THE SOLUTION: As was shown in Example 2.7, the overall Gibbs free energy for an electrochemical system is the sum of the free energy changes occurring at the individual electrodes. To illustrate this point, consider the reactions occurring in the chlorine electrolysis of Examples 2.6 and 2.7. At the anode we have numerical results for $G^o$ (kJ/mol):

$$2Cl^- \rightarrow Cl_2(g) + 2e \qquad +262$$

At the cathode:

$$2H_2O + 2e \rightarrow H_2(g) + 2OH^- \qquad +160$$

The overall reaction is:

$$2Cl^- + 2H_2O \rightarrow Cl_2(g) + H_2(g) + 2OH^- \qquad +422$$

Using Eq. (2.46), the open-circuit voltage from the overall free energy change is:

$$E^o = \frac{-\Delta G^o}{nF} = \frac{-422 \times 10^3}{2 \times 96500} = -2.187\,\mathrm{V}$$

This is identical with the value obtained in Example 2.6 and follows the convention that a positive free energy change indicates a reaction that has to be driven by an emf.

---

### 2.4.2.2. The Effect of Activity or Concentration on Open-Circuit Voltage

In chlorine electrolysis the open-circuit voltage, according to Eq. (2.47), will be:

$$E = E^o - \frac{RT}{nF} \ln \left[ \frac{a_{Cl_2} a_{H_2} a_{OH^-}^2}{a_{H_2O}^2 a_{Cl^-}^2} \right] \tag{2.64}$$

Equation (2.64) is made up of two expressions involving the standard electrode potentials which are often referred to as the Nernst equations:

$$(-E_A) = (-E_A^o) - \frac{RT}{nF} \ln \left[ \frac{a_{Cl_2}}{a_{Cl^-}^2} \right] \tag{2.65}$$

and

$$E_C = E_C^o - \frac{RT}{nF} \ln \left[ \frac{a_{H_2} a_{OH^-}^2}{a_{H_2O}^2} \right] \tag{2.66}$$

Equations (2.64), (2.65), and (2.66) enable you to calculate open-circuit voltages for practical reactor operating conditions. The problem with using these equations is the attainment of appropriate values of activities for the constituent species. A detailed treatment of activities is beyond the scope of this book; the reader should consult Ref. 30. We present only a brief discussion.

We defined the relative activity of a species $j$ by Eq. (2.44) in terms of its chemical potential $\mu_j$:

$$\mu_j = \mu_j^o + RT \ln (a_j)$$

For any species $j$ in a solution and in equilibrium with its vapor:

$$\mu_j^{\text{sol}} = \mu_j^{\text{vap}}$$

If the vapor behaves as an ideal gas it can be shown[31] that

$$a_j = p_j/P_j^o \quad (2.67)$$

where $p_j$ is the partial pressure of $j$ above the solution and $P_j^o$ is the vapor pressure of pure $j$.

We can define activity in terms of an activity coefficient $\gamma_j$:

$$a_j = \gamma_j x_j \quad (2.68)$$

where $x_j$ is the mole fraction of $j$.

An ideal solution is defined as one where $\gamma_j = 1$, in other words where the solution obeys Raoult's law.

Activities are dimensionless quantities since they are always related to a standard state $a_j^o$ which has a value of unity. For a solvent the standard state is taken as the pure liquid, or pure solid, at 1 atm pressure. For a solute the standard state is that at infinite dilution.

Problems arise when dealing with electrolytic solutions. Ideally, one would like to use activities of individual ions but this is not possible, since there is no way of measuring them. The activitity of, say, $Na^+$ in an electrically neutral solution will be quite different from that in a solution that is positively or negatively charged. We define the activities of an electrolyte in terms of its ions as follows. Consider a salt $MX$ dissociating according to the equation:

$$MX \rightarrow M^+ + X^- \quad (2.69)$$

If we denote the activity of the cation as $a_+$ and that of the anion as $a_-$, we can write for $a_{MX}$, the activity of the salt:

$$a_{MX} = a_+ a_- = a_\pm^2 \quad (2.70)$$

The quantity $a_\pm$, the geometric mean of $a_+$ and $a_-$, is called the mean activity of the ions. This definition is readily extended to more complex ions.[32] We can write again:

$$a_\pm = \gamma_\pm m_{MX} \quad (2.71)$$

where $m_{MX}$ is the molality of the salt. When dealing with the activities of electrolytes it is best to relate the former to molality (moles per unit weight of solution) rather than mole fraction. In Eq. (2.71), $\gamma_{\pm}$ becomes unity at infinite dilution.

Activity coefficients vary with electrolyte composition; often, necessary data are not available. We must then use molalities instead, although our answer may be very inaccurate. Some data on activity coefficients can be found in Ref. 33.

---

EXAMPLE 2.9. Open-Circuit Voltage During Water Electrolysis

THE PROBLEM: Calculate the open-circuit voltage for the electrolytic production of oxygen and hydrogen from a 6 molal potassium hydroxide electrolyte. The temperature is 378 K and the pressure, 15 atmospheres.

For thermodynamic data, see Table 2.5.

$p^{o}_{H_2O}$ and $p_{H_2O}$, the partial pressure of water vapor above pure water and above a KOH solution of molality $m$ respectively, can be calculated from:

$$\ln (p^{o}_{H_2O}) = 37.043 - \frac{6275.7}{T} - 3.4159 \ln T$$

where $T$ is the temperature in K.

$$\ln(p_{H_2O}) = 0.016214 - 0.13802m + 0.19330m^{1/2} + 1.0239 \ln (p^{o}_{H_2O})$$

THE SOLUTION: The constituent reactions are as follows. At the cathode:

$$2H_2O + 2e \rightarrow H_2(g) + OH^-$$

At the anode:

$$2OH^- \rightarrow 0.5O_2(g) + H_2O + 2e$$

TABLE 2.5. Thermodynamic Data

| Gas | $C_p$ (cal/mol K) |
|---|---|
| $H_2$ | $6.95 - 0.196 \times 10^{-3}T + 0.476 \times 10^{-6}T^2$ |
| $O_2$ | $8.27 + 0.258 \times 10^{-3}T - 1.877 \times 10^{5}T^{-2}$ |
| $H_2O$ | $1.593 - 3.883 \times 10^{-3}T + 6.335 \times 10^{-6}T^2$ |

Overall reaction:

$$H_2O \rightarrow H_2(g) + 0.5O_2(g)$$

From the thermodynamic data in Table 2.4 the overall standard free energy change is 237 kJ/mol. Hence the standard open-circuit voltage will be:

$$E^o = \frac{-237 \times 10^3}{2 \times 96500} = -1.228\,\mathrm{V}$$

We will assess the effect of temperature on voltage by both methods described in Section 2.4.2.1.

The first method: From the thermodynamic data of Example 2.6, $H^o = 285\,\mathrm{kJ/mol}$ and the temperature coefficient is:

$$\frac{\partial E^o}{\partial T} = \frac{\Delta H^o - \Delta G^o}{nFT} = \frac{(285 - 237) \times 10^3}{2 \times 96500 \times 298} = 8.346 \times 10^{-4}\,\mathrm{V/K}$$

From Eq. (2.55):

$$E^o_T = -1.228 + (T - 298) \times 8.346 \times 10^{-4}$$

at 378 K (105°C):

$$E^o = -1.161\,\mathrm{V}$$

The second method uses the more fundamental thermodynamic Eq. (2.63), which requires appropriate heat capacity data:

$$\Delta C_p = C_{p(H_2)} + 0.5C_{p(O_2)} - C_{p(H_2O)}$$

Substituting the thermodynamic data and simplifying:

$$\Delta C_p = 9.492 + 3.816 \times 10^{-3}T - 9.385 \times 10^4 T^{-2} - 5.859 \times 10^{-6}T^2$$

The appropriate integrals in Eq. 2.63 become:

$$\int_{T_1}^{T} \Delta C_p \mathrm{d}T = 9.492(T - T_1) + 1.908 \times 10^{-3}(T^2 - T_1^2) + 9.385 \times 10^3(1/T - 1/T_1) - 1.953 \times 10^{-6}(T^3 - T_1^3)$$

and

$$\int_{T_1}^{T} \frac{\Delta C_p}{T} dT = 9.492 \ln \frac{T}{T_1} + 3.816 \times 10^{-3}(T - T_1)$$

$$+ 4.693 \times 10^4 \left[\frac{1}{T_2} - \frac{1}{T_1^2}\right] - 2.930 \times 10^{-6}(T^2 - T_1^2)$$

With temperatures $T = 378$ K and $T_1 = 298$ K we get:

$$\int_{298}^{378} \Delta C_p dT = 9.492 \times 80 + 1.908 \times 10^{-3}(378^2 - 298^2)$$

$$+9.385 \times 10^3(1/378 - 1/298) - 1.953 \times 10^{-6}(378^3 - 298^3)$$

$$= 802.08 \text{ cal/mol}$$

and

$$\int_{298}^{378} \frac{\Delta C_p}{T} dT = 9.492 \ln \frac{378}{298} + 3.816 \times 10^{-3} \times 80$$

$$+ 4.693 \times 10^4 \left[\frac{1}{378^2} - \frac{1}{298^2}\right] - 2.930 \times 10^{-6}(378^2 - 298^2)$$

$$= 2.20 \text{ cal/mol}$$

Substituting in Eq. (2.63) with the appropriate units:

$$\Delta G^o_{378} = \left[1 - \frac{378}{298}\right] 285 + \frac{378}{298} 273 + \frac{802.08 \times 10^{-3}}{4.187} + 378 \frac{2.20 \times 10^{-3}}{4.187}$$

$$= 224 \text{ kJ/mol}$$

$$E^o_{378} = \frac{-224 \times 10^3}{2 \times 96500} = -1.161 \text{ V}$$

There is no significant difference between the values obtained by the two methods.

The open-circuit voltage at 15 atmospheres is calculated from Eq. (2.47):

$$E = -1.161 + \frac{RT}{nF} \ln \left[\frac{a_{H_2}(a_{O_2})^{1/2}}{a_{H_2O}}\right]$$

The activities of hydrogen and oxygen (since they are in separate compartments) are given by:

$$a = P - p_{H_2O}$$

where $P$ is the total pressure and $p_{H_2O}$ the partial pressure of water vapor above the solution.

The activity of water $a_{H_2O}$ is given by Eq. (2.68):

$$a_{H_2O} = p_{H_2O}/p^o_{H_2O}$$

where $p^o_{H_2O}$ is the vapor pressure of pure water at the corresponding temperature.

The equations for $\ln(p^o_{H_2O})$ and $\ln(p_{H_2O})$ give these values:

$$\ln(p^o_{H_2O}) = 37.043 - \frac{6275.7}{378} - 3.4159 \ln 378 = 0.168$$

Hence $p^o_{H_2O} = 1.183$ atm, and:

$$\begin{aligned}\ln(p_{H_2O}) &= 0.016214 - 0.13802 \times 6 + 0.19330 \times 6^{1/2} + 1.0239 \times 0.168 \\ &= -0.166\end{aligned}$$

Hence:

$$p_{H_2O} = 0.847 \text{ atm}$$

Substituting these values into Eq. (2.47):

$$E = -1.161 + \frac{8.31 \times 378}{2 \times 96500} \ln \frac{(15 - 0.847)^{3/2} \times 1.183}{0.847} = -1.091 \text{ V}$$

Thus increasing the pressure from 1 atmosphere to 15 saves 70 mV and if on top of that we work at 105°C instead of 25°C the total saving will be 137 mV. Since in water electrolysis electricity costs are of overriding importance, the saving is worthwhile; this is why most water electrolysis processes work at elevated temperatures and pressures.

### 2.4.3. Parallel Reactions

An important application arising from a knowledge of reversible electrode potentials is the assessment of possible parallel side reactions. In typical electrolytes anions and cations are present, all of which can under suitable conditions be discharged. Evaluation of reversible electrode potentials from a thermodynamic viewpoint will give information about which reactions could occur in the cell. At a cathode it will be the reaction with the least negative potential and at the anode, with the least positive one. Electrolyte composition, (in particular, pH and concentration) and temperature will have a bearing on which reaction is preferred. It should also be remembered that the same ion can be involved in more than one reaction.

---

EXAMPLE 2.10. Parallel Reactions of Metal Deposition and Hydrogen Evolution

THE PROBLEM: Determine appropriate conditions of $pH$ and concentration of cobalt which favor its deposition from an aqueous solution of cobalt sulfate.

Although cobalt deposition proceeds via a complex reaction path involving deposition and discharge of hydroxide species, for this example we shall assume two reactions occurring at the cathode, namely;

$$Co^{2+} + 2e \rightarrow Co \qquad \text{and} \qquad 2H^+ + 2e \rightarrow H_2$$

The standard equilibrium potential for cobalt deposition $E^o_{Co}$ can be taken as $-0.277$ V.

THE SOLUTION: The equilibrium potential for cobalt at 25°C is:

$$E_{Co^{2+}} = E^o_{Co^{2+}} - \frac{RT}{2F} \ln \left[ \frac{a_{Co}}{a_{Co^{2+}}} \right] = -0.277 + 0.01283 \ln (a_{Co^{2+}})\ \text{V}$$

since $a_{Co} = 1$. The equilibrium potential for hydrogen evolution is:

$$E_{H^+} = 0 - \frac{RT}{2F} \ln \left[ \frac{a_{H_2}}{a^2_{H^+}} \right] = 0.02566 \ln (a_{H^+})\ \text{V}$$

since $a_{H^2} = 1$. As $pH = -\log(a_{H^+})$, we can write:

$$E_{H^+} = 0.02566 \times 2.303 \log(a_{H^+}) = -0.0591\, pH\ \text{V}$$

Figure 2.15 shows that the equilibrium potential for hydrogen evolution becomes more negative with increasing $pH$, while the equilibrium potential for cobalt deposition increases with cobalt concentration. At low

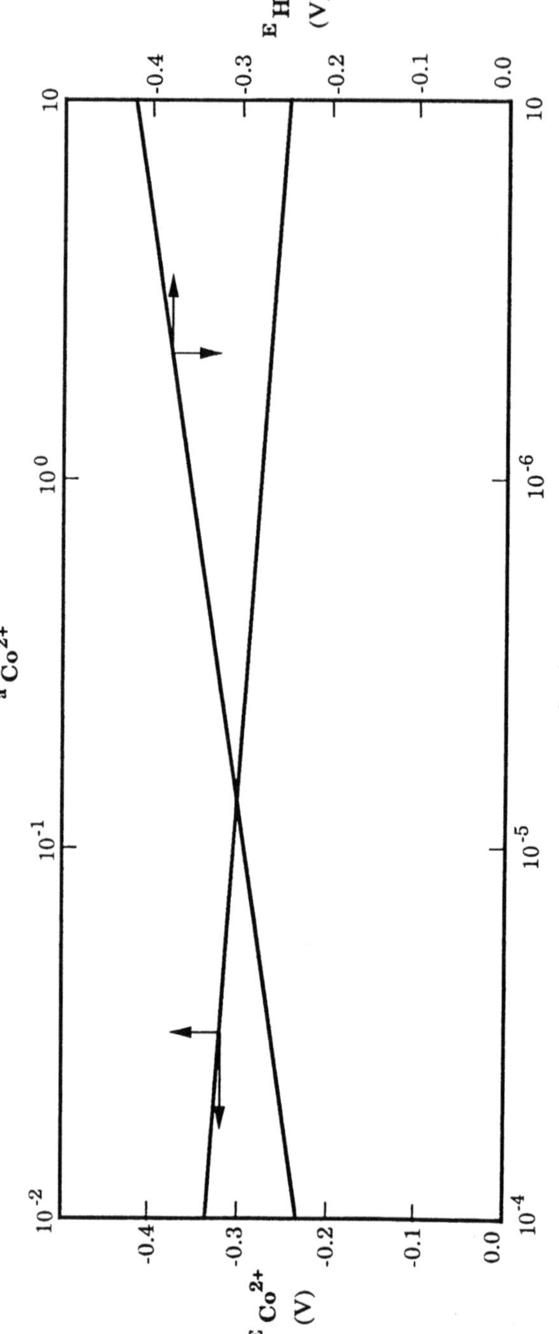


**FIGURE 2.15.** Equilibrium potentials for hydrogen evolution and cobalt deposition.

pH it is almost impossible to deposit cobalt. At a relatively high pH and a sufficiently high cobalt concentration, cobalt deposition becomes the favored reaction.

For example, at pH 4.5, $E_H = -0.266\,V$. Then for cobalt deposition to occur its activity must satisfy the condition (remembering that > means more positive) that $E_{Co^{2+}} > E_H$, or:

$$-0.277 + 0.01283 \ln(a_{Co^{2+}}) > -0.266\,V$$

This means that $a_{Co^{2+}}$ must exceed 2.36.

---

### 2.4.4. Ohmic Voltage Losses

In any synthesis where electricity costs are important it is vital to minimize ohmic voltage losses. In addition to those covered by Eq. 2.41, namely in the electrolyte and the diaphragm, one must add losses caused by resistance in the current leads or bus bars. Of these, electrolyte resistance is usually the most important; it is affected by several parameters.

#### 2.4.4.1. Electrolyte Resistance

Flow of current and voltage drop in the electrolyte are governed by Ohm's law:

$$V_{ohm} = I \times R \tag{2.72}$$

where the resistance $R$ can be defined in terms of the specific resistance or resistivity $\rho$ (ohm-m) or specific conductance or conductivity $\kappa$ (mho/m):

$$R = \rho l/A = l/\kappa A \tag{2.73}$$

where $A$ is the mean cross-sectional area in the direction of the current flow path $l$.

The assumption of a one-dimensional current flow is adequate for most electrolytic reactors that have narrow interelectrode spaces. Appropriate mean cross-sectional area $A$ and equivalent electrolyte length $l$ must be assigned. For electrodes of nonplanar cross section McMullin[34] describes procedures to estimate $A$ and $l$.

It is usual for electrolytic reactors, where the reaction rate is expressed in terms of current density, to denote ohmic losses in the electrolyte in the form:

$$V_{ohm} = i \times l/\kappa \tag{2.74}$$

For reactors with more than one electrolyte, the total voltage loss is:

$$V_{ohm} = i \sum \frac{l}{\kappa} \qquad (2.75)$$

In Eq. (2.75) the minimization of ohmic losses requires low current densities, low interelectrode spaces, and low electrolyte resistances. As we shall see in Chapter 6, current density is a key optimizing parameter in process economics. Usually a minimum value of the interelectrode gap is required, to comply with pumping and mass transfer requirements; mechanical considerations require a sufficiently large interelectrode distance in order to avoid electrical shorting and blockage of flow. The latter is important in gas evolution. This leaves conductivity, which can in principle be "maximized" independently—but in practice there are severe limitations because the conditions which optimize the yield of the product take precedence, particularly for a high-value product. For example it may be necessary to run the reaction in a solvent which imparts a low conductivity to the system. In reactors which accommodate diaphragms or membranes we may be able to maximize the conductivity of the electrolyte which contains the counterelectrode. Electrolyte conductivity usually increases almost linearly with temperature—but, again, the optimum temperature may be dictated by the nature of the reaction.

The effect of electrolyte composition on conductivity is complex. Typically, in relatively dilute solutions conductivity increases with concentration. Above a certain value of concentration the effect is reversed: typical conductivity maxima result. This is true of electrolyte solutions of strong electrolytes such as NaCl, KOH, and HCl; at temperatures approaching 100°C, they have conductivities of 100 mho/m at molarities of 4–6 M. Such conditions are typical of water and chlorine electrolyzers and mean that ohmic voltage losses are of the order of 10 mV/mm interelectrode gap/(kAm$^{-2}$).

There are ways to estimate conductivities of very dilute solutions of concentrations up to 0.2 M, with application in waste water treatment, although in most situations one has to rely on experimental data. Conductivity may be suppressed in solutions containing more than one species by the presence of a second species. Changing the $p$H of a salt solution can have a similar effect.

---

EXAMPLE 2.11. Ohmic Electrolyte Resistance between Two Concentric Cylindrical Electrodes

THE PROBLEM: Estimate the ohmic voltage loss in the electrolyte between two cylindrical electrodes of diameters 0.05 m and 0.065 m and length 0.2 m. The conductivity is 200 mho/m and the cell current is 200 amp.

THE SOLUTION: The system is shown in Fig. 2.16. To estimate the voltage loss the effective resistance of the interelectrode space is estimated from a current balance on a section of thickness $\delta r$. The current going into the segment will be $I$ and that going out will be $I + dI/dr$. Since the current in and out must be equal it follows that $dI/dr = 0$.

We now write Ohm's law in the form:

$$I = -\kappa A(dV/dr) = -\kappa 2\pi r L(dV/dr)$$

Differentiating:

$$\frac{dI}{dr} = \frac{d}{dr}[-\kappa 2\pi r L(dV/dr)] = 0$$

Integrating:

$$\kappa 2\pi r L(dV/dr) = \text{constant} = K_1$$

Integrating again between limits $r_1$ and $r_2$:

$$V = \frac{K_1 \ln[r_1/r_2]}{2\pi L\kappa}$$

From the boundary condition at $r_1$ or $r_2$ we get $K_1 = I$. Therefore the electrolyte resistance $R = \ln[r_1/r_2]/(2\pi L\kappa)]$. The voltage drop $V$ is:

$$\frac{200 \ln[0.065/0.05]}{2\pi \times 0.2 \times 200} = 0.209 \text{ V}$$

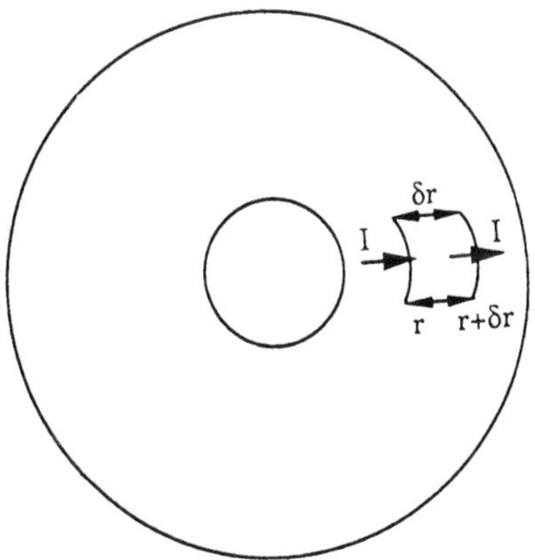


**FIGURE 2.16.** Voltage balance for a cell with cylindrical geometry. $I$ = current; $r$ = cell radius.

### 2.4.4.2. Gas Evolution

The evolution of a gas during an electrolytic reaction produces a dispersion of bubbles in the electrolyte. Since the bubbles have virtually zero electrical conductivity the current flow path becomes restricted and ohmic voltage losses become greater than those for the electrolyte. The usual way to deal with bubbles is to assign to the electrolyte an "effective" conductivity $\kappa_e$ which is typically correlated as a function of gas voidage $f_g$. There are a number of correlations of this type: one of the most reliable is the so-called Bruggeman equation:[35]

$$\kappa_e = \kappa(1 - f_g)^{3/2} \tag{2.76}$$

This correlation (and others) requires no knowledge of bubble size and is applicable to systems with an approximately uniform voidage distribution. Bubbles may cause a considerable effect by accumulating near the vicinity of the gas-evolving electrode and producing an electrolyte layer rich in gas bubbles, the "bubble curtain effect." This layer can be as thick as a few millimeters; if not controlled, it can be a major source of ohmic resistance. The usual way to meet this problem is to induce effective electrolyte circulation and/or to use perforated electrodes. The latter are used in chlorine electrolyzers in order to permit gas release at the rear, away from the potential field.

In designs where gas release can only be effected at the top of the electrolyzer there will be a variation of bubble fraction along the length of the flow channel. This gives rise to effective electrolyte resistance that varies in the direction parallel to the electrode, which will, among other things, influence current distribution. For very low current densities, a first estimate of the ohmic voltage loss in an electrolyzer with a flowing electrolyte can be made from the following expression[36] for its mean conductivity:

$$\frac{\kappa_e}{\kappa} = \frac{(J/2 + 1)}{(J + 1)^2} \tag{2.77}$$

where

$$J = \frac{\kappa}{\kappa_e} \frac{RT \times I}{2 \times nF \times V_g \times P \times W \times d}$$

$I$ being the current, $P$ the atmospheric pressure, $W$ the width of the electrode, $d$ the electrolyte gap, and $V_g$ the rise velocity of the bubbles (as predicted by Stokes's law) augmented by the electrolyte flow.

EXAMPLE 2.12. Electrolyte Voltage Losses during Gas Evolution

THE PROBLEM: Estimate the electrolyte voltage requirement for an electrolyzer with a single gas evolution reaction with a current efficiency of 100%. The rectangular electrodes are 0.1 m wide and 0.25 m long; the electrolyte gap is 0.01 m.

The cell current is 10 amperes. The following data are also available: $V_g = 0.002\,\text{m/s}$, $\kappa = 40\,\text{mho/m}$, $T = 298\,\text{K}$, $P = 10^5\,\text{N/m}^2$, and $n = 2$.

THE SOLUTION: The gas effect parameter $J$ in Eq. 2.77 is:

$$J = \frac{\kappa}{\kappa_e} \frac{8.31 \times 298 \times 10}{2 \times 2 \times 96500 \times 0.002 \times 10^5 \times 0.1 \times 0.01} = 0.32\kappa/\kappa_e$$

Denoting $\kappa/\kappa_e$ by $K$, Eq. 2.77 becomes:

$$1/K = (0.32K/2 + 1)/(0.32K + 1)^2$$

Rearrangement gives a quadratic equation:

$$0.06K^2 + 0.36K - 1 = 0$$

so that $K = 2.08 = \kappa/\kappa_e$. Hence:

$$\kappa_e = \kappa/2.08 = 40/2.08 = 19.2\ \text{mho/m}$$

Voltage loss:

$$\frac{I \times 1}{\kappa_e \times A} = \frac{10 \times 0.01}{19.2 \times 0.25 \times 0.1} = 0.21\ \text{V}$$

### 2.4.4.3. Diaphragm Resistance

Although one tries to use undivided cells, diaphragms are sometimes essential. The voltage loss due to their presence can be greater than 10% of the total voltage loss in the cell.

The effective conductivity $\kappa_d$ of a diaphragm saturated with electrolyte can be predicted from a knowledge of its hydraulic permeability $p$ and its hydraulic radius $m$, from the relationship:[37]

$$\frac{\kappa}{\kappa_d} = \frac{0.272m^2}{p} \tag{2.78}$$

The hydraulic radius is the diaphragm porosity $e$ divided by its specific surface area $a$. The permeability is defined by Darcy's law as:

$$p = \frac{\mathrm{e}^3}{ka^2(1 - \mathrm{e})} \tag{2.79}$$

where $k$ is a constant. Combining Eqs. (2.78) and (2.79) and writing $\mathrm{e}/a$ for $m$:

$$\frac{\kappa_d}{\kappa} = \frac{\mathrm{e}}{0.272k(1 - \mathrm{e})} \tag{2.80}$$

From Eq. (2.80) it is clear that the relative conductivity of a diaphragm is a function of its voidage. Other correlations of the type of Eq. (2.80) exist for specific materials, but it is safest to use manufacturers' data when possible.

#### 2.4.4.4. Resistance of Solid Conductors

This kind of ohmic resistance is associated with that of the bus bars, the conductors of electricity between the rectifier and the cell and between the cells themselves, and that of the electrodes. It is also important to avoid contact resistances. The resistances of electrode materials are varied: for example, copper has a low resistivity of $1.7 \times 10^{-2}$ ohm-m, titanium's is $46 \times 10^{-2}$ ohm-m, and graphite's is $700 \times 10^{-2}$ ohm-m. Voltage losses for all solid conductors can be estimated from Ohm's law.

The importance of low voltage losses is not so much the saving of electricity costs but the avoidance of bad current distribution. This factor, discussed further in Chapter 5, is particularly important for coated metal electrodes, e.g., the so-called dimensionally stable electrodes used in the chloralkali industry which consist of titanium covered by a thin layer (1–2 $\mu m$) of catalyst. Voltage losses are kept low by making sure that the titanium is sufficiently thick.

#### 2.4.4.5. Total Cell Voltage

Returning to Eq. 2.41, many interrelated parameters govern the required values of the open-cell voltage and ohmic resistances. We will mention the third component of the total cell voltage $V^C$, the overpotential or overvoltage—but polarization and electrode kinetics are the subject of Chapter 3.

It is practical to linearize overpotential relationships over a limited current density change about a mean current density $i'$:

$$\eta = \eta(i') + (i - i')\left[\frac{d\eta}{di}\right]_{i=1'} \tag{2.81}$$

Total cell voltage $V^C$ is:

$$V^C = A + Bi \tag{2.82}$$

where $A$ contains the open-cell voltage and the constant part of Eq. (2.81) and $B$ contains all ohmic resistance terms and the slope of the linear overpotential relationship.

---

EXAMPLE 2.13. Voltage Analysis of an Electrowinning Cell

THE PROBLEM: A divided cell is to be used for an electrowinning application. The anticipated range of current densities is 500–3000 A/m$^2$. Show how variations in the composition of the anolyte will affect the total cell voltage if the specific conductance of the anolyte is given by:

$$\kappa = 6 + 25c - 3c^2 \text{ mho/m}$$

where $c$ is the concentration of the anolyte.

The electrolyzer has the following characteristics: Catholyte gap = 0.01 m; anolyte gap = 0.003 m; diaphragm thickness = 0.0005 m; diaphragm relative conductivity $\kappa_d/\kappa = 0.2$; resistance of bus bars and electrodes = $0.4 \times 10^{-3}$ V/Am$^{-2}$; catholyte conductivity = 60 mho/m; open-circuit voltage = 2.19 V; cathode overvoltage $\eta_C = 0.45 + 0.0007i$ V; anode overvoltage $\eta_A = 0.70 + 0.0012i/c$ V.

Assume that the diaphragm contains catholyte; then its effective conductivity is $60 \times 0.2 = 12$ mho/m.

THE SOLUTION: The voltage balance is written according to Eq. (2.82):

$$V^C = 2.19 + 0.45 + 0.70 + 0.0007i + 0.0012i/c$$
$$+ i\left[\frac{0.01}{60} + \frac{0.0005}{12} + \frac{0.003}{6 + 25c - 3c^2}\right] \text{V}$$

Minimum cell voltage is obtained when $dV^C/dc = 0$.

Differentiating and setting the result equal to zero:

$$-\frac{0.0012}{c^2}-\frac{0.003\times(25-6c)}{(6+25c-3c^2)^2}=0$$

Rearranging and solving by iteration, we find that $c = 6.61\,\text{kmol/m}^3$.

The minimum cell voltage $V^C_{min}$ can be given as a function of current density:

$$V^C_{min} = 3.34 + 0.001165i$$

Hence $V^C_{min}$ will vary between 3.92 and 6.82 V for a current density range of 500–3000 A/m$^2$.

---

## 2.5. ENERGY BALANCES

In designing an electrolytic plant it is essential to know accurately the amount of energy consumed in order to supply required sources of heat and to have necessary heat exchanger capacity available for addition or removal of heat. A heat or energy balance has to be performed, using the laws of thermodynamics. Only the outlines can be discussed here: for further study, the reader is referred to a standard book of physical chemistry such as that by W. J. Moore.

### 2.5.1. Thermodynamic Relationships

The first law of thermodynamics states that for a closed system the enthalpy change $\Delta H$ is:

$$\Delta H = \Delta U + \Delta(PV) \tag{2.83}$$

where $\Delta U$ is the change in internal energy and $V$ the volume of the system.

Under constant volume conditions $\Delta(PV) = V\Delta P$, the work due to change of pressure. For electrochemical systems where changes in kinetic and potential energy are small:

$$\Delta H = Q' - W' - (-V\Delta P) \tag{2.84}$$

where $Q'$ is the heat absorbed by the system and $W'$ the work done by the system.

The useful work, i.e., work excluding the work due to the pressure change, is $-W' - (-V\Delta P)$; in an electrochemical system it equals the electrical work. Under reversible conditions it equals the free energy change $\Delta G$, and $Q = T\Delta S$. We have seen in Section 2.4.2.1 that $\Delta S$, the change in entropy, is given by Eq. (2.49), slightly rearranged here as $T\Delta S = \Delta H - \Delta G$. In an electrochemical reaction $\Delta G$ represents minimum electrical energy used to make the reaction proceed. In general $Q'$, the amount of heat absorbed by the system, is:

$$Q' = T\Delta S + nF(V^C - E) \tag{2.85}$$

Depending on the sign of $\Delta S$, $\Delta H$ can be smaller or larger than $\Delta G$. For example, in the electrolysis of water (see Example 2.9), the overall reaction:

$$H_2O \rightarrow H_2 + 0.5O_2 \tag{2.86}$$

is endothermic with $\Delta H = 285\,\text{kJ/mol}$ (S.T.P.). The difference between $\Delta H$ and $\Delta G$ is caused by change in entropy. Heat must be supplied or removed from the system to maintain this balance. Conditions for isothermal operation ($\Delta H = \Delta G$) are referred to as the thermoneutral voltage $E_{tn}$, defined by:

$$E_{tn} = \frac{\Delta H}{nF} \tag{2.87}$$

For water electrolysis this is approximately 1.48 V. Below 1.48 V heat and electricity must be supplied for hydrogen generation. Above 1.48 V, the typical operating condition, heat must be removed for isothermal operation. In water electrolysis, operating temperatures are high so as to capitalize on the lower open-circuit voltage ($E$ falls approximately by 0.08 V for an increase of 100 degrees) and the improved energy efficiency. This is favorable due to the relatively small increase in $\Delta H$ with temperature.

As we said at the beginning of this section, an energy balance of the process is essential. A major user of energy is the electrochemical reactor, which will be considered next.

### 2.5.2. Reactor Energy Balances

Inclusion of a substantial electrical term makes the energy balance, and therefore the calculation of the cooling or heating requirements, somewhat different from that of other chemical plants. Further consideration to the calculation of heat exchanger duties is given in Section 6.5.2.

### 2.5.2.1. Continuous Operation

To see how the electrical term fits into the energy balance let us look at a continuous electrolytic process (Fig. 2.17a) which comprises a cell, streams in and out (there may be more than two if a gas is evolved), a cooler (which may be an interstage cooler if it is followed by another reactor), and a power source.

An energy balance can be based on Eq. (2.85):

$$H_2 - H_1 = W + E'' - Q - r_1\Delta H_{R1} - r_2\Delta H_{R2} - \cdots \tag{2.88}$$

which relates the change in enthalpy flow between the inlet and outlet streams $H_2 - H_1$ with the duty on the coolers $Q$, the power supplied via the pumps W, the electrical energy supplied to the cell $E''$, the rate of reaction

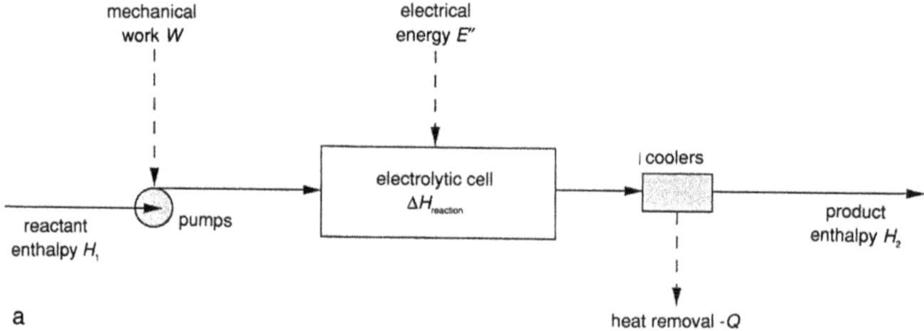


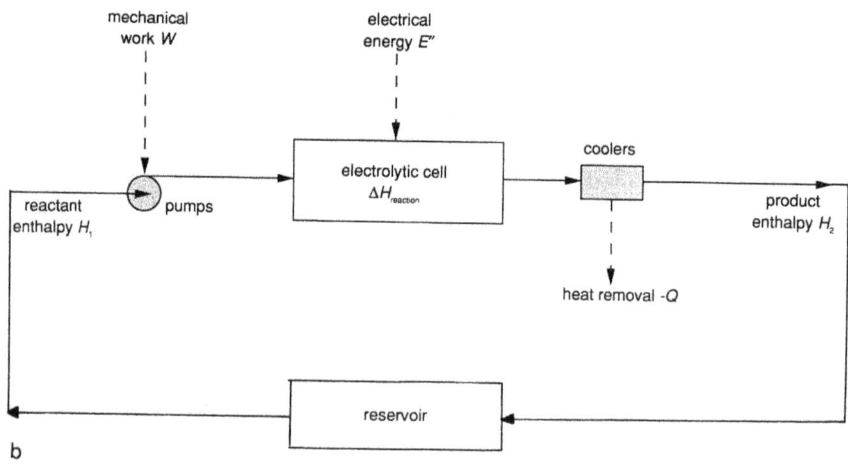


**FIGURE 2.17.** Process energy balance.

$r$ and the enthalpy of reaction per kmol $\Delta H_R$. $Q$ is the heat removed from the system and $W$ the work done on the system.

The electrical energy will be removed in the coolers, other than that required by the change in enthalpy due to reaction and temperature.

### 2.5.2.2. Batch Operation

Let us look at the cooling requirements for batch operation. The flow sheet in Fig. 2.17b now comprises a cell, a reservoir, a circulating pump, and of course a power supply.

Batch operation is essentially unsteady-state, so the energy balance is drawn up with respect to conditions obtaining at some instant $t$. If the reactor is run isothermally the cooling requirements at $t$ will be given by:

$$Q = W + E'' - r_1 \Delta H_{R1} - r_2 \Delta H_{R2} - \cdots \tag{2.89}$$

where the symbols have the same significance as before.

In Eq. (2.89) we have taken the value of $H_2 - H_1$ as zero. This is true when we take a datum line for the enthalpies. However, under those circumstances the values of $r_1 \Delta H_{R1}$ and $r_2 \Delta H_{R2}$ must be those for the temperature of the datum line and not of the reactor.

The average cooling rate of the batch $\bar{Q}$ will be given by integration with respect to time:

$$\bar{Q} = \frac{1}{t_b} \int_0^{t_b} Q \, dt \tag{2.90}$$

where $t_b$ is the total batch time.

---

EXAMPLE 2.14. Estimation of the Heat Load on the Heat Exchangers of an Organic Electrosynthetic Plant

THE PROBLEM: To get an idea of the size of heat exchangers required, an approximate energy balance has to be carried out for a plant producing compound $B$ according to the reaction:

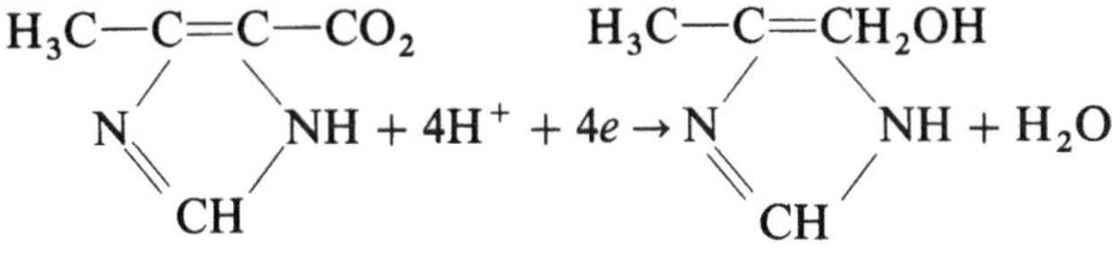

```
H3C—C=C—CO2            H3C—C=CH2OH
    /    \                  /    \
   N      NH + 4H+ + 4e → N       NH + H2O
    \\   /                  \\   /
     CH                      CH
  compound A             compound B
```

The plant operates batchwise, each batch having a three-day running time (Table 2.6).

THE SOLUTION: Using Eq. (2.88), the energy balance can be written as:

$$H_2 - H_1 = -Q + W + E'' - \Delta H_{reaction}$$

The reaction in the cell is mainly the breakdown of water according to the reaction:

$$H_2O \rightarrow H_2(g) + 0.5O_2(g)$$

and from the thermodynamic data in Table 2.4 (Example 2.6), $\Delta H_R$ for this reaction is $285 \times 10^3$ kJ/kmol. Since 1 kW = 3600 kJ/h, $\Delta H_{reaction}$ is:

$$\Delta H_{reaction} = \frac{32 \times 285 \times 10^3}{72 \times 3600} = 35.2 \text{ kW}$$

Since this is only a preliminary estimate of plant size, we shall add 10% for the cathodic reaction:

$$\Delta H_{reaction} = 35.2 + 3.5 = 38.7 \text{ kW}$$

$$E'' = I \times V^C = 3000 \times 8.3 \times 6 = 149.4 \text{ kW}$$

TABLE 2.6. Organic Electrosynthetic Plant Conditions

| | |
|---|---|
| *Process Conditions:* | |
| Feed composition | 25% v/v $H_2SO_4$ + 10% of compound A in aqueous solution |
| Current density | 3000 A/m$^2$ |
| Cell voltage | 6 V |
| Amount of water/batch | 32 kmol |
| Catholyte flow rate | 5.2 l/s |
| Specific heat of catholyte | 3 kJ/kg °c |
| Density of catholyte | 1.15 kg/l |
| Overall heat transfer coefficient | 0.57 kW/m$^2$ °C |
| Temperature | 55°C |
| Pressure | Atmospheric |
| *Plant Items:* | *Number:* |
| Filterpress-type cell with a total cathode area of 8.3 m$^2$ | 1 |
| 6-kW electrolyte pump | 2 |

Using 80% for the pumping efficiency gives $W$;

$$W = 0.8 \times 2 \times 6 = 9.6 \text{ kW}$$

Taking a datum for the enthalpy at 55°C, $H_1$ and $H_2$ are zero. Hence:

$$Q = W + E'' - \Delta H_{reaction} = 120 \text{ kW}$$

This duty will be met by inserting identical coolers with a duty $Q$ of 60 kW each in the catholyte and anolyte circuits.

The required heat transfer area of the cooler must be determined on the basis of the catholyte stream. The temperature rise $\Delta T$ in the catholyte stream (flow rate $U_L$) is:

$$\Delta T = \frac{Q}{\rho U_L C_p} = \frac{60}{1.15 \times 5.2 \times 3} = 3.3°\text{C}$$

The temperatures in the cooler (assuming cooling water is available at 25°C) are:

Inlet catholyte $T_{iC} = 58.3°\text{C}$

Outlet catholyte $T_{oC} = 55.0°\text{C}$

Inlet cooling water $T_{iW} = 25.0°\text{C}$

Outlet cooling water $T_{oW} = 35.0°\text{C}$

Using the basic heat transfer relationship:[38]

$$Q = hA\Delta T_m$$

where $h$ is the overall heat transfer coefficient, $A$ the heat transfer area of the cooler, and $\Delta T_m$ the mean temperature difference.

The log mean temperature difference $\Delta T_{lm}$ in the cooler (assuming countercurrent flow) is:

$$\Delta T_{lm} = \frac{(T_{oC} - T_{iW}) - (T_{iC} - T_{oW})}{\ln[(T_{oC} - T_{iW})/(T_{iC} - T_{oW})]} = \frac{30 - 23.3}{\ln(30/23.3)} = 26.5°\text{C}$$

Hence,

$$A = Q/(U_L \times \Delta T_{lm}) = 60/(0.57 \times 26.5) = 3.97 \approx 4 \text{ m}^2$$

An identical cooler will be used for the anolyte stream.

### 2.5.2.3. Energy Balance for Aluminum Production[39]

Worldwide, the electrolytic production of aluminum is the second largest consumer of electrical energy for electrochemical processes. It is a molten salt process at high temperature ($\approx 1250\,K$). Consumable anodes are used and the overall reaction is:

$$2Al_2O_3 + 3C \rightarrow 4Al + 3CO_2 \tag{2.91}$$

The use of carbon anodes is primarily due to the lower open-circuit voltage of Eq. 2.91 compared to the straightforward reduction of $Al_2O_3$ to aluminum and oxygen. Equation (2.91) represents an oversimplified view of aluminum production; a number of side reactions occur. One of the more important is the formation of CO:

$$2Al_2O_3 + 6C \rightarrow 4Al + 6CO \tag{2.92}$$

and arises from the recombination of anodic and cathodic products. Anode gas from the process contains some 10–30% CO with 90–70% $CO_2$. Current efficiency of aluminum production is typically 90%.

If we refer to the schematic diagram of an aluminum cell (Figure 2.18) the need for an energy balance will become apparent. Aluminum deposited in liquid form on the carbon cathodes at the base of the cell and the molten electrolyte must be maintained in liquid form at 1250 K. A crust of frozen

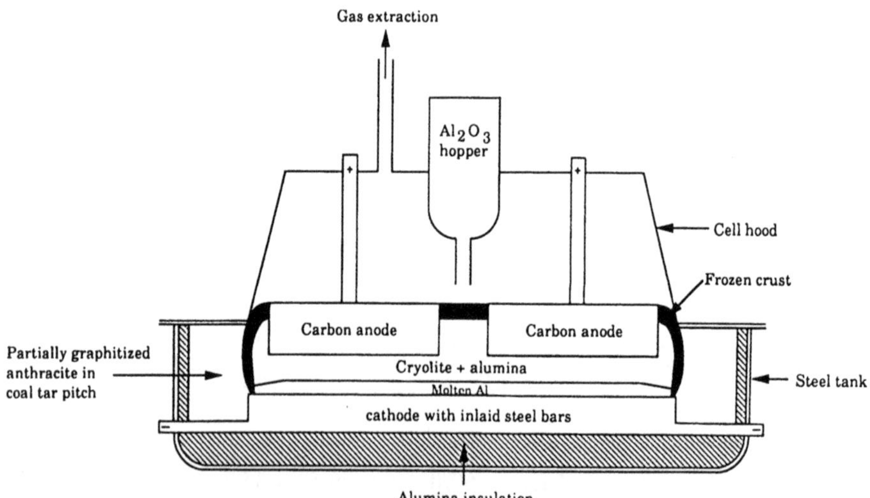


**FIGURE 2.18.** Cell for the extraction of aluminum (after Pletcher[39]).

electrolyte must also be formed to protect the cell housing from erosion. The carbon electrodes serve as a source of internal ohmic heat generation and as a means of heat conduction out of the cell. Overall heat conduction through the cell walls is crucial in determining their thickness, which must be such that with a temperature difference, $\Delta T$, of 950 K (1250 − 300) between the temperatures inside and outside the cell, the quantity of heat $nFV^C - \Delta H$ flows from the cell. This balance of heat is one of the essential factors which determine the construction of a cell for aluminum production and is illustrated in Example 2.15.

---

EXAMPLE 2.15. Heat Balance for an Aluminum Cell

THE PROBLEM: Determine the wall thickness that will keep the fused salt temperature at 1250 K.

Equations (2.91) and (2.92) are assumed to be the only ones in the cell and are rewritten as:

$$Al_2O_3 + xC \rightarrow 2Al + yCO_2 + 2CO \tag{2.93}$$

The standard free energies of Eqs. (2.91) and (2.92) are 6535 kJ/mol and 608.0 kJ/mol at 1250 K. The free energy change for Eq. (2.93) depends on the relative amounts of $CO_2$ and CO produced (Table 2.7).

(In this and some previous worked examples specific heat data in the literature are in non-SI units. We will convert to SI units only at a later stage).

THE SOLUTION: First, calculation of the cell voltage $V^C$. $V_{ohm}$, the voltage drop through the electrolyte, is:

$$V_{ohm} = \frac{10^5 \times 0.04}{28 \times 200} = 0.71\ \text{V}$$

The cell voltage is:

$$V^C = 1.2 + 0.5 + 0.1 + 0.5 + 0.5 + 0.71 = 3.51\ \text{V}$$

There is an additional voltage loss in the external electrical conductors of the order of 0.1 to 0.2 V, but this does not contribute to generation of heat in the cell.

TABLE 2.7. Aluminum Cell Heat Balance: Physical Data and Process Conditions

| Species | Molar mass | $H^o_{298}$ (kcal/mol) | $C_p$ (kcal/mol K) |
|---|---|---|---|
| Physical Data: | | | |
| $Al_2O_3$ | 101.94 | −399.0 | |
| Al | 26.97 | 0 | solid: $4.80 + 3.22 \times 10^{-3}T$<br>liquid 7.00 |
| C | 12.00 | 0 | |
| $CO_2$ | 44.01 | −94.0 | 10.6 |
| CO | 28.01 | −26.4 | $6.60 + 1.20 \times 10^{-3}T$ |
| Melting point of aluminum | | 933 K | |
| $\Delta H$ melting | | 2.52 kcal/mol | |
| Thermal conductivity of graphite | | 18.4 kJ/h m K | |
| Thermal conductivity of insulating alumina | | 1.46 kJ/h m K | |
| Equilibrium potential for reaction (2.93) | | −1.2 V | |
| Specific electrical conductivity of melt allowing for anode gas bubbles | | 200 mho/m | |
| Process Conditions: | | | |
| Current | $10^5$A | | |
| Current efficiency | 90% | | |
| Anode overvoltage | 0.5 V | | |
| Cathode overvoltage | −0.1 V | | |
| Ohmic voltage losses: | | | |
| Anode + connectors + contacts | | +0.5 V | |
| Cathode + lining + collector bars + contacts | | −0.5 V | |
| Anode spacing | | 0.04 m | |
| Temperature of melt | | 1250 K | |
| Temperature of air surrounding the cell | | 298 K | |

Second, the energy balance is written as:

$$Q = V^C \times I \times t - \Delta H$$

where:

$$\Delta H = \lambda \Delta H^o_{298} + \int_{298}^{1250} n_j C_p \, dT$$

$\lambda$ being the number of moles of alumina reacting per hour in the cell.

With a $CO/CO_2$ ratio of 0.25, Eq. (2.93) becomes:

$$Al_2O_3 + (5/3)C \rightarrow 2Al + (4/3)CO_2 + (1/3)CO$$

Hence:

$$\Delta H^o_{298} = 2H^o_{Al} + 4/3H^o_{CO_2} + 1/3H^o_{CO} - H^o_{Al_2O_3} - 5/3H^o_C$$

Since $10^5 I = 3600 \times 10^5$ coulombs (ampere seconds) per hour, with a current efficiency of 90% and the requirement of 6F per mole of $Al_2O_3$ we have:

$$\lambda = \frac{0.9 \times 3600 \times 10^5}{6 \times 0.965 \times 10^5} = 560 \text{ moles/h}$$

Therefore $n_{Al_2O_3} = 560$, $n_C = 933$, $n_{Al} = 1120$, $n_{CO_2} = 747$, and $n_{CO} = 187$.

Now:

$$\Delta H^o_{298} = 560(-4/3 \times 94.0 - 1/3 \times 26.4 + 399) = 148{,}325 \text{ kcal/h}$$

The term:

$$\int_{298}^{1250} n_j C_p \, dT$$

is evaluated in the following way. For $CO_2$:

$$n_j \int_{298}^{1250} C_p \, dT = 747[10.6(120 - 298)] = 7538 \text{ kcal/h}$$

For CO the integral becomes:

$$187[6.60(1250 - 298) + 0.00102(1250^2 - 298^2)/2] = 1316 \text{ kcal/h}$$

For Al, $\Delta H$ melting must be included and the integral becomes:

$$1120[4.8(933 - 298) + 0.00322(933^2 - 298^2)/2 + 2520 + 7(1250 - 933)]$$
$$= 10{,}132 \text{ kcal/h}$$

Hence:

$$\Delta H = 148{,}325 + 7538 + 1316 + 10{,}132 = 1.673 \times 10^5 \text{ kcal/h}$$
$$= 7.00 \times 10^5 \text{ kJ/h}$$

The electrical energy supplied to the cell is:

$$V^C \times I \times t = 3.51 \times 10^5 \times 3600 \times 10^{-3} = 12.64 \times 10^5 \text{ kJ/h}$$

Hence:

$$Q = (12.64 - 7.00) \times 10^5 = 5.64 \times 10^5 \text{ kJ/h}$$

This is how much heat has to be transferred per hour to the surroundings so that the temperature of the fused salt remains at 1250 K. Heat is transferred through the cell walls as well as through the crust of solidified melt at the surface of the bath. To illustrate the method we shall restrict our calculation to one of the walls—the remaining calculations would be similar.

Third, calculation of the cell floor thickness: We shall consider heat transfer through the floor of the cell and assume that air circulates below the floor by natural convection, as is true in modern aluminum plants. Starting from the premise that 14% of the total heat is transferred through the floor (this corresponds approximately to the situation in practice), the heat flow will be:

$$0.14 \times 5.64 \times 10^5 = 0.79 \times 10^5 \text{ kJ/h}$$

Since the floor area is 28 m$^2$, the heat flux $Q_F$ will be:

$$Q_F = (0.79 \times 10^5)/28 = 2820 \text{ kJ/h m}^2$$

Heat transfer through the floor of the cell is shown schematically in Fig. 2.19. The heat is transferred by conduction through the graphite and insulating material (heat flux $Q_c$), by conduction and natural convection from the steel mantle to the air (heat flux $Q_n$), and by radiation from the steel mantle to the air (heat flux $Q_r$). At steady state:

$$Q_F = Q_c = Q_n + Q_r = 2820 \text{ kJ/h m}^2 \tag{2.94}$$

The temperature of the melt $T_M$ is 1250 K and that of the air $T_A$ is 298 K. These are determined by process conditions. In writing equations for the three heat fluxes we leave $T_W$, the temperature of the steel mantle, an unknown quantity which will be calculated later.

First, assuming any temperature differences in the steel mantle to be negligible, the heat flux due to conduction through the carbon cathode and

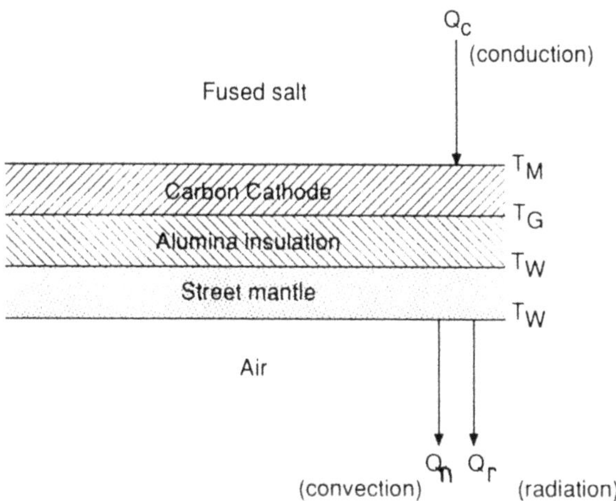


**FIGURE 2.19.** Temperatures during heat transfer through the floor of an aluminum cell. $T_M$ = temperature of alumina melt; $T_G$ = temperature of boundary between graphite cathode and solid alumina; $T_W$ = temperature of steel mantle of aluminum cell; $Q_c$ = heat flux by conduction; $Q_n$ = heat flux by convection, $Q_r$ = heat flux by radiation.

the insulating alumina is:

$$Q_c = k_G \frac{(T_M - T_G)}{d_G} = k_I \frac{(T_G - T_W)}{d_I} = h(T_M - T_W) \tag{2.95}$$

where $k_G$ and $k_I$ are the thermal conductivities of the graphite and insulating alumina, $d_G$ and $d_I$ their respective thicknesses, and $h$ the overall heat transfer coefficient. The thickness of the graphite is governed by the need to minimize ohmic resistance in order to avoid too much dissipation of electrical energy. A typical value for $d_G$ is 0.4 m.

From Eq. (2.95) one obtains the usual relationship for the overall heat transfer coefficient:

$$\frac{1}{h} = \frac{d_G}{k_G} + \frac{d_I}{k_I} \tag{2.96}$$

Second, we can write:

$$Q_n = h\Delta T = h(T_W - T_A) \tag{2.97}$$

For laminar natural convection of air at a horizontal plate heated from above, the heat transfer coefficient is given by:

$$h = 2.107(\Delta T/L)^{1/4}\ \text{kJ/h m}^2\ \text{K} \tag{2.98}$$

where $L$ is a characteristic length. In our case, assuming a square cell, $L = (28)^{1/2} = 5.3$ m. Combining Eqs. (2.97) and (2.98):

$$Q_n = 2.107(T_W - 298)^{5/4}/5.3^{1/4}\ \text{kJ/h m}^2 \tag{2.99}$$

Third, heat transfer by radiation obeys the relationship:

$$Q_r = 16.72[(T_W^4 - T_A^4)/100] = 16.72[(T_W^4 - 298^4)/100]\ \text{W/m}^2 \tag{2.100}$$

The unknown temperature of the outer cell wall $T_W$ can be obtained by combining Eqs. (2.94), (2.99), and (2.100) and determining $T_W$ by trial and error:

$$2820 = 1.389(T_W - 298)^{5/4} + \frac{0.1672}{3.6 \times 10^6}(T_W^4 - 298^4)$$

$T_W$ is 472 K, which is now used to calculate $d_I$ from Eq. (2.95) and (2.96):

$$d_I = k_I\left[\frac{T_M - T_W}{Q_c} - \frac{d_G}{k_G}\right] = 1.46\left[\frac{1250 - 472}{2890} - \frac{0.4}{18.4}\right] = 0.371\ \text{m}$$

This is the required thickness of the cell floor. Similar calculations can be made for the remaining walls of the cell. The energy consumption of aluminum production is of the order of 15 kWh per kg of aluminum. Since 60% of this energy is dissipated as heat, aluminum production does not demonstrate the best of energy utilization.

---

## 2.6. CONCLUDING REMARKS

Using thermodynamic analysis of electrochemical systems we have been able to perform energy accountancy on processes, to calculate parameters such as open-circuit voltage, and to get an estimate of heat exchanger size. A phenomenon which often becomes of overriding importance is polarization, a rate process involving charge transfer. This, combined with the rate of chemical steps and the rate of mass transfer discussed at the beginning of this chapter, enables us to produce a reaction model as described in Chapter 3.

## REFERENCES

1. Sharpe, G. J., 1967, *Fluid Flow Analysis*, Heinemann.
2. Treybal, R. E., 1980, *Mass Transfer Operations*, McGraw-Hill, New York, p. 107.
3. Crank, J., 1956, *The Mathematics of Diffusion*, Oxford University Press, p. 2.
4. Treybal, p. 68.
5. Sharpe, p. 262.
6. Rousar, I., Micka, K., and Kimla, A., 1986, *Electrochemical Engineering I*, Elsevier Scientific Publishing Co., New York, parts A–C, p. 27.
7. Pickett, D. J., 1979, *Electrochemical Reactor Design*, Elsevier Scientific Publishing Co., New York, p. 121.
8. Pickett, p. 131.
9. Coulson, J. M., and Richardson, J. F., 1977, *Chemical Engineering*, Vol. 1, Pergamon Press, p. 300.
10. Ross, T. K., and Wragg, A. A., 1965, "Electrochemical mass transfer studies in annuli," *Electrochim. Acta*, **10:** 1093–1106.
11. Taylor, G. I., 1923, "Stability of a viscous liquid contained between two cylinders," *Philos. Trans. R. S. London*, **A223**: 289–343.
12. Schlichting, H., 1968, *Boundary Layer Theory*, McGraw-Hill (1968).
13. Eisenberg, M., Tobias, C. W., and Wilke, C. R., 1954, "Ionic mass transfer and concentration polarisation at rotating electrodes," *J. Electrochem. Soc.*, **101**: 306–320.
14. Barz, R., Bernstein, C., and Vielstich, W., "On the investigation of electrochemical reactions by the application of turbulent hydrodynamics," in *Advances in Electrochemistry and Electrochemical Engineering*, Vol. 13 (H. Gerischer and C. W. Tobias, eds.), John Wiley & Sons, New York, pp. 261–353.
15. Rousar et al. p. 265.
16. Fahidy, T. Z., 1985, *Principles of Electrochemical Reactor Analysis*, Elsevier Scientific Publishing Co., New York, p. 81.
17. Wragg, T., 1977, "Applications of limiting diffusion current technique in chemical engineering," *Chem. Eng.*, 39–44, 49.
18. Goodridge, F., Pearse, B., Plimley, R. E., and Taama, W. M., unpublished.
19. Youn, J. H., 1988, *Mass Transfer Studies for the Design of a Thin-Gap Electrolytic Cell*, Ph.D. Thesis, University of Newcastle upon Tyne, U.K.
20. Letord-Quemere, M. M., Coeuret, F., and Legrand, J., 1988, "Mass transfer at the wall of a thin channel containing an expanded turbulence promoting structure," *J. Electrochem. Soc.*, **135**: 3063–3067.
21. Carlson, L., Sandegren, B., and Simonsson, D., 1983, "Design and performance of a modular, multi-purpose electrochemical reactor," *J. Electrochem. Soc.*, **130:** 342–346.
22. Goodridge, F., Mamour, G. M., and Plimley, R. E., 1986, "Mass transfer rates in baffled electrolytic cells," *Int. Chem. Eng. Symp. Ser.* **98**: 61–71.
23. Robertson, P. M., Schwager, F., and Ibl, N., 1975, "A new cell for electrochemical processes," *J. Electroanal. Chem.*, **65:** 883–900.
24. Lopez-Cacicedo, C. L., 1975 "Recovery of metals from rinse waters in chemelec electrolytic cells," *Trans. Inst. Met. Finish.*, **55:** 74–77.
25. Nassif, D. V., 1980, *Mass Transfer Studies in Fluidised Bed Cells*, Ph.D. thesis, University of Newcastle upon Tyne, U.K.
26. Coulson and Richardson, vol. 2, p. 233.
27. Stephan, K., and Vogt, H., 1979, "A model for correlating mass transfer data at gas evolving electrodes," *Electrochim. Acta*, **24:** 11–18.
28. Lu, R. and Alkire, R. C., 1989, "Mass transfer in parallel plate electrolysers with two phase liquid–liquid flow," *J. Electrochem. Soc.*, **131:** 1059–1067.

29. Moore, W. J., 1972, *Physical Chemistry*, Longmans, p. 67.
30. Moore, pp. 233, 300.
31. Moore, p. 237.
32. Moore, p. 440.
33. Harned, H. S., and Owen, B. B., 1950, *The Physical Chemistry of Electrolytic Solutions*, Reinhold Publishing.
34. MacMullin, R. B., 1963, "The problem of scale-up in electrolytic processes," *Electrochem. Technol.*, **1:** 5–17.
35. Bruggeman, D. A. G., 1935, "Berechnung verschiedener physikalischen Konstanten von heterogenen Substanzen," *Ann. Phys.*, **24:** 636–642, 665–679.
36. Pickett, p. 349.
37. MacMullin, R. B., and Muccini, G. A., 1956, "Characteristics of porous beds and structures," *Am. Inst. Chem. Eng. J.*, **2:** 393–403.
38. Coulson and Richardson, p. 167.
39. Pletcher, D., 1984, *Industrial Electrochemistry*, Chapman & Hall, p. 115.

# Chapter 3

# Rate Processes and Reaction Models

## 3.1. RATE PROCESSES

As will be shown in Chapter 6, process modeling is concerned with the construction of a mathematical description of a chemical process in terms appropriate to a specific purpose. To achieve this, it is necessary to obtain a reaction model which expresses the rate of reaction as a function of relevant variables. This requires, ideally, knowledge of the individual reaction steps, the rate at which they occur, and the mass transport rates for the species in the reaction. But often, detailed quantitative kinetic data are not available and not easy to determine. Semiempirical relations may have to be used.

To avoid having to refer continually to other texts, and to ensure that the reader appreciates that the techniques used in the formation of reaction models are common to all branches of reaction engineering, this chapter begins by introducing some basic concepts of reaction kinetics. (For a more detailed treatment the reader should consult Refs. 1–4.) We will then derive some simple kinetic relationships and discuss reaction models and experimental methods of obtaining numerical values for the kinetic constants involved. The chapter concludes by giving two examples of reaction models, one for the reduction of hexavalent uranium to a four-valent state and the other for electrosynthesis of *p*-anisidine from nitrobenzene.

### 3.1.1. Elementary and Overall Reactions

For a reaction to occur the reacting molecules must come into contact or, in other words, collide. We can visualize $A$ colliding with $B$ to give $C$:

$$A + B \rightarrow C \tag{3.1}$$

or two molecules of $A$ colliding to give $B$:

$$A + A \rightarrow B \tag{3.2}$$

We regard such elementary reactions as real events.

It is improbable that an elementary reaction will require the collision of four, five, or more molecules at one instant with the correct orientation to produce product. Reactions of the type

$$4H^+ + 2I^- + 2NO_2 \rightarrow I_2 + 2NO + 2H_2O \tag{3.3}$$

are thought to progress through a number of stages or steps, each step being an elementary reaction. The summation of these steps, which might give rise to an equation such as Eq. (3.3), is called the overall reaction.

### 3.1.2. Kinetics of Elementary Reactions

#### 3.1.2.1. Velocity Constant

Let us consider the elementary reaction:

$$aA + bB + cC \rightarrow \text{products} \tag{3.4}$$

Since an elementary reaction takes place in a single step, $a$ molecules of $A$, $b$ molecules of $B$, and $c$ molecules of $C$ must collide simultaneously. The probability of such an event is proportional to $c_A^a c_B^b c_C^c \cdots c_j$, where $c_j$ represents the concentration of a given species $j$. We can therefore write:

$$r_A = k_A c_A^a c_B^b c_C^c \tag{3.5}$$

where the constant of proportionality $k_A$ is called the rate constant or velocity constant of the reaction with reference to component $A$. The quantity $r_A$ is called the rate of reaction with respect to $A$; for homogeneous reactions it can be defined as the rate of consumption of reactant $A$ per unit volume of reaction mixture.

It is unlikely that an elementary reaction will involve more than three molecules: in general, $a + b + c \leqslant 3$.

The exponents in the rate law for an elementary reaction can be determined from the stoichiometric coefficients of the chemical equation. However, this is not true of overall reactions, the rate laws for which must be determined empirically.

### 3.1.2.2. Reversibility and Irreversibility

If a reaction can proceed to an equilibrium in which reactants and products are present the reaction is said to be reversible. It has been found that the rate of such reactions must be expressed as the difference between two terms, each of which contains a velocity constant. Theory regards the positive and negative terms as the individual rates of the forward and backward processes respectively. For an elementary reaction:

$$A + B \rightleftharpoons C + D \tag{3.6}$$

$$r_1 = kc_A c_B \tag{3.7}$$

$$r_{-1} = k' c_C c_D \tag{3.8}$$

so that the net rate of reaction $r$ will be:

$$r = r_1 - r_{-1} = kc_A c_B - k' c_C c_D \tag{3.9}$$

At equilibrium, since $r = 0$, it follows from Eq. (3.9) that $r_1 = r_{-1}$, or:

$$kc_A c_B = k' c_C c_D \tag{3.10}$$

giving:

$$K = (c_C c_D)/(c_A c_B) = k/k' \tag{3.11}$$

where $K$ is the equilibrium constant for the reaction. At this point (see Section 3.1.5.2), kinetics and thermodynamics are in agreement.

All reactions are to some degree reversible, although the equilibrium constant may be so large that the presence of reactants at equilibrium is negligible, and the reaction in effect goes to completion. Mathematically this condition is expressed by:

$$1/K \rightarrow 0 \qquad \text{or} \qquad k'/k \rightarrow 0 \tag{3.12}$$

The latter condition is fulfilled when $k' \rightarrow 0$, implying that no back reaction, or hardly any, occurs. Such reactions, appropriately, are called irreversible.

### 3.1.2.3. The Effect of Temperature on Reaction Rate

The rate of most chemical reactions increases rapidly with temperature; "slow" reactions can be transformed into "fast" reactions by heat. The effect was first explained and quantified in terms of activation energy. Later on (Section 3.1.4), when discussing the transition state theory, we shall use the concept of free energy of activation instead, but for now let us see how the simpler concept of energy of activation accounts for the effect of temperature on reaction rates.

The arrangements of atoms in a stable molecule corresponds to a minimum of potential energy. As indicated in Fig. 3.1, in an elementary chemical reaction the atoms of the reacting molecules corresponding to one potential energy minimum are rearranged to a different configuration, that of the products, which corresponds to another potential energy minimum. Somewhere between these minima there is a potential energy maximum representing an energy barrier which the reactants have to overcome if the chemical transformation is to take place. In order to cross this energy barrier the reactants acquire an energy $E_1$ and for the reverse reaction to take place the products require an energy $E_2$. These energies are the activation energies for the forward and reverse reaction respectively, and are related to the enthalpy of reaction $\Delta H_R$:

$$E_1 - E_2 = \Delta H_R \tag{3.13}$$

From thermodynamic considerations we know that the equilibrium constant $K$ is related to $\Delta H_R$:

$$d(\ln [K])/dT = \Delta H_R/(RT^2) = (E_1 - E_2)/(RT^2) \tag{3.14}$$

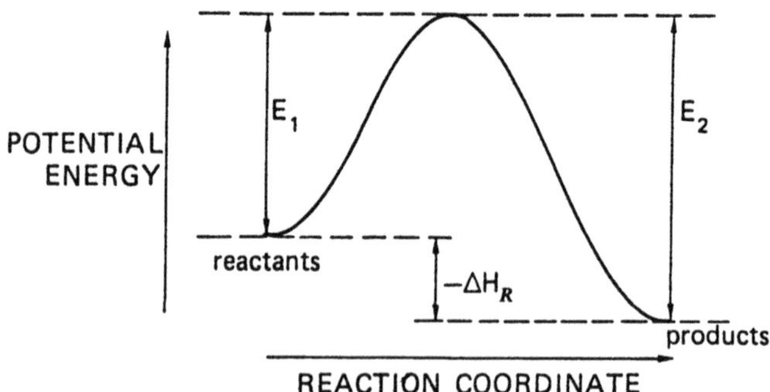


**FIGURE 3.1.** Plot of potential energy along the reaction coordinate (collision theory).

where $T$ is the absolute temperature. As we have seen in Section 3.1.2.2, for an elementary reaction $K = k/k'$. Therefore:

$$d(\ln [k])/dT - d(\ln [k'])/dT = (E_1 - E_2)/(RT^2) \tag{3.15}$$

From Eq. (3.15) we might speculate that:

$$d(\ln [k])/dT = E_1/(RT^2) + J \quad \text{and} \quad d(\ln [k'])/dT = E_2/(RT^2) + J \tag{3.16}$$

Agreement with observation is best when the constant $J$ is given the value zero and with this adjustment Eq. (3.16) integrates to give the Arrhenius equation,[5] generally as:

$$k = A' \exp(-E/RT) \tag{3.17}$$

$A'$ is called the frequency or pre-exponential factor, and $E$ is the activation energy. It follows that $E$ in Eq. (3.17) is an experimentally determined quantity, derived from a plot of $\ln [k]$ against $1/T$.

Many examples of "Arrhenius plots" are found in the literature, but for an electrochemical flavor the one shown in Fig. 3.2, taken from a paper by Kreysa and Medin, [6] refers to the chemical step in the indirect electrochemical oxidation of *p*-methoxytoluene to *p*-methoxybenzaldehyde using a $Ce^{4+}/Ce^{3+}$ redox couple. This reaction has an activation energy obtained from the slope of Fig. 3.2:

$$\text{slope} = -E/R = \frac{\ln k_1 - \ln k_2}{(1/T_1 - 1/T_2)} = \frac{5.19 - 4.49}{0.2 \times 10^{-3}} = -3490$$

Hence:

$$E = 3490 \times R = 3490 \times 8.314 = 2.9 \times 10^4 \text{ kJ/(K kmol)}$$

The frequency factor $A'$ is obtained from:

$$A' = \frac{k_1}{\exp[-E/RT]} = \frac{179.5}{\exp[-2.9 \times 10^4/8.314 \times 312]}$$
$$= 1.29 \times 10^7$$

The rate constant $k$ is:

$$k = 1.29 \times 10^7 \exp[-E/RT]$$

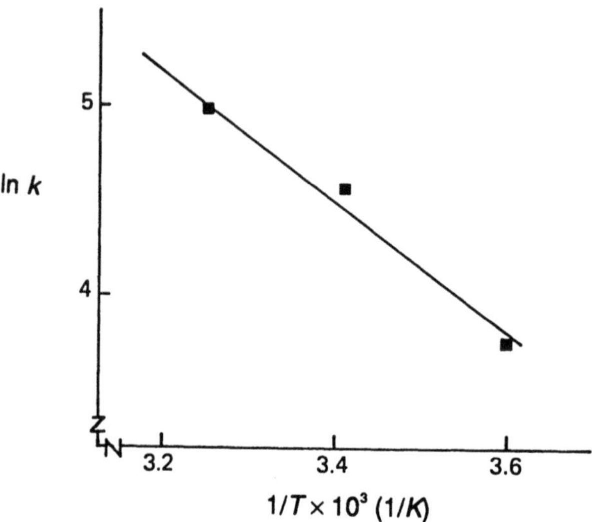


**FIGURE 3.2.** Arrhenius plot.

### 3.1.3. Reaction Mechanisms and Rate Laws

Consider the overall reaction representing the electrochemical oxidation of propylene in an undivided cell with an aqueous sodium bromide electrolyte:[7]

$$CH_3CH{=}CH_2 + H_2O \rightarrow CH_3\overset{\overset{\large O}{\diagup\;\diagdown}}{CH{-}CH_2} + H_2 \tag{3.18}$$

Even though only two molecules are apparently involved in Eq. (3.18), the reaction comprises at least five steps, even if the formation of any by-products is ignored:

- Electrochemical reaction at the anode:

$$2Br^- \rightarrow Br_2 + 2e \tag{3.19}$$

- Chemical equilibrium "reaction":

$$Br_2 + H_2O \rightleftharpoons HOBr + H^+ + Br^- \tag{3.20}$$

- Chemical reaction:

$$CH_3CH{=}CH_2 + HOBr \rightarrow CH_3CH(OH)CH_2(Br) \qquad (3.21)$$

- Electrochemical reaction at the cathode:

$$2H_2O + 2e \rightarrow 2OH^- + H_2 \qquad (3.22)$$

- Chemical reaction:

$$CH_3CH(OH)CH_2(Br) + OH^- \rightarrow CH_3\overset{O}{CH{-}CH_2} + Br^- + H_2O \qquad (3.23)$$

Every electrochemical reaction contains a number of steps (see Section 3.1.5.1) which in order to get a reaction model, must be taken into account. The chemical reactions are envisaged as actual events, i.e., elementary reactions. If we add Eqs. (3.19) to (3.23) we obtain the overall reaction of Eq. (3.18). This describes the chemistry only in terms of stable species whose consumption or production varies significantly with time. A mechanism, on the other hand, will also include reactive intermediates, which will be present in such minute quantities that sometimes they are virtually undetectable.

Reaction mechanisms may be developed to explain rate laws, but often they are developed in parallel, one helping the other. In translating a mechanism into a rate law a useful tool is Bodenstein's steady state approximation (SSA) or stationary state hypothesis.[8] This approximation assumes that after a very short interval of time any reactive intermediate, which because of its reactivity will only be present in negligible proportions, will have its rate of decay equal to its rate of production, i.e., it will reach a steady concentration on a vanishingly small time scale. If this did not happen the amount of the intermediate would build up to measureable proportions and it would become an "intermediate product." It is assumed at any instant that $dc_R/dt = 0$, where $R$ is the reactive intermediate.

It has already been stated (Section 3.1.2.2) that although all reactions are to some extent reversible, some for practical purposes go to completion. To obtain a completely general rate law, however, it is necessary to postulate all the elementary reactions of a mechanism as reversible.

#### 3.1.3.1. Steady State Approximation and the Development of Rate Laws

To demonstrate the use of SSA, take the hypothetical overall reaction:

$$2A + 2B \rightarrow X \qquad (3.24)$$

for which the following mechanism is suggested:

$$A + B \rightleftharpoons C \tag{3.25}$$

$$C + B \rightleftharpoons D \tag{3.26}$$

$$A + D \rightleftharpoons X \tag{3.27}$$

Since $C$ and $D$ do not appear in the overall reaction we shall assume they are reactive intermediates. To get a general rate law we shall assume that the three steps are reversible. For an enclosed reacting system of constant volume we can write for each component, using the same symbolism as for Eq. (3.6):

$$dc_A/dt = r_{-1} - r_1 + r_{-3} - r_3 \tag{3.28}$$

$$dc_B/dt = r_{-1} - r_1 + r_{-2} - r_2 \tag{3.29}$$

$$dc_C/dt = r_1 - r_{-1} + r_{-2} - r_2 \tag{3.30}$$

$$dc_D/dt = r_2 - r_{-2} + r_{-3} - r_3 \tag{3.31}$$

$$dc_X/dt = r_3 - r_{-3} \tag{3.32}$$

where subscripts 1 to 3 denote Eqs. (3.25) to (3.27). The reaction rates are not referred to specific components because all the stoichiometric coefficients are unity. Since, by definition, all the steps in a mechanism are elementary the rates of reaction can be replaced by the following rate laws:

$$r_1 = k_1 c_A c_B \tag{3.33}$$

$$r_{-1} = k'_1 c_C \tag{3.34}$$

$$r_2 = k_2 c_B c_C \tag{3.35}$$

$$r_{-2} = k'_2 c_D \tag{3.36}$$

$$r_3 = k_3 c_A c_D \tag{3.37}$$

$$r_{-3} = k'_3 c_X \tag{3.38}$$

Combining Eqs. (3.28) to (3.38), we have:

$$dc_A/dt = k'_1 c_C - k_1 c_A c_B + k'_3 c_X - k_3 c_A c_D \tag{3.39}$$

$$dc_B/dt = k'_1 c_C - k_1 c_A c_B + k'_2 c_D - k_2 c_B c_C \tag{3.40}$$

$$dc_C/dt = k_1 c_A c_B - k'_1 c_C + k'_2 c_D - k_2 c_B c_C \tag{3.41}$$

$$dc_D/dt = k_2 c_B c_C - k'_2 c_D + k'_3 c_X - k_3 c_A c_D \tag{3.42}$$

$$dc_X/dt = k_3 c_A c_D - k'_3 c_X \tag{3.43}$$

The rate of the overall reaction based on product $X$ is $r_X$, which under the prescribed conditions equals $dc_X/dt$, as given by Eq. (3.43). It is sometimes possible for academic electrochemists to determine experimentally concentrations of very short-lived reactive intermediates, but for us it is preferable to eliminate $c_D$, the concentration of the reactive intermediate, by expressing it in terms of the concentrations of $A$, $B$, and $X$. This can be done by solving simultaneous Eqs. (3.37) to (3.43). Alternatively the task can be simplified by applying the SSA, which allows us to equate $dc_C/dt$ and $dc_D/dt$ to zero and to obtain $r_X$ by simple algebra:

From Eq. (3.41):

$$c_C = \frac{k_1 c_A c_B + k'_2 c_D}{k'_1 + k_2 c_B} \tag{3.44}$$

Substituting for $c_C$ in (3.42):

$$0 = \frac{k_2 k_1 c_A c_B^2 + k_2 k'_2 c_B c_D}{k'_1 + k_2 c_B} - k'_2 c_D - k_3 c_A c_D + k'_3 c_X \tag{3.45}$$

$$c_D = \frac{k_2 k_1 c_A c_B^2 + k'_1 k'_3 c_X + k_2 k'_3 c_B c_X}{k'_1 k'_2 + k'_1 k_3 c_A + k_2 k_3 c_A c_B} \tag{3.46}$$

Substituting in Eq. (3.43) and simplifying:

$$dc_X/dt = \frac{k_1 k_2 k_3 c_A^2 c_B^2 - k'_1 k'_2 k'_3 c_X}{k'_1 k'_2 + k'_1 k_3 c_A + k_2 k_3 c_A c_B} \tag{3.47}$$

In an enclosed system $r_X$ can be equated to $dc_X/dt$. After simplification, Eq. (3.47) becomes:

$$r_X = \frac{\boldsymbol{a} c_A^2 c_B^2 - \boldsymbol{b} c_X}{1 + \boldsymbol{c} c_A + \boldsymbol{d} a_A c_B} \tag{3.48}$$

where ***a***, ***b***, ***c***, and ***d*** are lumped kinetic constants.

What at first sight looks like a simple overall reaction may lead to a complex rate law. Note that at equilibrium, when $r_X$ is zero, Eq. (3.48) is consistent with the condition following from Eq. (3.24):

$$\frac{c_X}{c_A^2 c_B^2} = K \tag{3.49}$$

#### 3.1.3.2. The Rate-Determining Step

Equation (3.48) represents the most general expression which emerges from the suggested mechanism; it contains all the rate constants and makes no assumption about irreversibility. The drawback is that it requires us to know six rate constants.

What would happen, however, if we were also told that $k_2' = 0$ and $k_2$ is very small compared with $k_1$, $k_1'$, $k_3$, and $k_3'$? We will use Eq. (3.47) for our argument since it presents the rate constants individually.

First, by inserting $k_2' = 0$ in Eq. (3.47);

$$r_X = (k_1 k_2 c_A^2 c_B^2)/(k_1' c_A + k_2 c_A c_B) \tag{3.50}$$

Second, since $k_1' \gg k_2$, Eq. (3.50) reduces to:

$$r_X = (k_1 k_2 c_A c_B^2)/k_1' = k_X c_A c_B^2 \tag{3.51}$$

where $k_X = k_1 k_2 / k_1'$.

What is the physical meaning of the conditions we have inserted? First, by making $k_2' = 0$ we postulate that the step in Eq. (3.26) is in practice irreversible. Second, by making $k_2$ much smaller than the remaining velocity constants we postulate that the step in Eq. (3.26) is much slower than the others. In fact the restrictions we have placed on $k_2$ and $k_2'$ have characterized the step in Eq. (3.26) as slow and irreversible; we have made it the rate-determining step.

A rate-determining step is one which is essentially irreversible and has a small forward velocity constant compared to the velocity constants of the other steps. All the steps preceding a rate-determining step must be reversible with small equilibrium constants, otherwise the concentration of reactive intermediates will build up. The concentration of the latter formed by reactions prior to the slow step will have their equilibrium values; the speed of the back reaction rather than the rate of the slow step maintains their concentrations at the required low level.

Once a rate-determining step has been identified, the derivation of a rate law is relatively simple and does not require overt use of the SSA. Thus in the present example we use Eqs. (3.25) to (3.27):

$$A + B \rightleftharpoons C \qquad \text{(fast)}$$

$$C + B \rightarrow D \qquad \text{(slow)}$$

$$A + D \rightleftharpoons X \qquad \text{(fast)}$$

Since the steps before the rate-determining step are at equilibrium, we have:

$$c_C/(c_A c_B) = K_1 \tag{3.52}$$

Furthermore, since the speed of the rate-determining step controls the net forward rate of all later steps, it follows that the rate at which $B$ is consumed by Eq. (3.26) will equal the rate at which $X$ is produced by Eq. (3.27):

$$(r_B)_2 = r_X \tag{3.53}$$

Since, however:

$$(r_B)_2 = k_2 c_C c_B \tag{3.54}$$

it follows that

$$r_X = k_2 c_C c_B \tag{3.55}$$

A combination of Eqs. (3.52) and (3.55) yields:

$$r_X = k_2 K_1 c_A c_B^2 \tag{3.56}$$

or

$$r_X = k_X c_A c_B^2 \tag{3.57}$$

which is identical with the equation we derived previously.

Use of a rate-determining step changed a complex rate law, for which an order could not be defined, into a relatively simple law of order three. It is for this reason that rate laws which can be assigned integral orders are so often encountered.

### 3.1.4. Transition State Theory

We shall mention the ideas of transition state theory,[9] since the kinetic relationships derived in Section 3.1.5 will be interpreted according to this theory.

Consider a conventional elementary chemical reaction:

$$X{-}Y + Z \rightarrow X + Y{-}Z \tag{3.58}$$

where $X$, $Y$, and $Z$ are not necessarily atoms. The reaction is thought to take place via a transition state or activated complex:

$$\underset{\text{reactants}}{X{-}Y + Z} \rightleftharpoons X \cdots Y + Z \rightleftharpoons \overrightarrow{X \cdots Y \cdots Z} \rightarrow X + Y \cdots Z \rightarrow \underset{\text{products}}{X + Y{-}Z}$$

The transition state is represented by $X \cdots Y \cdots Z$, in whose structure the old bond between $X$ and $Y$ is partly broken and that between $Y$ and $Z$ partly formed. If we regard each step between reactants on the left and products on the right as a point along a reaction coordinate, a plot of the minimum energy associated with each configuration can appear (Fig. 3.3). Strictly speaking, we should label the transition state in Fig. 3.3 with an arrow, $\overrightarrow{X \cdots Y \cdots Z}$, indicating that it is moving over the energy hump from left to right. This distinguishes it from the transition state associated with the back reaction:

$$\underset{\text{products}}{X{-}Y + Z} \leftarrow X \cdots Y + Z \leftarrow \overleftarrow{X \cdots Y \cdots Z} \leftrightharpoons X + Y \cdots Z \leftrightharpoons \underset{\text{reactants}}{X + Y{-}Z}$$

Figure 3.3 is almost identical with Fig. 3.1; $E_1$ and $E_2$ are again (see Section 3.1.2.3) the activation energies of the forward and backward reactions, respectively.

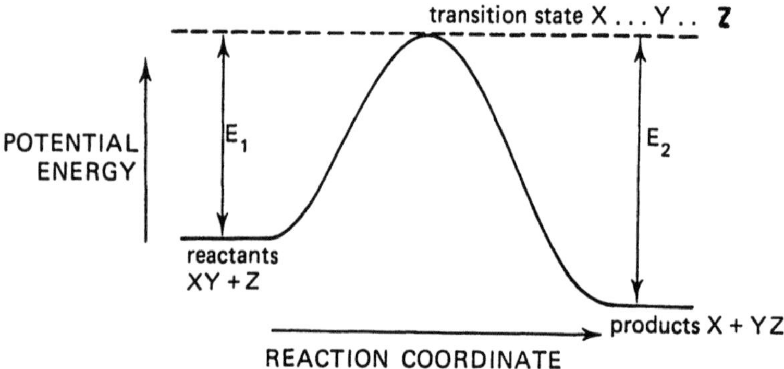


**FIGURE 3.3.** Plot of potential energy along the reaction coordinate (transition state theory).

In applying transition state theory (also called the theory of absolute reaction rates) it is postulated that the relevant activated complex and the reactants are in equilibrium. For an elementary bimolecular reaction:

$$A + B \rightleftharpoons C \rightarrow E + F \tag{3.59}$$

we have:

$$c_C^*/(c_A c_B) = K^* \tag{3.60}$$

where $c^*$ denotes the activated complex and $K^*$ the equilibrium constant which relates its concentration to those of the reactants. Furthermore, the rate of reaction is:

$$r = Ac_C^* \tag{3.61}$$

where $A$ is the frequency with which the activated complex is transformed into products.

Eliminating $c_C^*$ between Eqs. (3.60) and (3.61):

$$r = AK^* c_A c_B \tag{3.62}$$

so the velocity constant $k$ is:

$$k = AK^* \tag{3.63}$$

By analogy with the thermodynamics of stable species, $K^*$ is expressed in terms of a free energy of activation $G^*$, which is the difference between the free energy of the reactants and that of the activated complex, in standard states. Hence we can write:

$$k = A \exp(-\Delta G^*/RT) \tag{3.64}$$

where $R$ is the universal gas constant and $T$ the absolute temperature. A comparison between Eqs. (3.17) and (3.64) can be found in Ref. 10.

### 3.1.5. Derivation of Some Kinetic Relationships

We shall use rate processes, to derive quantitative relationships for the rate of electrode processes, which are examples of heterogeneous reactions.

TABLE 3.1. Mechanism for the Discharge of Hydrogen Ions

| Reaction | Type of reaction |
|---|---|
| $H_3O^+$ (electrolyte) $\rightarrow H_3O^+$ (near cathode) | Transport from bulk of electrolyte |
| $H_3O^+ + e \rightarrow H + H_2O$ | Charge transfer and dehydration |
| $H \rightarrow H$(ads) | Adsorption |
| 2H(ads) $\rightarrow H_2$(ads) | Chemical reaction |
| $H_2$(ads) $\rightarrow H_2$(gas) | Desorption |

#### 3.1.5.1. Steps in an Electrode Process

In its simplest form, an electrode process may proceed through all or some of the following steps:

1. The electroactive particle is transferred to the electrode surface from the bulk solution.
2. The electroactive particle is adsorbed on the electrode surface.
3. Electrons are transferred between the electrode and reactant.
4. The reacted particle suffers desorption and chemical reaction, in either order, then is transported back into the bulk solution. Or, the reacted particle becomes an adatom, then is incorporated into the electrode surface.

Take for example the discharge of hydrogen ions. One possible scheme is shown in Table 3.1.

For the discharge of a metal such as zinc we might expect the pattern in Table 3.2.

The overall reaction rate at an electrode will be shown to be directly proportional to the current density. It was mentioned in Chapter 1 that it is possible to relate the linear potential dependent portion of a polarization

TABLE 3.2. Discharge of Zinc Ions

| Reaction | Type of reaction |
|---|---|
| $Zn^{++}$ (electrolyte) $\rightarrow Zn^{++}$ (near cathode) | Transport from bulk of electrolyte |
| $Zn^{++} + 2e \rightarrow Zn$ | Charge transfer |
| Zn $\rightarrow$ Zn(adatom) | Adatom |
| Zn(adatom) $\rightarrow$ Zn(lattice) | Incorporated in the electrode surface |

curve (Fig. 1.4) to a situation where the rate-determining step is a charge transfer reaction. The horizontal portion, which we referred to as the limiting current, arises when the rate of reaction is governed by a mass transfer step; it is independent of the electrode potential. The former extreme is called charge transfer, activation, or kinetic control; the latter, diffusion or transport control. We shall derive quantitative relationships for the simplest example of these two extremes. Then we shall briefly examine the case when both types of controls have to be considered.

### 3.1.5.2. Charge Transfer, Activation, or Kinetic Control

As we have seen in Section 3.1.2.3, activation energy $E$ is experimentally derived. To illustrate the effect of an applied electric potential on the rate of reaction we used $G^*$, the free energy of activation, in Fig. 3.4 instead of $E$. Consider now the reversible step:

$$O + ne \rightleftharpoons R \tag{3.65}$$

where the negative sign represents a forward reduction reaction and the positive sign a backward oxidation reaction. The chemical forward and backward rates, $r_-$ and $r_+$ respectively, are given by (see Section 3.1.4):

$$r_- = A_- c_O \exp(-\Delta G^*_- / RT) \tag{3.66}$$

$$r_+ = A_+ c_R \exp(-\Delta G^*_+ / RT) \tag{3.67}$$

where subscripts $O$ and $R$ refer to the oxidized and reduced species.

When an electrochemical reaction is carried out, $\Delta G^*$ can be influenced by the imposed electric field, which is a source of potential energy. See Fig. 3.4, where we apply a cathodic field and the value of $\Delta G^*_-$ is reduced and that of $\Delta G^*_+$ increased. The fraction of the total electrode potential $E'$ affecting $\Delta G^*_-$ is $\alpha E'$; that affecting $\Delta G^*_+$ is $(1 - \alpha)E'$, where $\alpha$ is the transfer coefficient. Translated into energy per mole these potentials become $\alpha E'nF$ and $(1 - \alpha)E'nF$, where $F$ is the Faraday constant. The velocities of the forward and backward electrochemical reactions are:

$$r_- = A_- c_O \exp\{-(\Delta G^*_- + \alpha E'nF)/RT\} \tag{3.68}$$

$$r_+ = A_+ c_R \exp\{-[\Delta G^*_+ - (1 - \alpha)E'nF]/RT\} \tag{3.69}$$

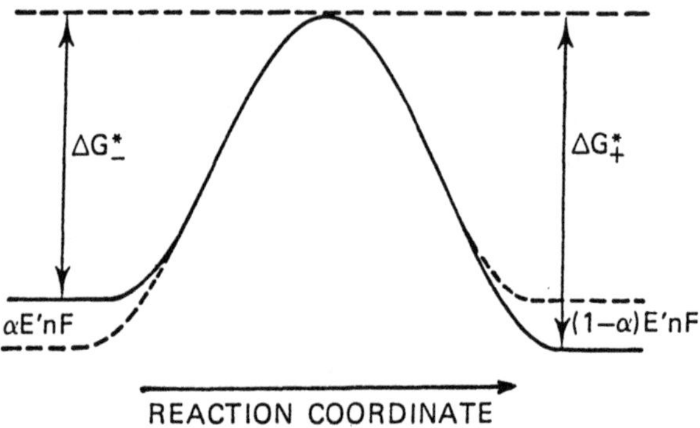


**FIGURE 3.4.** Effect of applied field on the free energy of activation.

Replacing $A_- \exp(-\Delta G^*_-/RT)$ and $A_+ \exp(-\Delta G^*_+/RT)$ by rate constants $k_-$ and $k_+$ respectively, we get:

$$r_- = k_- c_O \exp(-\alpha E'nF/RT) \tag{3.70}$$

$$r_+ = k_+ c_R \exp[(1-\alpha)E'nF/RT] \tag{3.71}$$

When values of $E'$ are inserted in Eqs. (3.70) and (3.71), attention must be given to its sign. Here the potential $E'$ is negative and the addition of $\alpha E'nF$ represents a reduction in the free energy of activation for the forward reaction (Fig. 3.4).

Reaction velocity $r$, in moles per unit area of electrode in unit time, can be converted directly to current density $i$ by multiplying by $nF$:

$$i_- = k_- c_O nF \exp(-\alpha E'nF/RT) \tag{3.72}$$

$$i_+ = k_+ c_R nF \exp[(1-\alpha)E'nF/RT] \tag{3.73}$$

Equations (3.72) and (3.73) imply that current flows in and out of an electrode simultaneously and that the amount of reaction occurring is a net effect:

$$i_c = i_- - i_+ \tag{3.74}$$

At or near equilibrium, i.e., under reversible conditions, $i_- = i_+$ and $E' = E$. From Eqs. (3.72) and (3.73):

$$k_- c_O \exp(-\alpha EnF/RT) = k_+ c_R \exp[(1-\alpha)EnF/RT] \tag{3.75}$$

Or, taking logarithms:

$$\ln[k_-] + \ln[c_O] - \alpha EnF/RT = \ln[k_+] + \ln[c_R] + (1-\alpha)EnF/RT \tag{3.76}$$

Solving Eq. (3.76) for $E$ we get:

$$\begin{aligned} E &= (RT/nF)\ln[k_-/k_+] + (RT/nF)\ln[c_O/c_R] \\ &= E^o + (RT/nF)\ln[c_O/c_R] \end{aligned} \tag{3.77}$$

$E^o$ is called the standard electrode potential and Eq. (3.77) is the well-known Nernst equation, usually derived from thermodynamic arguments. It should be pointed out, however, that a rigorous derivation requires the use of activities in Eq. (3.77) rather than concentrations. For further reading see, e.g., Ref. 11.

The value of $i_-$ or $i_+$ at equilibrium is called the exchange current density ($i_o$). Thus:

$$i_o = k_- nFc_O \exp[-\alpha EnF/RT] = k_+ nFc_R \exp[(1-\alpha)EnF/RT] \tag{3.78}$$

Since the overpotential $\eta$ is defined as[12] $\eta = E' - E$, then:

$$i_- = k_- nFc_O \exp[-\alpha E'nF/RT] = k_- nFc_O \exp[-\alpha(E+\eta)nF/RT] \tag{3.79}$$

and

$$\begin{aligned} i_+ &= k_+ nFc_R \exp[(1-\alpha)E'nF/RT] \\ &= k_+ nFc_R \exp[(1-\alpha)(E+\eta)nF/RT] \end{aligned} \tag{3.80}$$

It follows from Eqs. (3.78), (3.79), and (3.80) that:

$$i_- = i_o \exp[-\alpha\eta nF/RT] \tag{3.81}$$

and

$$i_+ = i_o \exp[(1-\alpha)\eta nF/RT] \tag{3.82}$$

Hence:

$$i_c = i_- - i_+ = i_o\{\exp[-\alpha\eta nF/RT] - \exp[(1-\alpha)\eta nF/RT]\} \tag{3.83}$$

Equation (3.83) is called the Butler–Volmer equation in the absence of concentration gradients.

If the charge transfer step is rate-controlling, conditions at the electrode interface will be far from equilibrium except at very low current densities, i.e., $i_- \gg i_+$ and therefore $i_c \approx i_-$. For this "Tafel approximation" we have:

$$i_c = i_o \exp[-\alpha\eta nF/RT] \tag{3.84}$$

or

$$-\eta = -(RT/\alpha nF)\ln[i_o] + (RT/\alpha nF)\ln[i_c] \tag{3.85}$$

It follows that for conditions far removed from equilibrium a plot of cathodic overpotential against $\ln[i_c]$ (Fig. 3.5) will give a straight line of slope $(RT/\alpha nF)$ and intercept $\ln[i_o]$ on the logarithmic coordinate. This kind of plot is called a Tafel curve and the slope is referred to as the Tafel slope.[13] For anodic reactions the slope is $[RT/(1-\alpha)nF]$. Based on a $\log_{10}$ scale the Tafel slope is $2.303RT/\alpha nF$. Slope value can be used as a diagnostic tool. Thus at 298 K and assuming $\alpha$ to be 0.5, a slope of 120 mV/decade indicates a one-electron charge transfer step.[14]

For very small values of $\eta$ (indicating that the electrode is operating near equilibrium), Eq. (3.83) can be linearized to:

$$(i_c/i_o) = (-\eta nF/RT) \tag{3.86}$$

The greater the value of $i_o$, the greater the net current that can be supported

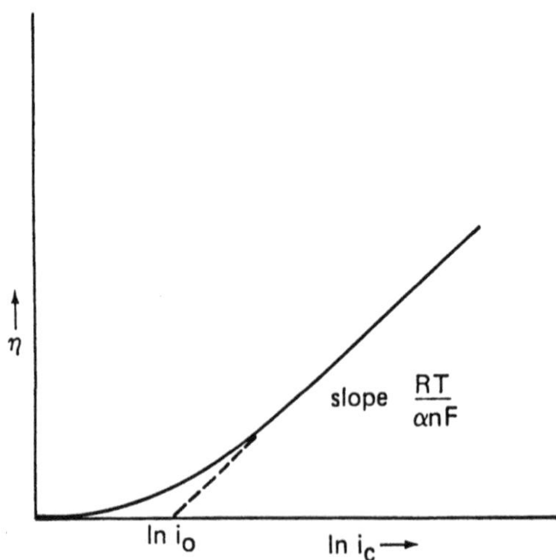


**FIGURE 3.5.** Typical Tafel plot.

by the electrode under reversible conditions. The corollary is that electrode reactions characterized by high exchange currents are less likely to show charge transfer control.

---

EXAMPLE 3.1. Determination of the Constants of a Tafel-Type Equation

THE PROBLEM: The polarization data in Table 3.3 have been obtained for the hydrogen evolution reaction on a rotating disk electrode. Calculate the constants of the resulting Tafel-type equation.

THE SOLUTION: First we write the Tafel-type equation (see Section 3.2.1 in general and Eq. (3.148), in particular):

$$i_H = k_H \exp(-b_H E')$$

Taking logarithms:

$$\ln(i_H) = \ln(k_H) - b_H E'$$

Substituting extreme values and subtracting (remembering that $E'$ is negative):

$$\ln(1000) - \ln(10) = b_H(772 - 669)$$

Hence, $b_H = 0.045$.

From $1000 = k_H \exp(0.045) \times 772)$ we get:

$$k_H = 8 \times 10^{-13}$$

The required equation is:

$$i_H = 8 \times 10^{-13} \exp(-0.045E')$$

TABLE 3.3. Polarization Data, Hydrogen Evolution Reaction

| Current density, $i_H$ ($A/m^2$) | Electrode potential (mV) |
|---|---|
| 10 | 669 |
| 100 | 720 |
| 1000 | 772 |

The value of $k_H$ is highly sensitive to small changes in $b_H$. A 2% change in $b_H$ results in a 50% change in $k_H$.

---

### 3.1.5.3. Diffusion or Mass Transport Control

We have seen in Chapter 2 that if the Reynolds number is not too high, the resistance to mass transfer to and from an electrode is confined to a hydrodynamic laminar layer adjacent to the surface of the electrode across which ions may be transferred either by diffusion or electrolytic migration. Resistance to mass transfer by diffusion causes concentration gradients to be formed (Fig. 3.6). The concentration profiles, normally curved, have been linearized as explained in Chapter 2. It is assumed that the bulk of the electrolyte is so well stirred that concentration is uniform. Let us use the simple reduction in Eq. (3.65) when considering activation control:

$$O + ne \rightleftharpoons R$$

If $c_O$ and $c_R$ are the concentrations of $O$ and $R$ in the bulk of the electrolyte, the electrode potential corresponding to reversible (i.e., equilibrium) conditions will be given by Eq. (3.87):

$$E = E^o + (RT/nF) \ln [c_O/c_R] \tag{3.87}$$

When finite currents are passed, conditions cease to be reversible, but since the controlling step is the diffusional mass transfer of species to and from the electrode we may assume that other steps are only insignificantly

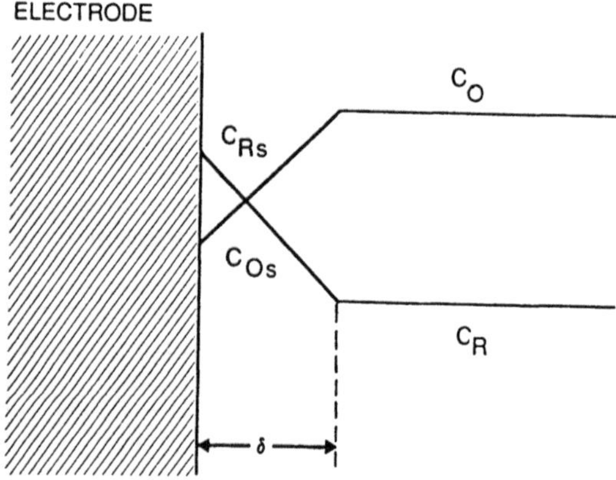


**FIGURE 3.6.** Concentration profiles during mass transfer control.

displaced from the reversible state. Then the electrode will take up the reversible potential corresponding to $c_{Os}$ and $c_{Rs}$, the concentrations of $O$ and $R$ respectively near its surface:

$$E' = E^o + (RT/nF) \ln [c_{Os}/c_{Rs}] \tag{3.88}$$

The overpotential $\eta$ is:

$$\eta = E' - E = (RT/nF) \ln [(c_{Os}c_R)/(c_{Rs}c_O)] \tag{3.89}$$

Current density is:

$$i = nFN_O \tag{3.90}$$

where $N_O$ is the flux of $O$ to the electrode surface due to diffusion and electrolytic migration.

The diffusional flux of $O$, $N_D$, will according to Section 2.3.1 be given by:

$$N_D = k_L(c_O - c_{Os}) \tag{3.91}$$

where $k_L$ is the individual liquid phase mass transfer coefficient (as explained in Section 2.3.1 and at greater length in Ref. 15.

Migration of a charged species under the influence of a potential or electric field can add to the diffusional flux. This addition due to migration[16] will be $(t^+ i)/nF$, where $t^+$ is the transport number for $O$. Hence the total flux $N_O$ is:

$$N_O = k_L(c_O - c_{Os}) + (t^+ i)/nF \tag{3.92}$$

Then:

$$i = nFN_O = nFk_L(c_O - c_{Os}) + t^+ i \tag{3.93}$$

or

$$i = [nFk_L(c_O - c_{Os})]/(1 - t^+) \tag{3.94}$$

The maximum current density which can be applied to the electrode, then is reached when $c_{Os}$ is zero. This is called the limiting current density $i_{lim}$:

$$i_{lim} = (nFk_L c_O)/(1 - t^+) \tag{3.95}$$

or

$$c_O = [i_{lim}(1 - t^+)]/k_L nF \tag{3.96}$$

From Eqs. (3.94) and (3.96):

$$c_{Os} = c_O - i(1 - t^+)/(k_L nF) = [(i_{lim} - i)(1 - t^+)/(k_L nF) \tag{3.97}$$

From Eqs. (3.96) and (3.97):

$$c_{Os}/c_O = (i_{lim} - 1)/i_{lim} \tag{3.98}$$

and substituting into Eq. (3.89):

$$\eta = \frac{RT}{nF} \ln \left\{ \left[ \frac{i_{lim} - i}{i_{lim}} \right] \frac{c_R}{c_{Rs}} \right\} \tag{3.99}$$

If diffusion of product plays no part in the controlling step, as in metal deposition or gas evolution, this equation reduces to:

$$\eta = (RT/nF) \ln [(i_{lim} - 1)/i_{lim}] \tag{3.100}$$

or

$$i = i_{lim}(1 - \exp [\eta nF/RT]) \tag{3.101}$$

A plot of Eq. (3.101) is shown in Fig. 3.7.

For an uncharged electroactive species or in the presence of an excess of supporting electrolyte (the usual case in industrial practice), $t^+ = 0$ and Eq. (3.95) becomes:

$$i_{lim} = nFk_L c_O \tag{3.102}$$

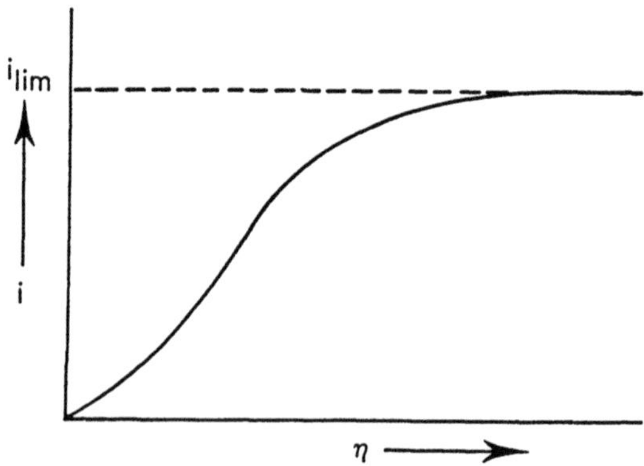


**FIGURE 3.7.** Plot of Eq. (3.101).

These equations have been derived for a cathodic reaction for which $\eta$ is always negative. The corresponding derivation for an anodic reaction results in a change of sign, for example in the power of the exponential in Eq. (3.101).

#### 3.1.5.4. Charge Transfer and Diffusion Control Combined

Often charge transfer and mass transfer will proceed at comparable rates: we will develop some relationships in this section. Consider again the cathodic reaction, Eq. (3.65). At conditions removed from equilibrium, $i_c$ will be given by a modified form of Eq. (3.84):

$$i_c = i_o(c_{Os}/c_O) \exp[-\alpha\eta nF/RT] \tag{3.103}$$

The difference between Eq. (3.84) and Eq. (3.103) arises from the fact that $i_o$ has been defined on the basis of bulk concentrations which had been identical to those at the cathode. The net cathodic current can also be expressed in terms of Eq. (3.94) with $t^+ = 0$:

$$i_c = nFk_L(c_O - c_{Os}) \tag{3.104}$$

From Eq. (3.104):

$$c_{Os}/c_O = 1 - i_c/(nFk_Lc_O) \tag{3.105}$$

Substituting Eq. (3.105) in Eq. (3.103):

$$i_c = i_o[1 - i_c/(nFk_Lc_O)] \exp[-\alpha\eta nF/RT] \tag{3.106}$$

giving:

$$i_c = \frac{i_o \exp[-\alpha\eta nF/RT]}{1 + (i_o \exp[-\alpha\eta nF/RT])/(nFk_Lc_O)} \tag{3.107}$$

or

$$i_c = \frac{1}{\dfrac{1}{nFk_Lc_O} + \dfrac{1}{i_o \exp[-\alpha\eta nF/RT]}} \tag{3.108}$$

Equation (3.108) shows that the current density is governed by two resistances in series: one due to mass transfer, the other to charge transfer. We shall use this type of equation again on reaction models. As $\eta$ tends to infinity in a cathodic sense the system becomes wholly diffusion-controlled and Eq. (3.108) reduces to Eq. (3.102).

Under conditions where diffusion and charge transfer play comparable roles, Eq. (3.108) may be re-expressed:

$$i_o/i_c = i_o/i_{lim} + 1/\exp[-\alpha\eta nF/RT] \tag{3.109}$$

or

$$\exp[\alpha\eta nF/RT) = \frac{i_o}{i_{lim}}\left[\frac{i_{lim}}{i_c} - 1\right] \tag{3.110}$$

Thus:

$$\eta = \frac{RT}{\alpha nF}\ln\frac{i_o}{i_{lim}} + \frac{RT}{nF}\ln\left[\frac{i_{lim}}{i_c} - 1\right] \tag{3.111}$$

When $i_c \ll i_{lim}$, which describes conditions when diffusion can be expected to play a minor role, Eq. (3.111) approximates to:

$$\eta = \frac{RT}{\alpha nF}\ln\frac{i_o}{i_{lim}} + \frac{RT}{\alpha nF}\ln\frac{i_{lim}}{i_c} = \frac{RT}{\alpha nF}\ln\frac{i_o}{i_c} \tag{3.112}$$

or

$$-\eta = -(RT/\alpha nF)\ln[i_o] + (RT/\alpha nF)\ln[i_c] \tag{3.113}$$

As one would expect, Eq. (3.113) is identical with the Tafel Eq. (3.85).

---

EXAMPLE 3.2. Polarization Curve for Deposition of Cadmium

THE PROBLEM: The equation representing the electrolytic deposition of cadmium is:

$$Cd^{2+} + 2e \rightleftharpoons Cd$$

The polarization data in Table 3.4 were obtained with the help of a rotating disc electrode at a temperature of 298 K.

The concentration of cadmium in a solution containing an excess of supporting electrolyte is 120 gmol/m³. Determine the parameters of the Tafel equation for this reaction.

TABLE 3.4. Polarization Data, Cadmium Deposition

| $E'$ (V) | $i_c$ ($A/m^2$) | $E'$ (V) | $i_c$ ($A/m^2$) |
|---|---|---|---|
| −0.2 | 1.5 | 0.3 | 214 |
| −0.1 | 2.16 | 0.4 | 228 |
| 0.0 | 12.5 | 0.5 | 250 |
| 0.1 | 59.3 | 0.6 | 320 |
| 0.2 | 156.5 | | |

THE SOLUTION: To determine whether Tafel characteristics are appropriate, we plot $\ln[i_c]$ versus $E'$ as shown in Fig. 3.8. Between potentials of −0.1 and 0.1, linearity exists and hence Tafel behavior. The plot also exhibits a limiting current plateau and thus Eq. (3.108)

$$i_c = \frac{1}{\dfrac{1}{nFk_L c_O} + \dfrac{1}{i_o \exp[-\alpha\eta nF/RT]}}$$

is appropriate.

Also, $i_o$ is given by Eq. (3.78):

$$i_o = k_- nFc_O \exp[-\alpha EnF/RT]$$

Equation (3.108) becomes:

$$i_c = \frac{c_O}{\dfrac{1}{nFk_L} + \dfrac{1}{nFk_- \exp[-\alpha nFE'/RT]}}$$

From the observed limiting current plateau:

$$\ln[i_{lim}] = 5.44$$

That is:

$$i_{lim} = 230\,\text{A/m}^2$$

Then the group:

$$1/nFk_L = c_O/i_{lim} = 120/230 = 0.52\,\text{gmol/mA}$$

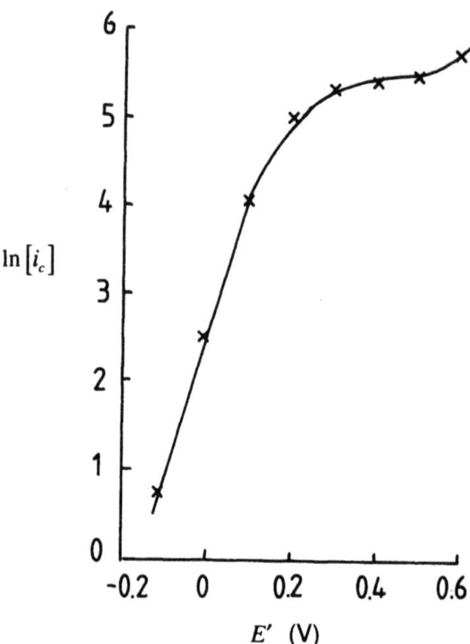


**FIGURE 3.8.** Plot of $\ln[i_c]$ versus $E'$.

The equation for $i_c$ can be rearranged:

$$nFk_-\left[\frac{c_O}{i_c} - \frac{1}{nFk_L}\right] = \exp\frac{\alpha nFE'}{RT}$$

Thus a plot of $\ln\left[\frac{c_O}{i_c} - \frac{1}{nFk_L}\right]$ against $E'$ would give a straight line with a slope of $\alpha nF/RT$ and an intercept of $-\ln[nFk_-]$. Although the Tafel part of the polarization curve in Fig. 3.8 appears to be well defined, the group $1/nFk_L = 0.52$ represents 26% of $c_O/i_c$ at $E' = 0.1$. Figure 3.9 demonstrates the plot of $\ln\left[\frac{c_O}{i_c} - \frac{1}{nFk_L}\right]$ against $E'$. From it we find the slope:

$$-\alpha nF/RT = -(4.01 - 0.41)/(-0.1 - 0.1) = 18.0$$

The reciprocal of the slope based on logarithms to the base 10 will be $2.303/18.0 = 0.128$ *V*/decade. This is close enough to 0.12 V/decade[14] to indicate that a one-electron charge transfer step is rate-determining.

When $E' = 0$ the log term is 2.21, i.e., $1/nFk_- = 9.08$. Hence:

$$k_- = 1/9.08nF = 1/(9.08 \times 2 \times 0.965 \times 10^5) = 5.7 \times 10^{-7}\ \mathrm{m/s}$$

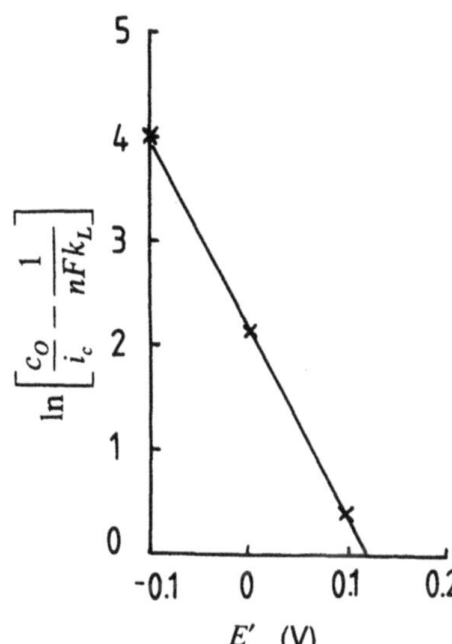


**FIGURE 3.9.** Plot of $\ln[c_O/i_c - 1/nFk_L]$ versus $E'$.

### 3.1.5.5. Charge Transfer with Adsorption on the Electrode

When the electrode acts as a source or sink for electrons with the reactant sitting in the outer Helmholtz plane (see Section 1.4.4), reaction rate is not usually dependent on the nature of the electrode material. For some reactions to proceed, however, the reactant or an intermediate must be adsorbed on the electrode surface. Then the nature of the electrode material often has a significant effect on the reaction rate by virtue of its ability or lack of ability to provide the necessary sites. We shall first discuss briefly the phenomenon of adsorption and then derive an example of kinetic expressions when adsorption is taken into account.

(i) Adsorption. Adsorption is physical or chemical. The physical adsorption of gases stems from their tendency to condense on flat surfaces: the standard free energy of physical adsorption is therefore akin to that of liquefaction, normally 8–20 MJ/kmol. In chemical adsorption, however, bonds are formed and the standard free energy of adsorption is of the same magnitude as that for chemical reactions. Chemical adsorption is restricted to a monolayer, but physical adsorption is not. Specific adsorption at an electrode falls under the heading of chemical adsorption, and it is convenient[17] to think of the standard free energy of adsorption as composed of two parts, a "specific" part $G_{sp}$ which is independent of potential and therefore specific to the chemical system, and a "nonspecific" part $G_{nsp}$ which is dependent on the electrode potential $E'$. Thus:

$$G_{ads} = G_{sp} + G_{nsp}(E') \tag{3.114}$$

When $G_{sp} \gg G_{nsp}$ the surface concentration of the adsorbed species can be obtained from an adsorption isotherm such as Langmuir's.

An adsorption isotherm relates the surface concentration of an adsorbed species to the variables which determine it. Langmuir's adsorption isotherm has its roots in the chemical adsorption of gases according to the expression:

$$A\,(\text{gas or solution}) + M\,(\text{site}) \rightleftharpoons A \cdots M\,(\text{adsorbed}) \tag{3.115}$$

where $M$ (site) represents a vacant site on the surface and $A \cdots M$ (adsorbed) an occupied site. Regarding the above as an elementary chemical reaction, Langmuir wrote:

$$r_a = k_a c_A S_M \tag{3.116}$$

$$r_d = k_d S_{AM} \tag{3.117}$$

where $r_a$ and $r_d$ represent the rate of adsorption and desorption, and $S_M$ and $S_{AM}$ the surface concentrations of unoccupied and occupied sites. If the concentration of the total number of available sites is $S_{MAX}$, then $S_M = S_{MAX} - S_{AM}$ and:

$$r_a = k_a c_A (S_{MAX} - S_{AM}) \tag{3.118}$$

$$r_d = k_d S_{AM} \tag{3.119}$$

Dividing Eqs. (3.118) and (3.119) by $S_{MAX}$:

$$r_a = k_a c_A (1 - \theta) \tag{3.120}$$

$$r_d = k_d \theta \tag{3.121}$$

where $\theta$ means $S_{AM}/S_{MAX}$ and is called the fractional coverage of the surface. Solving for $\theta$, we obtain the isotherm:

$$\theta = k_a c_A/(k_d + k_a c_A) = b c_A/(1 + b c_A) \tag{3.122}$$

where $b$, the adsorption coefficient, is the ratio of rate constants $k_a/k_d$.

Langmuir's analysis, by analogy with chemical reactions, implies that the standard free energy of the adsorption process remains constant and is independent of the number of sites already occupied. Sometimes this assumption is significantly in error and it is advisable to try the Temkin adsorption isotherm,[18] which postulates that the standard free energy varies linearly with fractional coverage.

(ii) Reaction rates in the presence of adsorption. Let us now apply the concepts discussed in Section (3.1.5.5.i) to an example. Consider the two-step process:

$$A_1 + M + e \underset{k_{-1}}{\overset{k_1}{\rightleftharpoons}} A_2 \cdots M \tag{3.123}$$

$$A_2 \cdots M + e \underset{k_{-2}}{\overset{k_2}{\rightleftharpoons}} A_3 + M \tag{3.124}$$

The rate of production per unit area of electrode of $A_3$ and that of the adsorbed species $A_2$, denoted by $dA_3/dt$ and $dA_2/dt$ *are:*

$$dA_2/dt = r_1 - r_{-1} - r_2 + r_{-2} \tag{3.125}$$

$$dA_3/dt = r_2 - r_{-2} \tag{3.126}$$

where subscripts 1 and 2 refer to Eqs. (3.123) and (3.124) respectively. Note that the adsorption process in 1 is from left to right, but in the reverse direction in 2. Replacing the rate terms in Eq. (3.125) and (3.126) by expressions of the form of Eqs. (3.70) and (3.71) gives:

$$\begin{aligned} dA_2/dt = {} & k_1(1-\theta)A_1 \exp(-\alpha f\ E') - k_{-1}\theta \exp[(1-\alpha)fE'] \\ & -k_2\theta \exp(-\alpha fE') + k_{-2}(1-\theta)A_3 \exp[(1-\alpha)fE'] \end{aligned} \tag{3.127}$$

$$dA_3/dt = k_2\theta \exp(-\alpha fE') - k_{-2}(1-\theta)A_3 \exp[(1-\alpha)fE'] \tag{3.128}$$

where $A_1$ and $A_3$ represent the concentrations of species $A_1$ and $A_3$ in the electrolyte near the surface of the electrode, and $f$ is $nF/RT$. Applying the steady state approximation (see Section 3.1.3.1) to the rate of formation of $A_2$:

$$\begin{aligned} 0 = {} & k_1A_1 \exp(-\alpha fE') - k_1\theta A_1 \exp(-\alpha fE') - k_{-1}\theta \exp[(1-\alpha)fE'] \\ & - k_2\theta \exp(-\alpha fE') + k_{-2}A_3 \exp[(1-\alpha)fE'] \\ & - k_{-2}\theta A_3 \exp[(1-\alpha)fE'] \end{aligned} \tag{3.129}$$

or

$$\theta = \frac{k_1A_1\exp(-\alpha fE') + k_{-2}A_3\exp[(1-\alpha)fE']}{k_1A_1\exp(-\alpha fE') + k_{-1}\exp[(1-\alpha)fE'] + k_2\exp(-\alpha fE') + k_{-2}A_3\exp[(1-\alpha)fE']} \tag{3.130}$$

Using Eq. (3.130) to eliminate $\theta$ from Eq. (3.128) and bringing the expression onto a common denominator:

$$\frac{dA_3}{dt} = \frac{k_1k_2A_1\exp(-2\alpha fE') - k_{-1}k_{-2}A_3\exp[2(1-\alpha)fE']}{k_1A_1\exp(-\alpha fE') + k_{-1}\exp[(1-\alpha)fE'] + k_2\exp(-\alpha fE') + k_{-2}A_3\exp[(1-\alpha)fE']} \tag{3.131}$$

If the reaction in Eq. (3.123) is rate-controlling we may take limits $k_{-1} \to 0$ and $k_2$, $k_{-2} \to \infty$. With these values Eq. (3.131) is simplified as follows, noting that $k_2/k_{-2} = 1/K_2$, where $K_2$ is an adsorption coefficient:

$$\frac{dA_3}{dt} = \frac{k_1(1/K_2)A_1\exp(-\alpha fE')}{(1/K_2) + A_3\exp(fE')} \tag{3.132}$$

and Eq. (3.130) becomes:

$$\theta = \frac{A_3\exp(fE')}{(1/K_2) + A_3\exp(fE')} \tag{3.133}$$

Taking the extreme cases of high and low surface coverage:

First, $\theta \to 0$ implies from Eq. (3.133) that $k_2/k_{-2} \gg A_3\exp(fE')$ and so Eq. (3.132) becomes:

$$dA_3/dt = k_1A_1\exp(-\alpha fE') \tag{3.134}$$

and

$$i = 2Fk_1A_1\exp(-\alpha fE') \tag{3.135}$$

Second, $\theta \to 1$ implies that $k_2/k_{-2} \ll A_3\exp(fE')$ and so Eq. (3.132) becomes:

$$dA_3/dt = k_1(k_2/k_{-2})(A_1/A_3)\exp[-(1+\alpha)fE'] \tag{3.136}$$

and

$$i = 2Fk_1(k_2/k_{-2})(A_1/A_3)\exp[-(1+\alpha)fE'] \tag{3.137}$$

The two extreme cases reduce to:

$$i = \text{constant } A_1 \exp(-\alpha fE') \tag{3.138}$$

$$i = \text{constant } (A_1/A_3)\exp[-(1+\alpha)fE'] \tag{3.139}$$

### 3.1.6. Electrocatalysis

Just as in chemical catalysis, electrocatalysis provides a reaction path which lowers the energy of activation (see Fig. 3.1) and hence increases the rate of reaction, i.e., the current density for a given overvoltage. We can distinguish between the situation when a species in solution acts as a catalyst and a process where the reactive intermediate has to be adsorbed on the electrode. We have already met an example of the former in Section 3.1.3, namely the "catalysis" of the epoxidation of propylene by means of bromide ions.

Pletcher, in his interesting book, *Industrial Electrochemistry*,[19] demonstrates reasoning that can sometimes lead to a decision on the most probable reaction path when an adsorption step is involved. Electrocatalysis is of major importance industrially. Use of dimensionally stable anodes in the chlorine industry[20] is a good example.

## 3.2. REACTION MODELS

In the context of electrode reactions, a reaction model is an expression for the dependence of current density on reactant concentration, electrode potential, and mass transfer coefficient. One or two other variables may also be important, such as temperature or $pH$. A reaction model is different from a reaction mechanism in that all that is needed for the model is an expression for the current density as a function of several parameters. Often it is not necessary to consider some very short-lived reaction intermediates which are an essential part of a reaction mechanism. As we shall see in the two examples given at the end of this chapter, it is often necessary to group reaction steps and resulting products together; otherwise the complexity of the reaction model will detract from its usefulness.

We will derive a typical expression for a reaction model and then consider ways of obtaining the necessary numerical values for the constants and variables in the expression. The chapter concludes with two examples of reaction models, one for an inorganic and the other for an organic.

### 3.2.1. General Considerations

Let us look at the simple reaction:

$$A + ne \rightarrow B \tag{3.140}$$

As shown in Fig. 3.10, for the reaction to occur $A$ has to be transported to the electrode, charge transfer possibly via intermediates occurs resulting in product $B$, which is then transported back into the bulk of the electrolyte. The rate of diffusional transport of $A$ to the electrode (see Section 2.3) is:

$$dQ_A/dt = k_L S(c_A - c_{As}) \tag{3.141}$$

where $Q_A$ is the number of kmols of $A$ transported to the electrode surface, $t$ the time, $k_L$ the mass transfer coefficient, $S$ the electrode surface area, and $c_A$ and $c_{As}$ the concentration of $A$ in the bulk and near the surface.

The equation giving $r_A$, the rate of reaction of $A$ at the electrode surface, has the form:

$$r_A = Sf(c'_{As'} E', \ldots) \tag{3.142}$$

where the functional relationship $f$ describes the reaction kinetics in terms of $c'_{As'}$ (the surface concentration of $A$), $E'$ (the electrode potential), and any other parameter required to define the kinetics. In practice $c_{As}$ and $c'_{As}$ are equated or an adsorption isotherm is used.

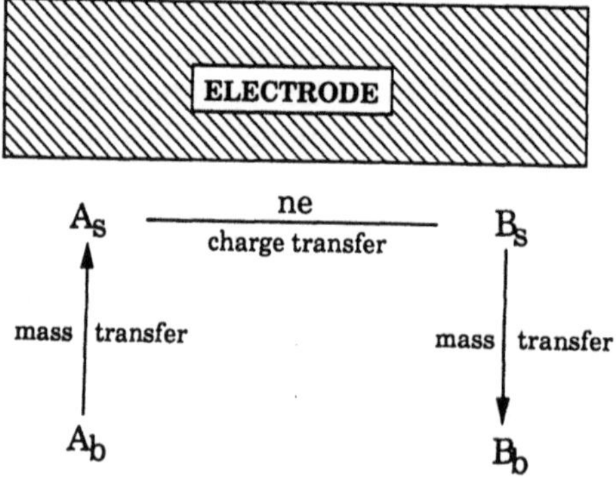


**FIGURE 3.10.** Simple reaction mechanism.

At steady state:

$$r_A = dQ_A/dt \tag{3.143}$$

Assume for simplicity's sake that $c_{As} = c'_{As'}$. Converting (Eqs. 3.141) and (3.142) to current densities:

$$i = n\mathfrak{F}k_L(c_A - c_{As}) \tag{3.144}$$

and

$$i = n\mathfrak{F}f(c_{As}, E', \ldots) \tag{3.145}$$

Equating (3.144) and (3.145) enables us to eliminate surface concentrations and to obtain a single expression for the current density in terms of the parameters mentioned.

Equation (3.144) enables one to make a quick estimate of the order of magnitude of the limiting current density. In industrial equipment $k_L$ usually ranges from $10^{-5}$ to $10^{-4}$ m/s. Since under complete mass transfer control $c_{As} = 0$, for $n = 1$ the expected limiting current density for, say, a $10^{-2}$ M solution lies between 10 and 100 A/m$^2$ assuming a current efficiency of 100%.

Let us develop a typical expression for a reaction model by assuming that we have a primary reaction $A \rightarrow B$ which is governed by a simple "Tafel-type" expression (see also Section 3.2.1.1). Equation (3.145) becomes:

$$i_A = n\mathfrak{F}k'_A c_{As} \exp(-b_A E') = k_A c_{As} \exp(-b_A E') \tag{3.146}$$

where $i_A$ is the partial current density for the formation of product $B$ and $b_A$ the slope of the polarization curve obtained by plotting $\ln[i_A]$ against $E'$. Also, $k'_A$ and $k_A$ are reaction rate constants, where $k_A = n\mathfrak{F}k'_A$. Although we are talking about "Tafel-type" curves or behavior, Eq. (3.146) is not identical with the usual Tafel Eq. (3.84) used in electrochemical science texts or papers. Instead it is of the form of Eq. (3.72), $k_A$ in Eq. (3.146) being equal to $n \times 10^3 Fk_-$. The reason for the use of Eq. (3.146) in reaction modeling is that many industrial systems are not reversible and hence the use of $i_o$, the exchange current density, and of $\eta$, the overpotential, is not possible.

Eliminating $c_{As}$ from Eqs. (3.144) and (3.146):

$$i_A = \frac{c_A}{\dfrac{1}{n\mathfrak{F}k_L} + \dfrac{1}{k_A \exp[-b_A E']}} \tag{3.147}$$

The primary reaction may be accompanied by an independent secondary reaction, often associated with the decomposition of the solvent or the discharge of a secondary ion. The secondary reaction for aqueous cathodic processes is, typically, hydrogen evolution. A simple Tafel relationship is assigned to the secondary reaction because mass transfer plays only a minor role in the discharge process; the concentration term is also omitted on the assumption that the solvent composition will change insignificantly during electrolysis. The partial current density for hydrogen evolution $i_{\mathrm{H}}$ is:

$$i_{\mathrm{H}} = k_{\mathrm{H}} \exp[-b_{\mathrm{H}}E'] \tag{3.148}$$

To get a reaction model we need numerical values for the "kinetic constants" $k_A$, $b_A$, $k_{\mathrm{H}}$, and $b_{\mathrm{H}}$ as well as a value for $k_L$, the mass transfer coefficient. Experimental methods of obtaining these values are described in Section 3.2.2.

Consider Eq. (3.147) a little more. Like Eq. (3.108) it contains two resistances to current flow, operating in series. The first, $1/(n\mathfrak{F}k_L)$, is due to mass transfer; the second, $1/(k_A \exp[-b_A E'])$, arises from the charge transfer process. The importance of the latter can be highlighted by putting numerical constants into Eq. (3.147) and (3.148). The values are taken from an organic reduction process requiring two electrons for the primary reaction where virtually the only by-product is hydrogen. Figure 3.11 shows the dramatic dependence of current efficiency, Eq. (1.22), on relatively small changes in the value of $k_L$ when mass transfer is not very efficient.

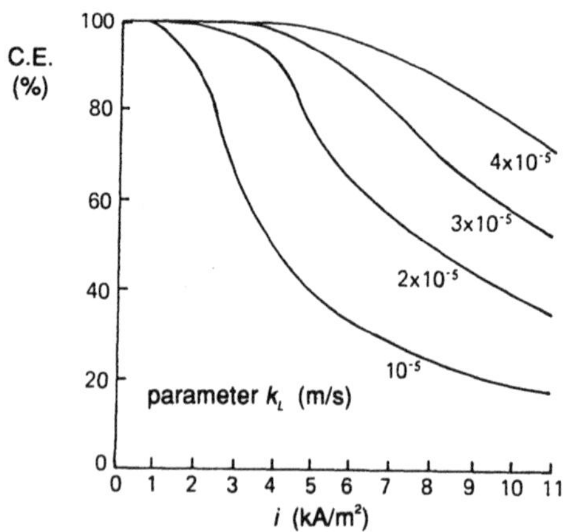


**FIGURE 3.11.** Effect of mass transfer on current efficiency.

EXAMPLE 3.3. Calculation of Mass Transfer Rates and Electrode Potential to Obtain a Required Conversion

THE PROBLEM: In a synthesis using an aqueous electrolyte, the reaction occurs according to the equation:

$$A + 2e \rightarrow B$$

A parallel secondary reaction is the evolution of hydrogen:

$$2H^+ + 2e \rightarrow H_2$$

For the synthesis to be economically viable a conversion of $A$ of 95% is required and the reaction must occur with a current density of at least 100 A/m$^2$ and a current efficiency of 95%.

Determine the required mass transfer rate and electrode potential if the reaction kinetics are given by:

$$i_A = 2.1 c_{As} \exp[-16E']$$

$$i_H = 0.126 \exp[-12E']$$

The initial concentration of $A$ is 200 gmol/m$^3$.

THE SOLUTION: The reaction model can be expressed using Eqs. (3.147) and (3.148):

$$i = i_A + i_H = \frac{c_A}{\dfrac{1}{nFk_L} + \dfrac{1}{k_A \exp[-b_A E']}} + k_H \exp[-b_H E']$$

The final concentration of $A = (1 - 0.95) \times 200 = 10$ gmol/m$^3$. For a C.E. of 95%:

$$i_H = 0.05 \times 100 = 5\ A/m^2$$

Hence, $5 = 0.126 \exp[-12E']$ and $E' = -0.307$.

Now $i_A = 0.95 \times 100 = 95\,\text{A/m}^2$, so for a final concentration of $A$ of $10\,\text{gmol/m}^3$:

$$95 = \frac{10}{\dfrac{1}{2 \times 0.965 \times 10^5 \times k_L} + \dfrac{1}{2.1\exp[16 \times 0.307]}}$$

Solving this equation gives $k_L = 5.09 \times 10^{-5}\,\text{m/s}$.

We require an electrode potential of $-0.307\,\text{V}$ and a mass transfer coefficient of $5.09 \times 10^{-5}\,\text{m/s}$. Achieving these values should present no problem.

---

#### 3.2.1.1. Tafel-Type Behavior

In Section 3.2.1 it was stated that sometimes in a model a number of charge transfer steps have to be grouped together to obtain a sufficiently simple model. It is important to know whether the overall charge transfer process still obeys Eq. (3.146). Consider as an example the overall charge transfer reaction:

$$A + 3e \xrightarrow{i_A} B \tag{3.149}$$

Let us assume that Eq. (3.149) is made up of three steps:

$$A + e \underset{i_{-1}}{\overset{i_1}{\rightleftharpoons}} C \tag{3.150}$$

$$C + e \underset{i_{-2}}{\overset{i_2}{\rightleftharpoons}} D \tag{3.151}$$

$$D + e \xrightarrow{i_B} B \tag{3.152}$$

Equation (3.152) is the rate-determining step (r.d.s., see Section 3.1.3.2). Since from the steady state approximation (SSA) $i_1 = i_{-1}$, you can write:

$$k_1 c_A \exp(-\alpha f E') = k_{-1} c_C \exp[(1 - \alpha) f E'] \tag{3.153}$$

or

$$c_C \propto c_A \exp(-f E') \tag{3.154}$$

where

$$f = n\mathfrak{F}/RT$$

For the second step, $i_2 = i_{-2}$ and:

$$k_2 c_C \exp(-\alpha f E') = k_{-2} c_D \exp[(1-\alpha) f E'] \tag{3.155}$$

or

$$c_D \propto c_c \exp(-fE') \tag{3.156}$$

Hence:

$$c_D \propto c_A \exp(-2fE') \tag{3.157}$$

Finally, for the third step:

$$i_B = k_3 c_D \exp(-\alpha f E') \tag{3.138}$$

Therefore:

$$i_B \propto c_A \exp[-(2+\alpha) f E'] \tag{3.159}$$

Since $i_A = i_B$:

$$i_A = k c_A \exp[-(2+\alpha) f E'] \tag{3.160}$$

On repeating the derivation of Eq. (3.160) if step 2 or step 1 is rate-determining, you get the results shown in Table 3.5. Clearly all three mechanisms obey a Tafel-type relationship.

Table 3.6 lists a variety of mechanisms, a Yes indicating Tafel behavior.

TABLE 3.5. Tafel-Type Behavior for Three-Step Reactions

| | $A + 3e \xrightarrow{i_A} B$ | |
|---|---|---|
| $A + e \rightleftharpoons C$<br>$C + e \rightleftharpoons D$<br>$D + e \rightarrow B$ | $A + e \rightleftharpoons C$<br>$C + e \rightarrow D$<br>$D + e \rightleftharpoons B$ | $A + e \rightarrow C$<br>$C + e \rightleftharpoons D$<br>$D + e \rightleftharpoons B$ |
| $i_A = kc_A \exp[-(2+\alpha) f E']$ | $i_A = kc_A \exp[-(1+\alpha) f E']$ | $i_A = kc_A \exp(-\alpha f E')$ |
| | $i_A = kc_A \exp[-b_A E']$ | |

TABLE 3.6. Mechanisms Exhibiting and Not Exhibiting Tafel Behavior

| | Mechanism | Tafel behavior | $\theta \to 1$ | $\theta \to 0$ |
|---|---|---|---|---|
| (1) | $A_1 + e \to A_2$ (rate-determining step)<br>$A_2 + e \to A_3$ | Yes | — | — |
| (2) | $A_1 + e \rightleftharpoons A_2$<br>$A_2 + e \to A_3$ (rate-determining step) | Yes | — | — |
| (3) | $A_1 + A_2 \rightleftharpoons A_3$<br>$A_3 + e \to A_4$ (rate-determining step) | Yes | — | — |
| (4) | $A_1 + e \to A_2$ (rate-determining step)<br>$A_2 + A_3 \to A_4$ | Yes | — | — |
| (5) | $A_1 + M + e \rightleftharpoons A_2 \cdots M$<br>$A_2 \cdots M + e \to A_3 + M$<br>(rate-determining step) | No | Yes | Yes?[a] |
| (6) | $A_1 + M + e \to A_2 \cdots M$<br>(rate-determining step)<br>$A_2 \cdots M + e \rightleftharpoons A_3 + M$ | No | Yes? | Yes |
| (7) | $A_1 + M + e \to A_2 \cdots M$<br>(rate-determining step)<br>$2A_2 \cdots M \rightleftharpoons A_3 + 2M$ | Yes | — | — |
| (8) | $A_1 + M + e \rightleftharpoons A_2 \cdots M$<br>$2A_2 \cdots M \to A_3 + 2M$<br>(rate-determining step) | No | Not potential dependent | |

[a] "Yes?" indicates that the equation differs from $i = kc_A \exp[-b_A E']$ due to absorption; see for example Eq. (3.139).

Some mechanisms involving adsorption steps (see Section 3.1.5.5) do not obey Eq. (3.146). The reader can easily test other mechanisms this way.

---

### EXAMPLE 3.4. Interpretation of Polarization Data in Terms of a Reaction Model

THE PROBLEM: Steady state electrochemical measurements of the overall reaction

$$A + 2e \rightleftharpoons B$$

were carried out at a temperature of 298 K. The Tafel slope changed with electrode potential from 35 mV/decade at low potentials to 65 mV/decade at high potentials. Explain these findings in terms of a reaction model.

THE SOLUTION: The data indicate that adsorption influences the reaction. The values of the Tafel slope can be equated to the coefficient $b_A$ in Eq. (3.146) (see also Section 3.1.5.2):

$$2.303/b_A = 0.035\ \text{V/decade}$$

Now:

$$f = nF/RT = 96500n/(8.314 \times 298) = 39n$$

Hence:

$$b_A/f = 2.303/(39n \times 0.035) = 1.69n \approx 1.5n \text{ at } 0.035\ \text{V/decade}$$

And:

$$b_A/f = 1.69n \times 0.035/0.065 = 0.91n \approx 1.0n \text{ at } 0.065\ \text{V/decade}$$

Assuming a one-electron rate-determining charge transfer step, $b_A/f$ at the lower potential is approximately $1 + \alpha$. From Table 3.5 we see that this is true if the second step is rate-determining. Considering the possibility that adsorption should be taken into account because of change in the Tafel slope, we suspect that mechanism (5) in Table 3.6 could be appropriate.

A proposed model:

$$A_1 + M + e \rightleftharpoons A_2 \cdots M$$

Rate-determining step:

$$A_2 \cdots M + e \rightarrow A_3 + M$$

Starting from Eq. (3.131):

$$\frac{dA_3}{dt} = \frac{k_1 k_2 A_1 \exp(-2\alpha f E') - k_{-1} k_{-2} A_3 \exp[2(1-\alpha) f E']}{k_1 A_1 \exp(-\alpha f E') + k_{-1} \exp[1-\alpha) f E'] + k_2 \exp(-\alpha f E') + k_{-2} A_3 \exp[(1-\alpha) f E']}$$

and using the conditions:

$$k_{-2} \rightarrow 0,\ k_1 \rightarrow \infty,\ k_{-1} \rightarrow \infty$$

Eq. (3.131) becomes:

$$\frac{dA_3}{dt} = \frac{K_1 k_2 A_1 \exp(-2\alpha f E')}{K_1 A_1 \exp(-\alpha f E') + \exp[(1-\alpha) f E']}$$

where $K_1 = k_1/k_{-1}$.

From Eq. (3.130):

$$\theta = \frac{K_1 A_1 \exp(-\alpha f E')}{K_1 A_1 \exp(-\alpha f E') + \exp[(1-\alpha) f E']}$$

The condition $\theta \to 0$ implies:

$$K_1 A_1 \exp(-\alpha f E') \ll \exp[(1-\alpha) f E']$$

That is:

$$dA_3/dt = K_1 k_2 A_1 \exp[-(1+\alpha) f E']$$

The condition $\theta \to 1$ implies:

$$K_1 A_1 \exp(-\alpha f E') \gg \exp[(1-\alpha) f E']$$

That is:

$$dA_3/dt = k_2 \exp(-2\alpha f E')$$

With $\alpha = 0.5$ the exponential varies from $1.5fE'$ to $1.0fE'$, in line with the experimental data.

---

### 3.2.2. Experimental Methods of Obtaining Model Constants

#### 3.2.2.1. Methodology

As we have seen in Section 3.2.1, the expression for current density of an electrochemical reaction contains a mass transfer and a kinetic element, as shown in Eq. (3.147). Steady state polarization curves and preparative runs are needed to provide information, respectively, on the dependence of total current densities and of partial current densities on electrode potential

### 3.2.2.2. Experimental Equipment

This is not a book on experimental methods, but the determination of model constants is so essential to modeling technique that a relatively detailed treatment of the necessary experimental equipment is justified. For further information the reader should consult Ref. 21.

The three types of experimental cells commonly employed are shown in Fig. 3.12. The simplest is the undivided cell with a magnetic stirrer, but if catholyte and anolyte must be separated a magnetically stirred H-cell divided by a glass frit is used. If a membrane separator is required the frit is replaced by a flange to clamp it. Both cells have a stationary working electrode, but the third type, also of a H-type configuration, employs a rotating disk electrode (RDE)[22,23]

The RDE (already discussed in Section 2.3.4.2) preferred not only for obtaining polarization data but also (with a somewhat larger electrode area) for preparative runs, because it offers uniform, predictable, and reproducible mass transfer rates. It is important not to make the RDE too large, since (see Section 2.3.3.4) the advantage of a clearly defined hydrodynamic regime might be lost.

The reason why magnetic stirring is frequently used in the cells shown in Fig. 3.12 is its simplicity and the relative ease with which the cells may be closed for the purpose of determining amounts of gaseous products. It has been our experience, however, that magnetic stirrers result in unreproducible mass transfer rates and that a mechanical stirrer is preferable.

The arrows appearing in the cells in Fig. 3.12 indicate the positioning of the Luggin probes. These are connected, as already described in Chapter 2, via a salt bridge to a reference electrode suitable to a particular electrolyte.

### 3.2.2.3. Ohmic Correction to the Electrode Potential

Using a Luggin probe introduces an error into the measurement of potential due to the gap which exists between the Luggin tip and the electrode surface. Consider the diagram in Fig. 3.13. The electrode potential $E'$ is given by Eq. (1.14):

$$E' = E_m - E_s$$

while the potential actually measured ($V$) is:

$$V = E_m - E_{sl} \tag{3.161}$$

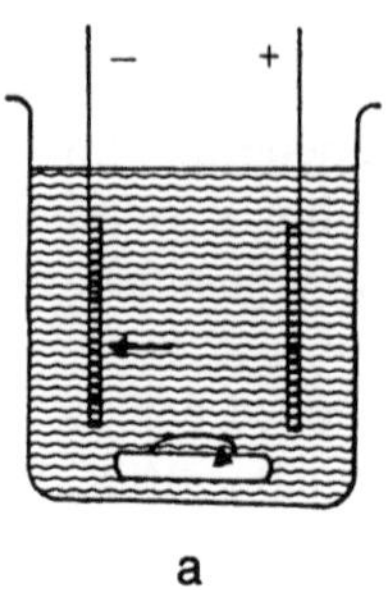


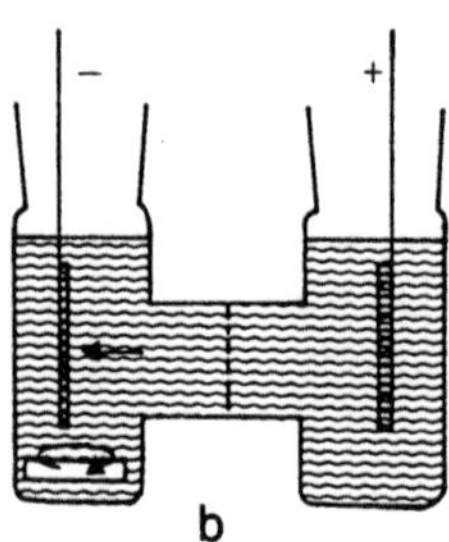


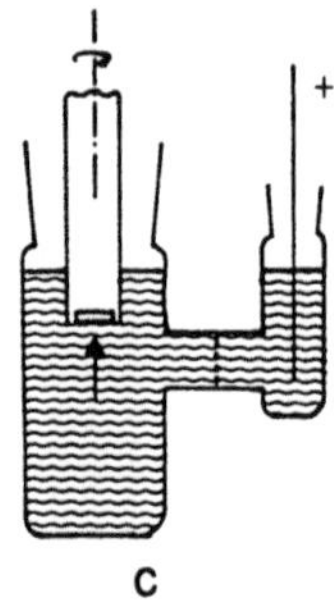


**FIGURE 3.12.** Laboratory cells.

It follows from Fig. 3.13 that:

$$iR = E_s - E_{sl} \tag{3.162}$$

where $R = \rho l/A$ (see Section 2.4.4.1), and that:

$$E' = V - iR \tag{3.163}$$

where $iR$ is the "ohmic correction."

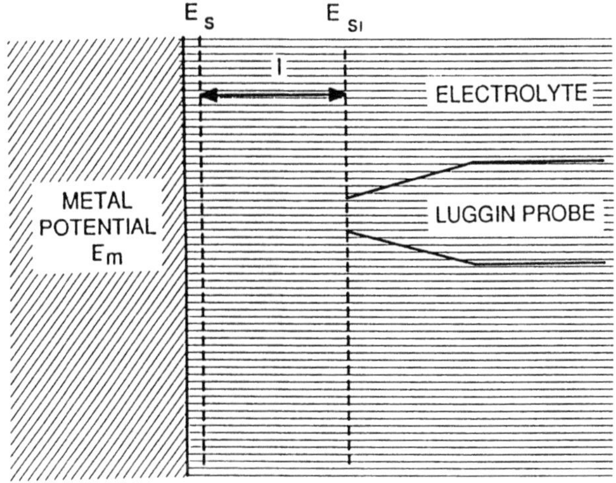


**FIGURE 3.13.** Ohmic error.

Electrochemical kinetics often work at not far from equilibrium conditions where current densities are relatively low. Also, a large number of fundamental studies are done in reasonably conducting electrolytes. These two factors plus with the fact that $l$ the distance between the Luggin tip and the electrode surface, is normally only 1–2 mm make the "ohmic correction" unimportant. In contrast, industrial practice aims at current densities that are as high as possible and electrolyte conductivity, in for example organic electrosynthesis, is often very low. If such a system is to be modeled, determining this ohmic drop is necessary. Figure 3.14 presents raw and corrected data for hydrogen evolution from a 99.999% lead cathode at 20°C. Although the electrolyte is a highly conducting aqueous solution of 0.5 M sulfuric acid, correction at 7000 A/m$^2$ amounts to 150 mV. In poorly conducting electrolytes (as in the example in Section 3.2.3.2) the correction can amount to a significant percentage of the actual electrode potential. Even in the more conductive uranium system (see Section 3.2.3.1) it was essential to apply an "ohmic correction" to the measured electrode potential.

#### 3.2.2.4. Methods of Determining the Ohmic Correction

The difficulty of measuring the gap between the Luggin and the electrode surface precludes the calculation of $iR$. Indirect methods have to be employed; three of them are described below.

(i) Hydrogen evolution methods. The fact that hydrogen evolution displays Tafel behavior up to reasonably high current densities (see Fig. 3.14) allows

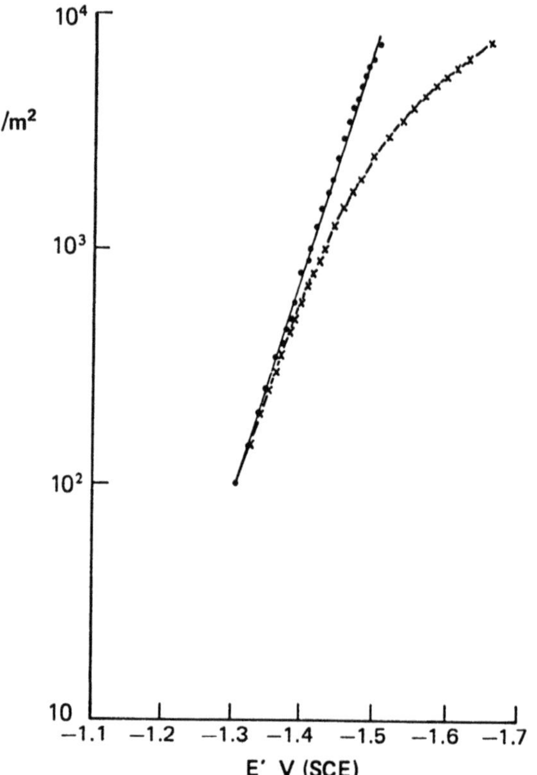


**FIGURE 3.14.** Ohmic correction for hydrogen evolution. $x$ = Raw data, ● = data corrected for ohmic drop.

it to be used as a means of determining the ohmic correction without special equipment.

After assembling a laboratory cell before experimentation, it is first filled with 0.5 M sulfuric acid and polarization data for hydrogen evolution are determined for current densities $i_1, i_2, \ldots, i_m$ and $ni_1, ni_2, \ldots, i_{nm}$ as indicated in Table 3.7, where $n$ is an integral multiple, say 10. The cell is washed out and the sulfuric acid replaced by the required working electrolyte, being careful to ensure that the relative positions of the Luggin probe and electrode are undisturbed. Using the notation of Table 3.7, the observed current densities and potentials are related:

$$2.303 \log [i] = a + bE_i' = a + b(V_i - iR) \tag{3.164}$$

TABLE 3.7. Polarization Data for the Hydrogen Evolution Method

| Current density $i$ | Observed potential $V_i$ | Current density $ni$ | Observed potential $V_{ni}$ | $\frac{V_{ni} - V_i}{n - 1}$ |
|---|---|---|---|---|
| $i_1$ | $V_{i1}$ | $ni_1$ | $V_{ni1}$ | $(V_{ni1} - V_{i1})/(n - 1)$ |
| $i_2$ | $V_{i2}$ | $ni_2$ | $V_{ni2}$ | $(V_{ni2} - V_{i2})/(n - 1)$ |
| $i_3$ | $V_{i3}$ | $ni_3$ | $V_{ni3}$ | $(V_{ni3} - V_{i3})/(n - 1)$ |
| . | . | . | . | . |
| . | . | . | . | . |
| $i_m$ | $V_{im}$ | $ni_m$ | $V_{nim}$ | $(V_{nim} - V_{im})/(n - 1)$ |

Also:

$$2.303 \log [ni] = a + b(V_{ni} - niR) \tag{3.165}$$

Subtracting Eq. (3.164) from Eq. (3.165) and rearranging:

$$(V_{ni} - V_i)/(n - 1) = Ri - \text{constant} \tag{3.166}$$

A plot of $(V_{ni} - V_i)/(n - 1)$ against $i$ will have a slope of $R$. This value, corrected for the change in conductivity, is directly transferable to the working electrolyte if the Luggin is not disturbed during the emptying and filling operations after the polarization observations of the sulfuric acid. The method is precluded if hydrogen evolution has a deleterious effect on the electrode material.

(ii) Interrupter method. The principle of this method is demonstrated in the electrode equivalent circuit in Fig. 3.15. When the current to a cell is switched off the electrode potential falls in two stages, a sudden drop and a gradual decay (Fig. 3.16). The gradual decay is associated with a resistive/capacitive model of the electrode process, while the sudden drop is associated with the purely resistive loads in the lead to the electrode ($R_L$ and $R_C$) and in the electrolyte gap $R_G$ between the electrode and the Luggin tip. Measurement of these drops in voltage is conveniently carried out with a storage oscilloscope. If resistive loads to the electrode are made negligible, the vertical portion of the oscilloscope trace in Fig. 3.16 is a direct measure of the ohmic correction at a given current density. Repeating the procedure at different current densities enables the ohmic nature of $R_G$ to be determined. The transient is obtained by a manual make-and-break of the circuit

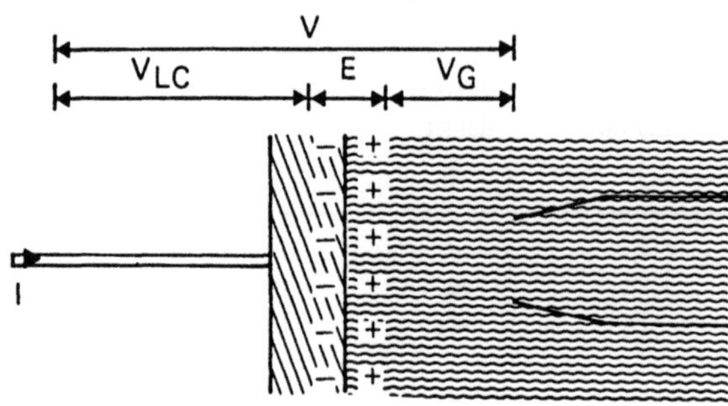


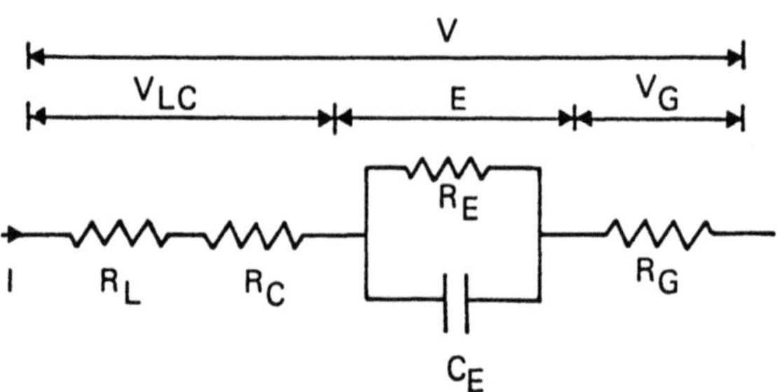


**FIGURE 3.15.** Electrode equivalent circuit.

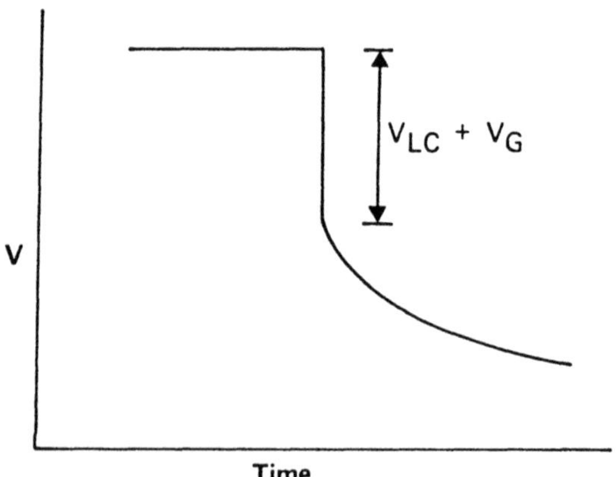


**FIGURE 3.16.** Potential–time trace.

using a logic switch. Equipment for using this interrupter technique is commercially available.

(iii) Frequency response methods. This method superimposes low-amplitude, high-frequency (about 1000 Hz) AC signals over the DC potential supplied to the electrode. The response of the resulting AC component of current to changes in the frequency is analyzed to give information on a variety of electrode parameters including the ohmic correction. For details of the principle and its application, consult Refs. 24 and 25.

### 3.2.3. Example of Reaction Models

#### 3.2.3.1. The Reduction of U(VI) to U(IV)

The reduction of hexavalent to tetravalent uranium is an important step in the reprocessing of spent nuclear fuels. Using an electrochemical technique instead of a chemical reducing agent avoids having to remove a radioactive chemical reagent. The model described below was developed to predict the performance of pilot plant cells. A model should enable one to describe an electrode-potential window (see below) for satisfactory operation of the cell.

The experimental results for the basic model of the system indicated in Fig. 3.17 were obtained with a titanium RDE. The electrolyte was aqueous nitric acid containing hydrazine, which stabilizes the U(IV) ion.

The primary reaction is:

$$\mathrm{U(VI)} + 2e \rightarrow \mathrm{U(IV)} \tag{3.167}$$

The transport of U(VI) from the bulk solution to the electrode surface is given by Eq. (3.141):

$$N_\mathrm{U} = dQ_\mathrm{U}/dt = Sk_L(c_\mathrm{U} - c_{\mathrm{U}s}) \tag{3.168}$$

where $N_\mathrm{U}$ is the number of kmoles of U(VI) reaching the electrode surface in unit time, $c_\mathrm{U}$ and $c_{\mathrm{U}s}$ are the concentrations of U(VI) in the bulk and near the electrode surface, and all other symbols are as defined in Eq. (3.141).

$R_\mathrm{U}$, the number of kmoles of U(VI) converted to U(IV) in unit time, is given by Eq. (3.146):

$$R_\mathrm{U} = Sk'_\mathrm{U}c_{\mathrm{U}s}\exp[-b_\mathrm{U}E'] \tag{3.169}$$

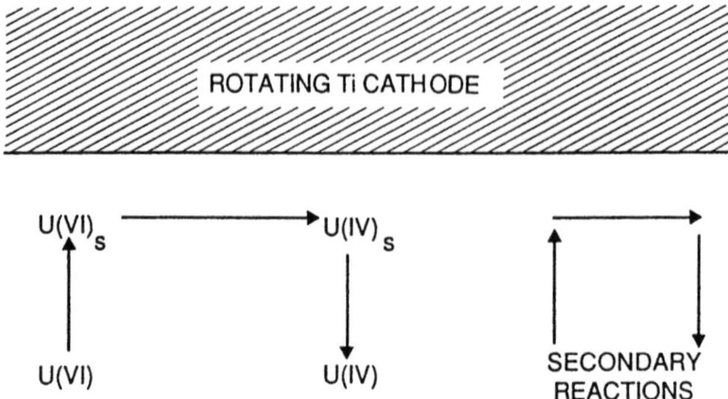


**FIGURE 3.17.** Model for U(VI) reduction.

Since accumulation of U(VI) at the surface of the electrode will be negligible compared with the rate terms $N_U$ and $R_U$, the two can be assumed equal and will be given the symbol $N$:

$$N = Sk_L(c_U - c_{Us}) = Sk'_U c_{Us} \exp[-b_U E'] \tag{3.170}$$

Eliminating $c_{Us}$ from Eq. (3.170):

$$N = \frac{Sc_U}{1/k_L + 1/[k'_U \exp(-b_U E')]} \tag{3.171}$$

The coulombic requirement $I_U$ for this reduction leads to:

$$2\mathfrak{F}N = I_U = \frac{Sc_U}{1/(2\mathfrak{F}k_L) + 1/(k_U \exp[-b_U E'])} \tag{3.172}$$

Hence:

$$i_U = I_U/S = \frac{c_U}{1/(2\mathfrak{F}k_L) + 1/(k_U \exp[-b_U E'])} \tag{3.173}$$

where $i_U$ is the partial current density for the reduction of U(VI). Note that Eq. (3.173) is identical with Eq. (3.147).

The secondary reactions (a number of complex reactions including the reduction of nitric acid and hydrogen evolution) are lumped together under the partial current density $i_S$ (see also Section 1.4.2), where:

$$i = i_U + i_S \tag{3.174}$$

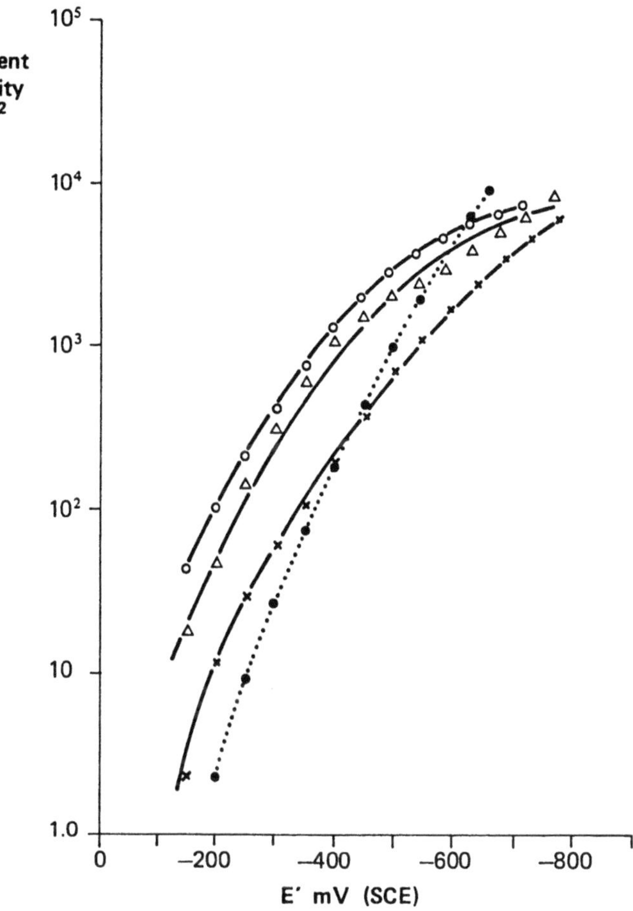


**FIGURE 3.18.** U(VI) polarization curves for 2.0 M $HNO_3$, 0.2 M $N_2H_4$. U(VI) concentrations: ○ = 0.8 M, Δ = 0.3 M, $x$ = 0.1 M, ● = background.

$i$ being the observed current density. As shown in Eq. (1.22):

$$\text{C.E.} = 100 i_U / i \qquad (3.175)$$

(i) Partial current density $i_U$. Equation (3.175) is used to convert the data in Figs. 3.18 and 3.19 to determine values of $i_U$. Referring to Eq. (3.173), $i_U/c_U$ will be dependent, for a given rotation speed, only on the electrode potential. A plot of $i_U/c_U$ against $E'$ should result in a unique curve. Such a plot (Fig. 3.20), within the error commonly encountered in this kind of work bears this out. The plateau of the curve represents the limiting current

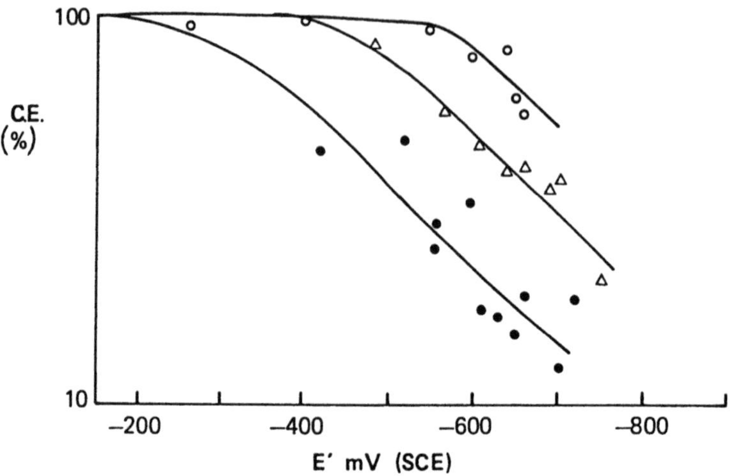


**FIGURE 3.19.** U(VI) current efficiency curves: RDE, 1000 rpm, 2.0 M $HNO_3$, 0.2 M $N_2H_4$. U(VI) concentrations: ○ = 0.8 M, Δ = 0.3 M, ● = 0.1 M.

density $i_{lim}$ for U(VI) reduction as expressed in Eq. (3.102). In accordance with this equation Fig. 3.20 suggests for $2\mathfrak{F}k_L$ a value of 6200 $A/m^2$ per kmol/$m^3$. This corresponds to an actual partial limiting current density of 6200 $A/m^2$ for a solution which is molar with respect to U(VI).

(ii) Kinetic constants $b_U$ and $k_U$. Rearrangement of Eq. (3.173) leads to:

$$c_U/i_U - 1/2\mathfrak{F}k_L = 1/k_U \exp[-b_U E'] \tag{3.176}$$

Taking logarithms:

$$\log_{10}[c_U/i_U - 1/2\mathfrak{F}k_L] = -\log_{10}[k_U] - b_U E'/2.303 \tag{3.177}$$

A plot of Eq. (3.177) for the present data, using $2\mathfrak{F}k_L$ determined in Section (3.2.3.1.i) from Fig. 3.20, is shown in Fig. 3.21. From the slope of Fig 3.21 $b_U$ is 16 $V^{-1}$ and $k_U$ is 5.0 A m/kmol, giving an equation for $i_U$, Eq. (3.173):

$$i_U = \frac{c_U}{1/(19 \times 10^7 k_L) + 1/(5.0 \exp[-16E'])} \tag{3.178}$$

(iii) The secondary reactions. Although the parallel secondary or background reactions involve complex reduction processes of nitric acid and hydrogen evolution, for the purpose of the present model they have been lumped together under the partial current density $i_S$. A background polar-

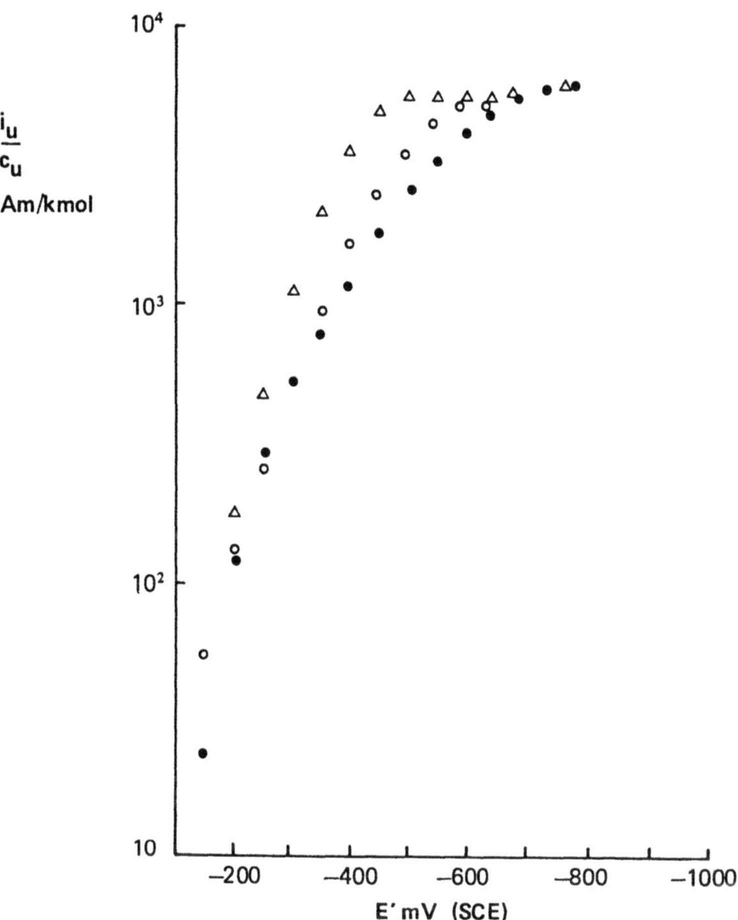


**FIGURE 3.20.** Effect of $c_U$ on $i_U$: RDE, 1000 rpm, 2.0 M $HNO_3$, 0.2 M $N_2H_4$. U(VI) concentrations: ○ = 0.8 M, Δ = 0.3 M, ● = 0.1 M.

ization curve is plotted in Fig. 3.18. Other values of $i_S$ can be obtained from the data in Fig. 3.18 using Eq. (3.174):

$$i_S = i - i_U \tag{3.179}$$

From the set of polarization data in Fig. 3.18, at higher potentials there is a marked suppression of the secondary reactions in the presence of U(VI). Secondary reactions are not truly independent, as the associated kinetics can, not surprisingly, depend on changes taking place at the electrode caused by the main reaction.

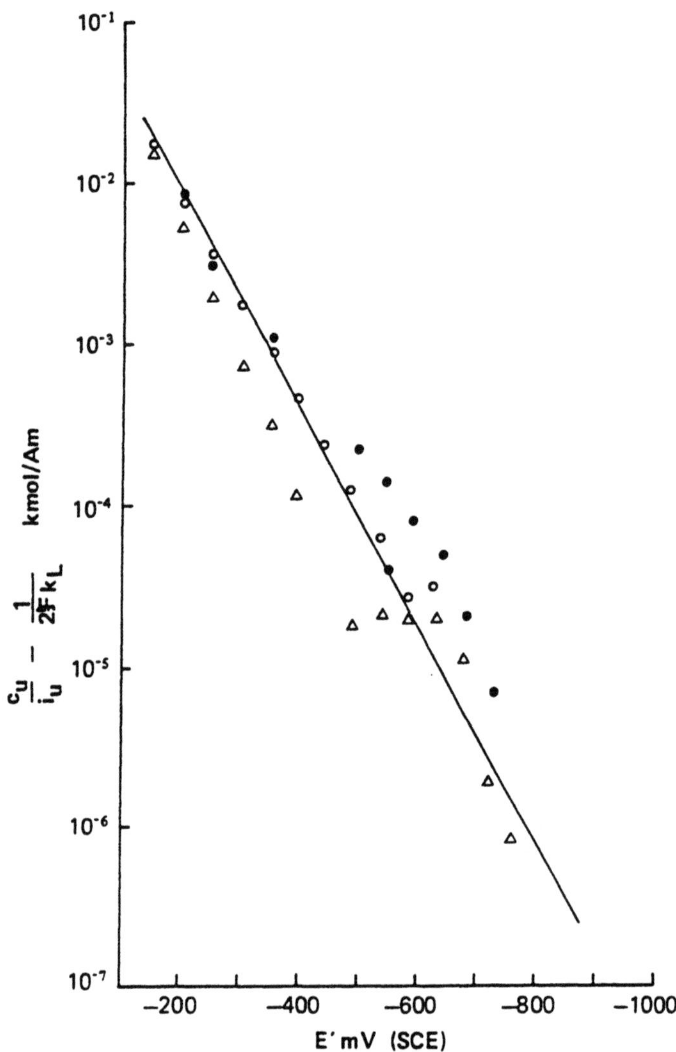


**FIGURE 3.21.** Plot for determining $b_U$ and $k_U$: RDE, 1000 rpm, 2.0 M $HNO_3$, 0.2 M $N_2H_4$. U(VI) concentrations: ○ = 0.8 M, Δ = 0.3 M, ● = 0.1 M.

At this point a warning should be issued: **Insufficient experimental work can be harmful to the predictive powers of a model.**

Looking at the model so far, the results hang together and make sense. The plot in Fig. 3.21 produces a straight line over a large potential range (700 mV). Again, 16 $V^{-1}$ for $b_U$ is not far removed from 19 $V^{-1}$, the value you would expect for a one-electron change rate-determining step. But further experimental work has indicated complexities which did not appear in the preliminary work just described. We will not go into details, but it was found that the nature of the titanium surface depends on the range of electrode potential used and the time needed to form the new surface. In simple terms, below $-500\,\text{mV}$ (vs. SCE) the reaction takes place on a catalytic oxide surface; above that value, hydride formation occurs which eventually damages the electrode, which is undesirable in an industrial context. The model finally produced uses equations similar to those derived above but in addition takes notice of the surface on which the reaction takes place. The apparent suppression of the secondary reactions is not caused by the presence of U(VI) but that at those higher potentials the reaction occurs on a hydride surface. Secondary reactions are not necessarily truly independent: their kinetics can depend on surface changes caused by the main reaction.

### 3.2.3.2. The Production of *p*-Anisidine

*p*-Anisidine,[26] like *p*-aminophenol,[27] can be prepared by the reduction of nitrobenzene in an acidic medium. The reaction is carried out in a methanol environment and (Fig. 3.22) proceeds via phenylhydroxylamine as an intermediate, which by direct rearrangement can lead to the formation of *p*-aminophenol as a small but significant by-product. For the purpose of the present model, products *p*- and *o*-anisidine and *p*-aminophenol are referred to collectively as *C*. The chemical yield is also decreased by the reduction of phenylhydroxylamine to aniline.

The reaction scheme is shown in Fig. 3.23. In contrast to the previous example, we will treat the two secondary reactions separately.

On the assumption of Section 3.2.3.1, namely that there is no accumulation of the reacting $A_s$, $B_s$, and $D_s$ and that the Tafel-type expression in Eq. (3.146) applies, one obtains a relationship for $i_A$, the current density, for the primary reaction:

$$i_A = \frac{c_A}{1/(4\mathfrak{F}k_L) + 1/(k_A \exp[-b_A E'])} \tag{3.180}$$

Assuming (1) the rate of diffusion of $B$ away from the electrode is faster than the rate of the chemical step $B \rightarrow C$; i.e., as a first approximation the buildup of $B$ in the bulk of the electrolyte is negligibly small (in practice $c_B$ represents 10 to 15% of $c_C$), and (2) Eq. (3.146) can also be used for $i_B$, the current density of the consecutive secondary reaction, an equation can be derived:

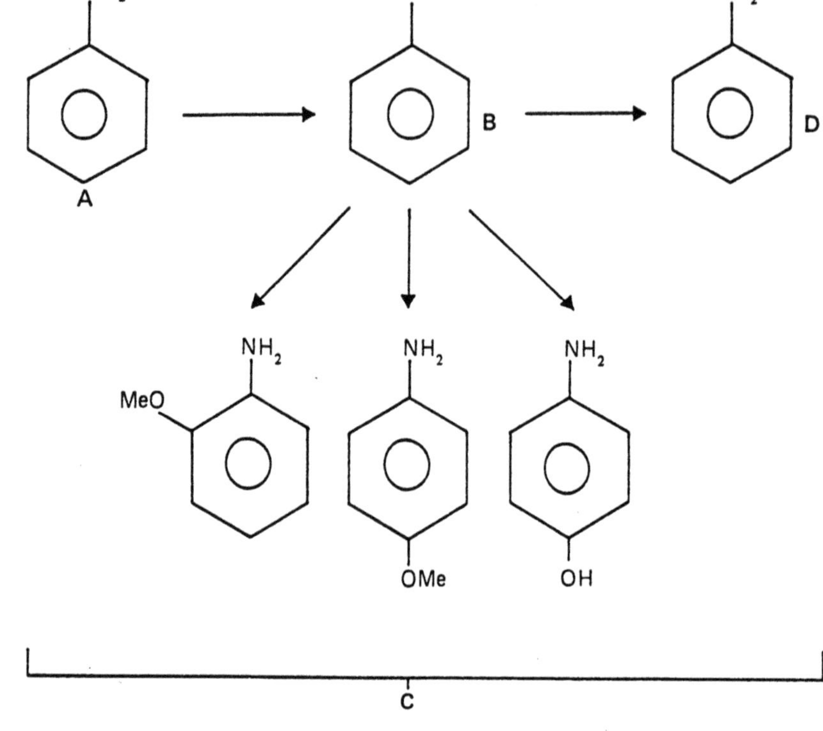


**FIGURE 3.22.** Reduction products of nitrobenzene in methanol.

$N_B$, the number of kmoles of $B$ diffusing away from the surface will be, with $c_B \approx 0$:

$$N_B = Sk_L c_{Bs} \tag{3.181}$$

$R_A$ and $R_B$, the number of kmoles of $A$ and $B$ converted to $B$ and $D$ respectively, will be:

$$R_A = Sr_A \qquad \text{and} \qquad R_B = Sr_B \tag{3.182}$$

With the previously mentioned Tafel behavior:

$$i_B = k_B c_{Bs} \exp(-b_B E') \tag{3.183}$$

A coulombic balance gives:

$$i_A = 4\mathfrak{F} r_A \qquad \text{and} \qquad i_B = 2\mathfrak{F} r_B \tag{3.184}$$

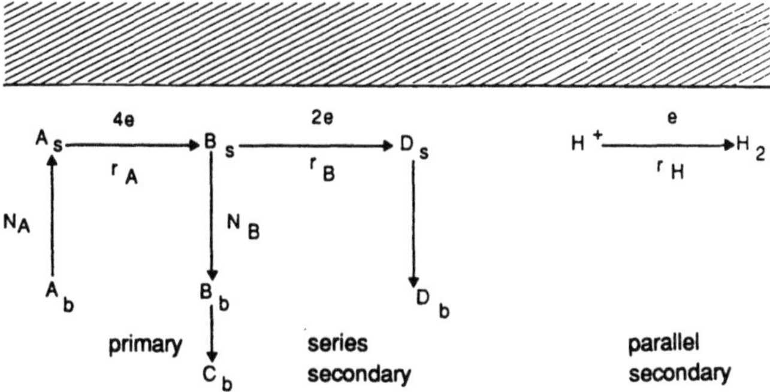


**FIGURE 3.23.** Reaction scheme for the production of *p*-anisidine.

Since accumulation of $B$ at the electrode surface is negligible,

$$R_A = N_B + R_B \qquad \text{or} \qquad Sr_A = Sk_L c_{Bs} + Sr_B \tag{3.185}$$

giving:

$$i_A/4 = k_L c_{Bs} + i_B/2 = k_L c_{Bs} + k_B c_{Bs} \exp(-b_B E')/2 \tag{3.186}$$

From Eq. (3.186):

$$c_{Bs} = \frac{i_A}{4\mathfrak{F}k_L + 2k_B \exp(-b_B E')} \tag{3.187}$$

and hence:

$$i_B = \frac{i_A k_B \exp[-b_B E']}{4\mathfrak{F}k_L + 2k_B \exp[-b_B E']} \tag{3.188}$$

Using the simple Tafel-type relationship $i_H = k_H \exp(-b_H E')$: one can write:

$$i = i_A + i_B + i_H \tag{3.189}$$

and hence:

$$i = \left[\frac{c_A}{\dfrac{1}{4\mathfrak{F}k_L} + \dfrac{1}{k_A \exp(-b_A E')}}\right]\left[\frac{4\mathfrak{F}k_L + 3k_B \exp(-b_B E')}{4\mathfrak{F}k_L + 2k_B \exp(-b_B E')}\right] + k_H \exp(-b_H E') \tag{3.190}$$

To produce a reaction model we must obtain numerical values for constants $k_A$, $k_B$, $k_H$, $b_A$, $b_B$, and $b_H$ (see Section 3.2.2.2). These are determined largely from preparative runs carried out galvanostatically in an H-cell at 55°C, the data obtained being presented in Table 3.8. The cathode is copper. A few polarization runs were carried out with a copper RDE; they are a source of Fig. 3.24. It can be seen that the background polarization, i.e., $i_H$, is a negligible fraction of the preparative current densities. We can therefore write:

$$i = i_A + i_B \tag{3.191}$$

(i) Determination of $b_A$. At low cathodic potentials, $1/(4\mathfrak{F}k_L) \ll 1/(k_A \exp[-b_A E'])$. Equation (3.180) reduces to:

$$i_A = k_A c_A \exp[-b_A E'] \tag{3.192}$$

From Table 3.8, at low electrode potentials amounts of aniline are very small, i.e., $i = i_A$. Hence:

$$i = k_A c_A \exp[-b_A E'] \tag{3.193}$$

TABLE 3.8. Preparative Results for the Production of Anisidine and Aniline

| Electrode potential $E'$ vs.SCE (mV) | Products (kmoles × $10^6$) | | | | | Current density $i$ ($A/m^2$) |
|---|---|---|---|---|---|---|
| | *o*-anisidine | 4e *p*-anisidine | *p*-aminophenol | 6e aniline | Total | |
| 216 | 1.46 | 6.22 | 1.76 | 0.28 | 9.72 | 250 |
| 258 | 1.36 | 6.30 | 1.65 | 0.43 | 9.74 | 770 |
| 260 | 1.23 | 5.75 | 1.48 | 0.39 | 8.85 | 700 |
| 262 | 1.40 | 6.27 | 1.72 | 0.22 | 9.61 | 500 |
| 268 | 1.26 | 5.77 | 1.70 | 0.66 | 9.39 | 900 |
| 274 | 1.50 | 6.66 | 1.94 | 0.19 | 10.29 | 250 |
| 276 | 1.48 | 6.65 | 1.81 | 0.18 | 10.12 | 500 |
| 278 | 1.24 | 6.22 | 1.61 | 0.60 | 9.67 | 365 |
| 291 | 1.12 | 5.60 | 1.65 | 0.88 | 9.25 | 900 |
| 311 | 0.84 | 3.74 | 1.11 | 2.05 | 7.74 | 1600 |
| 316 | 1.23 | 5.70 | 1.57 | 0.73 | 9.23 | 1200 |
| 317 | 1.19 | 5.50 | 1.49 | 0.92 | 9.10 | 1200 |
| 320 | 1.10 | 5.08 | 1.48 | 0.94 | 8.60 | 1140 |
| 324 | 1.17 | 5.46 | 1.46 | 1.20 | 9.29 | 1170 |
| 334 | 1.10 | 4.87 | 1.30 | 1.25 | 8.52 | 1450 |
| 338 | 1.14 | 5.38 | 1.41 | 1.58 | 9.51 | 1780 |
| 344 | 1.07 | 4.67 | 1.40 | 1.63 | 8.77 | 1450 |
| 348 | 1.08 | 5.29 | 1.49 | 2.71 | 10.57 | 2040 |
| 373 | 0.93 | 4.18 | 1.14 | 2.51 | 8.76 | 1930 |

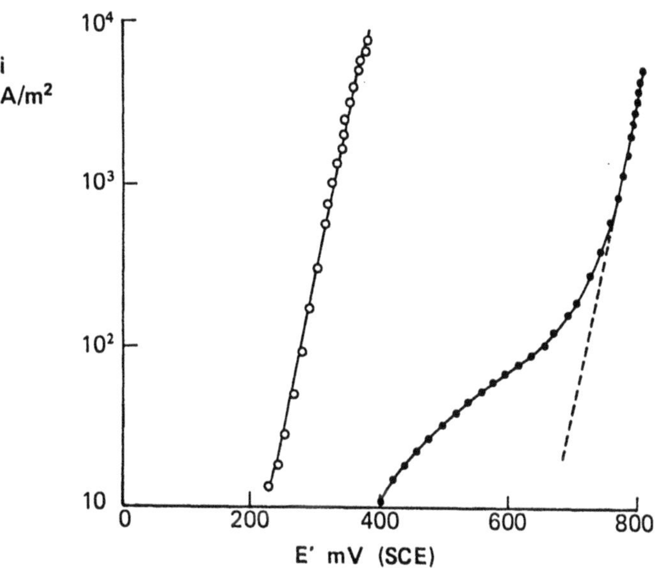


**FIGURE 3.24.** Polarization curves for the reduction of nitrobenzene in $CH_3OH/H_2SO_4$: RDE, 40 Hz, 55°C. ○ = 0.3 M $\phi NO_2$, ● = background.

or

$$\ln[i/c_A] = \ln[k_A] - b_A E' \tag{3.194}$$

As shown in Fig. 3.25, a plot of $\ln[i/c_A]$ against $E'$ of the results obtained with the RDE gives a straight line with a slope of $b_A = 0.0339\,\mathrm{mV}^{-1}$. Although an approximate value of $k_A$ could have been obtained by extrapolating Fig. 3.25, this was not appropriate, since the highly polished surface of the RDE was quite different from that of the copper electrode being used in prolonged synthesis.

(ii) Determination of $b_B$ and $k_B$. On dividing Eq. (3.180) by Eq. (3.188) we get:

$$\frac{i_A}{i_B} = \frac{4\mathfrak{F}k_L + 2k_B \exp[-b_B E']}{k_B \exp[-b_B E']} \tag{3.195}$$

or

$$\frac{i_A}{i_B} - 2 = \frac{4\mathfrak{F}k_L}{k_B}\exp(b_B E') \tag{3.196}$$

Taking logarithms:

$$\ln\left[\frac{i_A}{i_B} - 2\right] = \ln\left[\frac{4\mathfrak{F}k_L}{k_B}\right] + b_B E' \tag{3.197}$$

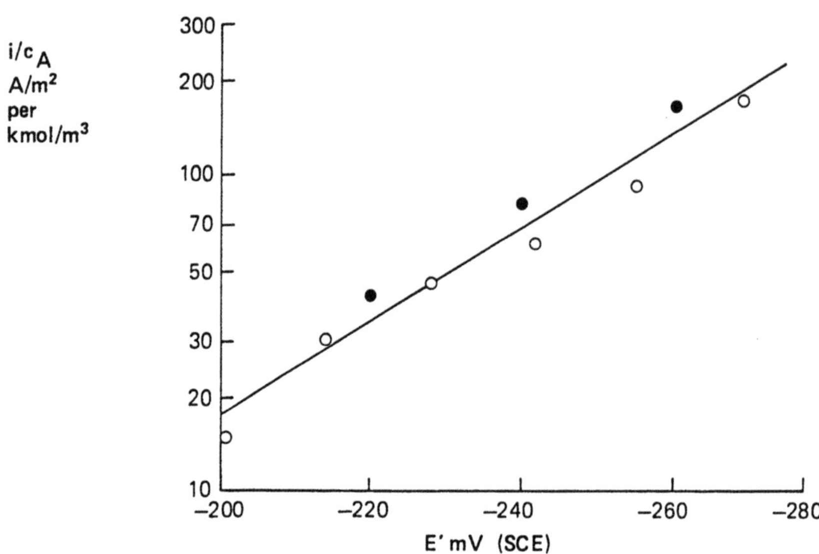


**FIGURE 3.25.** Determination of $b_A$: RDE, 40 Hz, 55°C. $\phi NO_2$ concentrations: ○ = 0.3 M, ● = 0.03 M.

A plot of $\ln[(i_A/i_B) - 2]$ against $E'$ will have a slope of $b_B$ and an intercept from which $k_B$ can be determined if $k_L$, the mass transfer coefficient of the cell, is known. In order to use Eq. (3.197), values of $i_A/i_B$ were calculated on the basis of the following argument (see Fig. 3.23). The main products require the transfer of four electrons, but aniline needs the transfer of six. Also from Fig. 3.23:

$$r_A = N_B + r_B \tag{3.198}$$

that is, $r_A$ = production rate of four electron products + production rate of aniline. Hence, $i_A = 4\mathfrak{F} r_A = 4\mathfrak{F} \times$ production rate of all products.

Since errors in the production rate of all products are unavoidable it is better to write: $i_A \propto 4\mathfrak{F} \times$ measured production rate of all products.

Similarly, $i_B \propto 2\mathfrak{F} \times$ measured production rate of aniline. Thus:

$$\frac{i_A}{i_B} = \frac{2 \times \text{measured production rate of all products}}{\text{measured production rate of aniline}}$$

A plot of Eq. (3.197) is presented in Fig. 3.26. The slope gives a value for $b_B$ of $0.019\,\text{mV}^{-1}$. The value of $k_L$ used to determine $k_B$ is $1.6 \times 10^{-5}$ m/s as found by the limiting current density technique discussed in Section 2.3.4 using copper

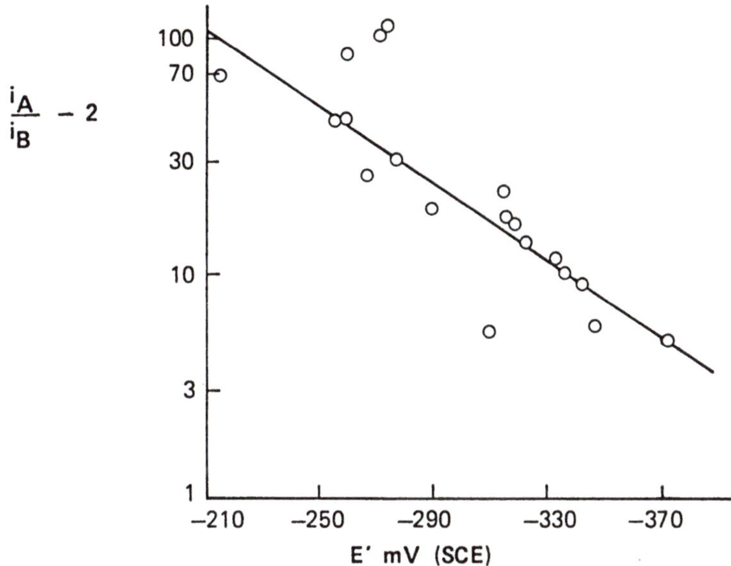


**FIGURE 3.26.** Determination of $b_B$ and $k_B$. Points calculated from data in Table 3.4.

deposition from aqueous acidic copper sulfate. The observed experimental value is adjusted for differences in viscosity and diffusivity when applied to the organic system. Then $k_B$ is 1.03 A/m$^2$ per kmol/m$^3$.

(iii) Determination of $k_A$. Equation (3.180) can be rearranged:

$$\frac{1}{Y} = \frac{1}{k_A c_A \exp[-b_A E']} = \frac{1}{i_A} - \frac{1}{4\mathfrak{F} k_L c_A} \tag{3.199}$$

This gives:

$$Y = k_A c_A \exp[-b_A E'] \tag{3.200}$$

From a plot of $Y$ against $\exp[-b_A E']$ (Fig. 3.27), $k_A$ is 0.217 A/m$^2$ per kmol/m$^3$. Values of $i_A$ are calculated from $i_B/i_A$ obtained from the preparative results (Section 3.2.3.2.ii).

(iv) Determination of $b_H$ and $k_H$. The kinetic constants for the hydrogen evolution reaction are determined from the background polarization curve (Fig. 3.24) which gives values for $b_H$ and $k_H$ of 0.037 mV$^{-1}$ and $4.9 \times 10^{-10}$ A/m$^2$. The value of $b_H$ is very sensitive to small changes in $k_H$.

(v) The final equations of the reaction model. Having determined appropriate values of the rate constants for the partial current densities of the reaction

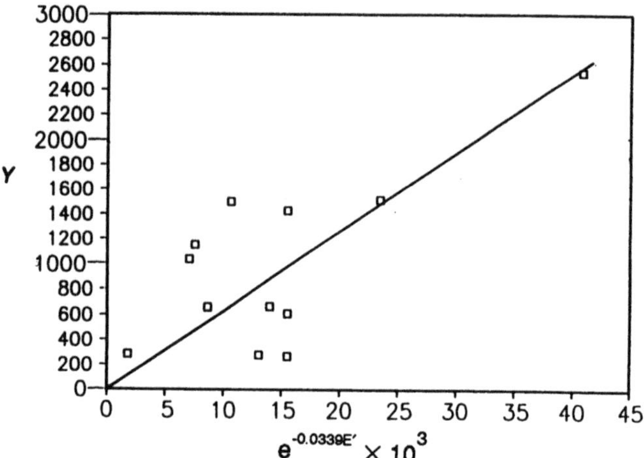


**FIGURE 3.27** Determination of $k_A$.

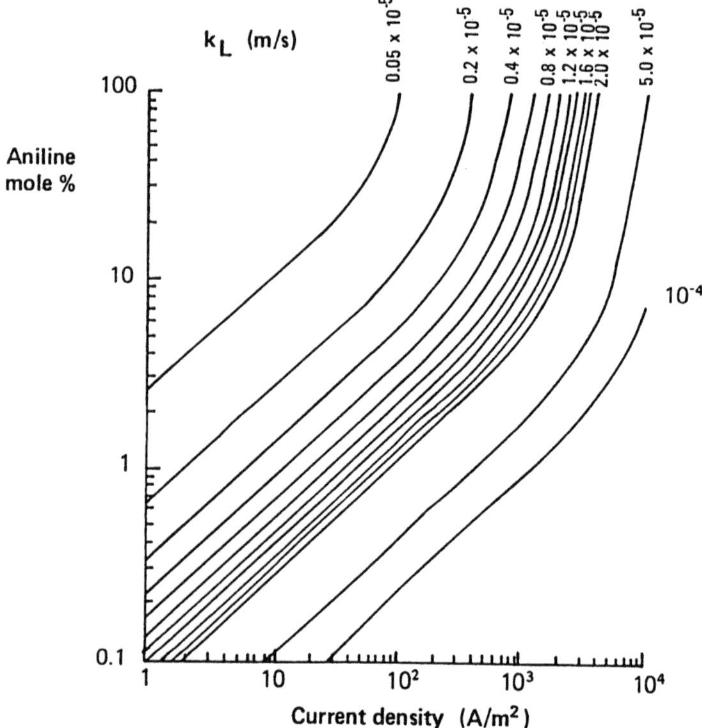


**FIGURE 3.28** By-product formation as a function of mass transfer rate.

model, we can fill in numerical values in Eqs. (3.180), (3.188), and (3.148):

$$i_A = \frac{c_A}{1/4\mathfrak{F}k_L + 1/0.22\exp[-0.034E']}$$

$$i_B = \frac{1.03 i_A \exp[-0.019E']}{4\mathfrak{F}k_L + 2.06\exp[-0.019E']}$$

$$i_{\mathrm{H}} = 4.9 \times 10^{-10}\exp[0.037E']$$

All current densities are in $A/m^2$ and $E'$ is in mV.

Using the model, Fig. 3.28 shows how strongly the relative production of the unwanted by-product aniline is dependent on the magnitude of the mass transfer coefficient.

## 3.3. CONCLUDING REMARKS

The two examples of reaction models presented in Section 3.2.3 should acquaint readers enough with the principle of setting up a reaction model that they can apply the principle to any synthesis. The difference between a reaction mechanism and a reaction model (mentioned in Section 3.2) should now be clear.

We emphasize again the need for reaction modeling. Returning to Fig. 3.28, the plot enables us to specify the required mass transfer characteristics of any cell intended to be used in a synthesis plant. A relatively simple model (if it is appropriate; some uncertainty in the numerical values of the model constants is not usually important) enables us to select the variables which govern performance. Experimentation, particularly at the expensive pilot plant stage, can be minimized and development costs significantly reduced. Development of a reaction model is an essential first step to obtaining a reactor model, which in turn is a must for process optimization.

## REFERENCES

1. Pilling M. J., 1975, *Reaction Kinetics*, Oxford Clarendon Press, Oxford Chemistry Series.
2. Moore, W. J., 1972, *Physical Chemistry*, Longman.
3. Barrow, G. M., 1979, *Physical Chemistry*, McGraw-Hill, New York.
4. Smith J. M., 1981, *Chemical Engineering Kinetics*, McGraw-Hill, New York.
5. Barrow, p. 681.
6. Kreysa, G., and Medin. H., 1986, "Indirect electrosynthesis of p-methyoxybenzaldehyde," *J. Appl. Electrochem.*, **16**: 757–767.

7. Goodridge, F., Harrison, S., and Plimley, R. E., 1986, "The electrochemical production of propylene oxide," *J. Electroanal. Chem.*, **214**: 283–293.
8. Barrow, p. 670.
9. Glasstone, S., Laidler, K. J., and Eyring, H., 1941, *The Theory of Rate Processes*, McGraw-Hill, New York.
10. Barrow, pp. 697–700.
11. Bard, A. J., and Faulkner, L. R., 1980, *Electrochemical Methods*, John Wiley & Sons, New York.
12. Barrow, p. 19.
13. Koryta, J., Dvorak, J., and Bohackova, V., 1970, *Electrochemistry*, Methuen & Co., New York, p. 257.
14. Thirsk, H. R., and Harrison, J. A., 1972, *A Guide to the Study of Electrode Kinetics*, Academic Press, p. 15.
15. Treyball, R. E., 1980, *Mass-Transfer Operations*, McGraw-Hill, New York, p. 47.
16. Rousar, I., Micka, K., and Kimla, A., 1986, *Electrochemical Engineering*, Vol. 1, Elsevier, pp. 23–27.
17. Fried, I., 1973, *The Chemistry of Electrode Processes*, Academic Press, p. 54.
18. Delahay, P., 1965, *Double Layer and Electrode Kinetics*, John Wiley & Sons, New York, p. 83.
19. Pletcher, D., 1982, *Industrial Electrochemistry*, Chapman & Hall, pp. 32–41.
20. Trasatti, S., and O'Grady, W. E., 1981, "Properties and applications of $RuO_2$-based electrodes," in *Advances in Electrochemistry and Electrochemical Engineering*, Vol. 12, (H. Gerischer and C. W. Tobias, eds.), John Wiley & Sons, New York, pp. 180–261.
21. Goodridge, F., and King, C. J. H., 1974, "Experimental methods and equipment," in *Techniques of Electroorganic Synthesis I*, (N. L. Weinberg, ed.), John Wiley & Sons, New York, pp. 7–147.
22. Albery, J., 1975, *Electrode Kinetics*, Clarendon Press, Oxford, pp. 49–55.
23. Thirsk and Harrison, pp. 83–95.
24. Harrison, J. A., and Small, C. E., 1980, "The automation of electrode kinetic measurements—Part 1: The instrumentation and fitting of the data using a library of reaction schemes," *Electrochim. Acta*, **25**: 447–452.
25. Harrison, J. A., 1982, The automation of electrode kinetic measurements—Part 7," *Electrochim. Acta*, **27**: 1113–1122.
26. Clark, J. M. T., Goodridge, F., and Plimley, R. E., "A reaction model for the electrochemical production of *p*-anisidine," *J. Appl. Electrochem.*, **18**: 899–903.
27. Lund, H., 1991, "Cathodic reduction of nitro and related compounds," in *Organic Electrochemistry, an Introduction and a Guide*, (Henning Lund and M. M. Baizer, eds.), Marcel Dekker, New York, pp. 401–432.

# Chapter 4

# Reactor Models

## 4.1. GENERAL CONSIDERATIONS

An electrolytic reactor or cell is a device in which one or more chemical species are transformed to alternative states with an associated energy change. There are many kinds of reactors; selection of an appropriate one is dealt with in Chapter 5. The form a reactor takes depends largely on the number of phases involved and the energy requirements. The size depends on the rate of reaction: we shall see that a reactor model can provide important answers about this rate. The reaction rate of an electrolytic reactor is defined by the possible current density, expressed in terms of reaction models in Chapter 3.

It is a unique feature of electrolytic reactors that the current and hence the reaction rate can readily be measured. The fact that the current causes the reaction provides us with an additional control parameter, the mode of current supply. Three modes can be used: galvanostatic, potentiostatic, and voltastatic, referring in turn to a constant current, a constant electrode potential, and a constant voltage. Because of the cost and nonavailability of large potentiostats, industrial practice is almost confined to a galvanostatic or less frequently voltastatic mode, using cheaper and more readily available DC rectifiers. Only when the reactant is very expensive and chemical yield is of overriding importance would one consider running an industrial reactor in potentiostatic mode.

Reactor models can be formulated in several ways depending on their purpose. For instance, if optimization is intended, we will be interested in costs, and the cost of a reactor as a function of its size, size in this context being measured by $S$, the surface area of the working electrode. Our model will therefore take the functional form:

$$S = f(i, X, \dots) \tag{4.1}$$

where $i, X, \dots$ are process variables which can influence $S$; $i$ is the operating current density and $X$ the fractional chemical conversion.

If we are interested in pilot plant experimentation the available area is usually fixed and our model will be expressed in terms of other dependent variables, e.g., current efficiency and batch time.

We will develop models for both purposes, starting with those primarily intended for design and optimization, and then see how they can be applied in notional examples.

As we shall see in Section 5.1.1.1, reactors can be classified as batch and continuous reactors, which in turn can be idealized as stirred-tank and plug-flow reactors. We shall not consider any nonideality of fluid flow behavior, since most industrial reactors exhibit only small deviations from ideality. One object of reactor design and operation is to ensure this.

It is impossible to cover every form of reactor design and operation with an appropriate reactor model. As with reaction models in Chapter 3, we can only try to provide some basic principles that can be applied to particular processes. To simplify the models we ignore dynamic behavior, apart from that inherent in batch operation, and we assume isothermal reactor operation. We also adopt "Tafel-type" kinetic expressions, but this—as demonstrated in Section 3.2.1.1—should not make any difficulty.

Even after these simplifications most reactor model equations involving charge and mass transfer resistances can only be solved numerically. For those not well versed in computational skills, we have included hints on how to solve some model equations.

## 4.2. THE BATCH REACTOR

A batch reactor is charged with reactant, the required conversion takes place, and the reactor is emptied. One consequence is that the concentration of the reactant and products in the reactor are a function of time. There are two types of electrochemical batch reactors: those without electrolyte recirculation (Fig. 4.1a), and those with recirculation (Fig. 4.1b). The former are suitable as laboratory devices for obtaining performance data as a function of electrode potential, reactant concentration, and hence conversion as described in Section 3.2.2.2. The reactor is often run in a potentiostatic

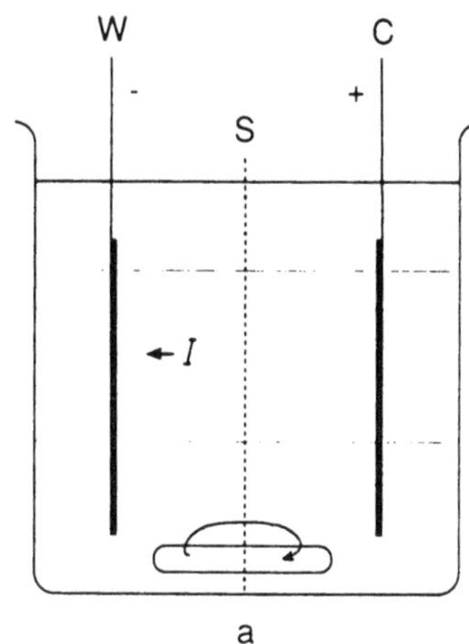


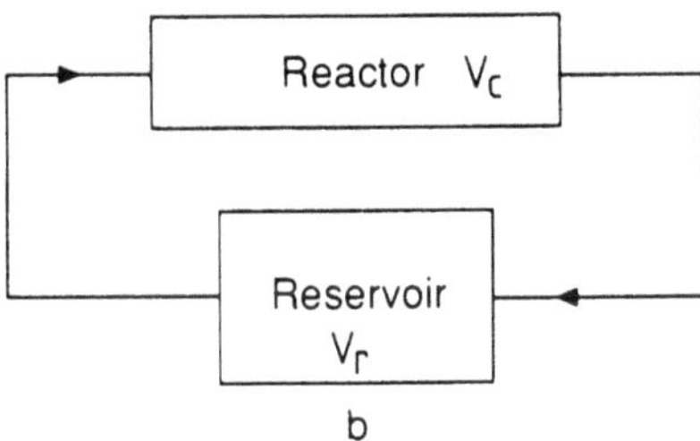


**FIGURE 4.1.** Schematic representation of batch reactors. (a) Without recirculation ($W$ = working electrode, $C$ = counter electrode, $S$ = separator); (b) with recirculation. ($V_c$ = reactor volume, $V_r$ = reservoir volume).

mode. Recirculation is often used in industry because the recycle via a reservoir has the advantage of providing a flexible batch volume while the electrolyte flow through the reactor results in good mass transfer characteristics. Even with a recycle, capacity is limited and for really large-scale throughput a continuous process is preferable. Reactors (with or without recycle) are designed to have as uniform a current distribution as possible (see Chapter 5); this condition will be assumed in deriving reactor models.

### 4.2.1. Batch Reactors Without Electrolyte Recycle

We will begin by deriving an integral relationship for $S$, the electrode area. Let us assume that a single cathodic reaction with 100% current efficiency occurs according to the equation:

$$A \xrightarrow[i_A = i]{ne} B \tag{4.2}$$

A charge balance for the conversion of $A$ gives:

$$i_A S = -n\mathfrak{F}V \frac{dc_A}{dt} \tag{4.3}$$

The kinetic expression implying Tafel-type behavior is Eq. 3.147 (see Section 3.2.1):

$$i = i_A = \frac{c_A}{\dfrac{1}{n\mathfrak{F}k_L} + \dfrac{1}{k_A \exp[-b_A E']}}$$

Eliminating $i_A$ from Eqs. (4.3) and (3.147):

$$\frac{Sc_A}{\dfrac{1}{n\mathfrak{F}k_L} + \dfrac{1}{k_A \exp[-b_A E']}} = -n\mathfrak{F}V \frac{dc_A}{dt} \tag{4.4}$$

Integrating Eq. (4.4) gives the required electrode area $S$:

$$S = \frac{n\mathfrak{F}V}{t_B} \int_{c_{Af}}^{c_{A0}} \left[\frac{1}{n\mathfrak{F}k_L} + \frac{1}{k_A \exp[-b_A E']}\right] \frac{dc_A}{c_A} \tag{4.5}$$

where $c_A^0$ and $c_A^f$ are the initial and final concentrations and $t_B$ is the "batch time."

Equations (3.147) and (4.5) constitute a reactor model, the former being the reaction model and the latter an expression for the required electrode area in terms of process parameters.

The evaluation of the integral depends on the mode of operation of the cell. Most often the mode is one of constant current supplied to a bipolar arrangement. This results in a current density which is nominally constant and for which $E'$, the electrode potential, will be a complex function of $c_A$. Direct integration is therefore not possible; a numerical technique must be employed using the expression for $i$ to evaluate appropriate values of $E'$. This is why the reaction model is an essential element of the reactor model.

The reactor model can also be expressed in terms of conversion. Remembering that the conversion $X$ is defined as the amount of $A$ reacted divided by the amount initially in the feed:

$$X_A = \frac{c_A^0 - c_A}{c_A^0} \tag{4.6}$$

and

$$dX_A = -\frac{dc_A}{c_A^0} \tag{4.7}$$

The model becomes:

$$S = \frac{n\mathfrak{F}V}{t_B}\int_0^{X_A^f}\left[\frac{1}{n\mathfrak{F}k_L} + \frac{1}{k_A \exp[-b_A E']}\right]\frac{dX_A}{1 - X_A} \tag{4.8}$$

and

$$i = \frac{c_A^0(1 - X_A)}{\dfrac{1}{n\mathfrak{F}k_L} + \dfrac{1}{k_A \exp[-b_A E']}} \tag{4.9}$$

In principle, equations for $S$ and $i$ can be combined to eliminate $E'$, producing an expression for the electrode area $S$ as a function of $X$, $i$, initial concentration of $A$, kinetic parameters, and run-time, as indicated by Eq. (4.1).

It is assumed for the sake of simplicity that the reaction in Eq. (4.2) occurs with a current efficiency of 100%. In practice this is rarely true. The second reaction is usually the parallel decomposition of the solvent, but as we saw when deriving reaction models in Chapter 3, simultaneous or consecutive reactions of species already present or produced by the primary reaction can occur. For now we will only extend the model to the situation where the solvent decomposes into an inert gas. We will therefore stay with Eq. (4.2); if it occurs in an aqueous medium, the second reaction will be hydrogen evolution which is kinetically controlled and unaffected by mass transfer:

$$A \xrightarrow[i_A]{ne} B$$

$$2H^+ \xrightarrow[i_H]{2e} H_2 \tag{4.10}$$

Equations (4.5) and (4.8) for the electrode area remain unchanged but the reaction model in terms of concentration will now be:

$$i = \frac{c_A}{\dfrac{1}{n\mathfrak{F}k_L} + \dfrac{1}{k_A \exp[-b_A E']}} + k_H \exp[-b_H E'] \tag{4.11}$$

and in terms of conversion:

$$i = \frac{c_A^0(1 - X_A)}{\dfrac{1}{n\mathfrak{F}k_L} + \dfrac{1}{k_A \exp[-b_A E']}} + k_H \exp[-b_H E'] \tag{4.12}$$

We return to Eq. (3.147), which as explained in Chapter 3 shows the current density, i.e., the rate of reaction, to be governed by two resistances in series, one due to mass transfer the other due to charge transfer. Dependence of reaction rate on mass and charge transfer can be expressed in a slightly different way. By inserting $k_{Af}$ in Eq. (3.147) for $k_A \exp[-b_A E']$ and rearranging the expression, one obtains:

$$i_A = \frac{c_A k_{Af}}{1 + Da_A} \tag{4.13}$$

where the dimensionless Damköhler number $Da_A = k_{Af}/n\mathfrak{F}k_L$ expresses the influence of mass transfer on the rate of reaction. Thus if $k_L$ is large $Da_A$ is small and mass transfer is not rate-controlling. Conversely, if $k_L$ is small compared to $k_{Af}$, $Da_A$ will be large and mass transfer will be a limiting factor.

As indicated in Section 4.1, reaction models which involve charge and mass transfer limitations have to be solved numerically; this should not produce undue difficulties. It is instructive to simplify and be able to get an analytical solution, thereby gaining a better understanding of the behavior of the system. We shall postulate that Eq. (4.2) is mass transfer–controlled, i.e., occurs at limiting current density. Equation (4.10) will always be kinetically controlled.

Combining Eqs. (4.3) and (4.13) and writing "$a$" for "$S/V$," the specific electrode area,

$$-\frac{dc_A}{dt} = \frac{ak_{Af}c_A}{n\mathfrak{F} + \dfrac{k_{Af}}{k_L}} \tag{4.14}$$

For a mass transfer–controlled reaction $k_{Af} \gg k_L$, and Eq. (4.14), after rearrangement, simplifies to:

$$\frac{dc_A}{c_A} = -ak_L\,dt \tag{4.15}$$

Integrating Eq. (4.15) between limits ($c_A = c_A^0$, $t = 0$) and ($c_A = c_A^f$, $t = t$), we get:

$$\ln \frac{c_A^f}{c_A^0} = -ak_L t \tag{4.16}$$

or

$$\frac{c_A^f}{c_A^0} = \exp[-ak_L t] \tag{4.17}$$

This is the classic exponential decay relationship for mass transfer in a stirred tank.

At any time total current density $i$ is:

$$i = i_A + i_{\mathrm{H}} \tag{4.18}$$

Current efficiency can be written as either:

$$\text{C.E.} = \frac{\int_0^t i_A\,dt}{\int_0^t (i_A + i_H)\,dt} \tag{4.19}$$

or

$$\text{C.E.} = \frac{c_B}{c_B + \dfrac{a}{\mathfrak{F}} \displaystyle\int_0^t \frac{i_{\mathrm{H}}}{2}\,dt} \tag{4.20}$$

---

EXAMPLE 4.1: Performance of a Stirred Batch Laboratory Cell Operating at a Limiting Current Density

THE PROBLEM: A batch laboratory reactor with an electrolyte volume of 700 $cm^3$ and an electrode area of 30 $cm^2$ is used to deposit a divalent metal from an aqueous solution in a potentiostatic mode. Initial concentration of the metal is 0.1 $kmol/m^3$. The reactor mass transfer coefficient has been measured as $3.3 \times 10^{-5}$ m/s. Hydrogen evolution occurs as a parallel reaction according to the equation $i_{\mathrm{H}} = k_{\mathrm{H}} \exp[-b_{\mathrm{H}} E']$, where $k_{\mathrm{H}} = 1.30 \times 10^{-4}$ $A/m^2$ and $b_{\mathrm{H}} = 12\ V^{-1}$. If the metal deposition is operated at its limiting current density at an electrode potential of $-0.9$ V (SCE), determine how conversion, total current density, and current efficiency vary with time, in a potentiostatic mode. What will be the current efficiency at the final metal concentration of 0.001 kmol/m$^3$? If the cell is operated galvanostatically at a current density of 637 A/m$^2$, calculate the change of conversion, electrode potential, and current efficiency with time. What will be the current efficiency at the final metal concentration?

THE SOLUTION: Applying Eq. (4.17), the conversion for potentiostatic and galvanostatic operation is:

$$1 - X_A = \frac{c_A^f}{c_A^0} = \exp\left[-\frac{30 \times 3.3 \times 10^{-5}}{700} t\right] = \exp[-1.414 \times 10^{-6} t]$$

The total current density $i$ for the potentiostatic mode will be from Eq. (4.18):

$$i = 2\mathfrak{F} k_L c_A^f + k_H \exp[-b_H E'] = 6369 c_A^f + 6.373 = 636.9(1 - X_A) + 6.373$$

For the galvanostatic mode we obtain $i_H$ as $i - [636.9(1 - X_A)]$ and hence $E'$ from $k_H \exp[-b_H E']$.

For the potentiostatic case the current efficiency is calculated from Eq. (4.20):

$$\text{C.E.} = \frac{c_B}{c_B + \dfrac{S \times i_H}{V \times 2 \times \mathfrak{F}} t}$$

$$= \frac{c_A^0 X_A}{c_A^0 X_A + \dfrac{30 \times 6.373}{700 \times 2 \times 0.965 \times 10^8} t}$$

$$= \frac{0.1 X_A}{0.1 X_A + 1.415 \times 10^{-9} t}$$

For the galvanostatic case, C.E. equals $i_A/i$.

The results are shown in Tables 4.1 and 4.2. Note the high current efficiency achieved at the cost of a rapidly diminishing current density.

For $c_A^f = 0.001$ kmol/m$^3$ we have:

$$1 - X_A = c_A^f / c_A^0 = 0.01 = \exp[-1.414 \times 10^{-6} t]$$

This gives a time of $4.6 \times 10^6$ s and the corresponding current efficiency is 93.8%.

TABLE 4.1. The Potentiostatic Case

| $t$ (s) | $X_A(-)$ | $i$ (A/m$^2$) | C.E. (%) |
|---|---|---|---|
| $10^4$ | 0.014 | 634 | 99.0 |
| $10^5$ | 0.13 | 560 | 98.9 |
| $2 \times 10^5$ | 0.25 | 484 | 98.9 |
| $10^6$ | 0.77 | 153 | 98.2 |
| $2 \times 10^6$ | 0.94 | 45 | 97.1 |
| $3 \times 10^6$ | 0.985 | 16 | 95.9 |

TABLE 4.2. The Galvanostatic Case

| $t$ (s) | $X_A(-)$ | $E'$ (V) | C.E. (%) |
|---|---|---|---|
| $10^4$ | 0.014 | −0.9 | 98.6 |
| $10^5$ | 0.13 | −1.11 | 87.0 |
| $2 \times 10^5$ | 0.25 | −1.17 | 75.0 |
| $10^6$ | 0.77 | −1.26 | 23.0 |
| $2 \times 10^6$ | 0.94 | −1.28 | 6.0 |
| $3 \times 10^6$ | 0.985 | −1.28 | 1.5 |

In Table 4.2 it is the current efficiency that decreases with conversion. For the final concentration of 0.001 kmol/m$^3$ the current efficiency will have dropped to 1% with the electrode potential virtually unchanged at −1.28 V.

### 4.2.2. Batch Reactors with Electrolyte Recycle

Electrolyte from a well-mixed reservoir of volume $V_r$ is rapidly recirculated through a reactor of volume $V_c$ (Fig. 4.1b). Analysis of this system is more complicated than that for the stirred batch reactor considered in Section 4.2.1, since the reactor may be assumed to have a plug-flow or stirred-tank configuration.

#### 4.2.2.1. Recirculation Through a Plug-Flow Batch Reactor

Let us see what happens with a plug-flow reactor and how it differs from the batch reactor model in Section 4.2.1. There will be no mixing or dispersion of the species along the reactor, only a gradual decrease in the concentrations of the reactants and an increase in those of the products as a result of the electrochemical reactions. At any plane a distance $x$ along the reactor in the direction of flow composition is assumed to be uniform, except at regions close to the electrode where mass transport takes place. Looking at Fig. 4.2 one can make a steady state material balance on a differential element $dx$. The change in concentration $dc_A$ for a flow rate $Q$ is related to the local current density $i_x$ according to:

$$-Q\,dc_{Ax} = \frac{i_{Ax}S}{n\mathfrak{F}L}\,dx \tag{4.21}$$

where $c_{Ax}$ is the concentration of $A$ at level $x$. By defining the time a unit section of fluid takes to progress from inlet to outlet as the residence time $\tau = x/u$ where the velocity $u = Q/A$ (Fig. 4.2), Eq. (4.21) transforms to:

$$i_{Ax}S = -n\mathfrak{F}V\frac{dc_{Ax}}{d\tau} \tag{4.22}$$

Equation (4.22) is analogous to Eq. (4.3), the equation for the batch reactor, except that overall partial current density $i_A$ is replaced by a local partial current density $i_{Ax}$. We shall return to Eq. (4.22) later but for now we assume a mass transfer limited reaction. Under those circumstances $i_{Ax} = n\mathfrak{F}k_L c_{Ax}$; substituting this quantity in Eq. (4.22) we get, after a slight rearrangement:

$$\int_{c_{A1}}^{c_{A2}} \frac{dc_{Ax}}{c_{Ax}} = -\int_0^{\tau} \frac{S}{V}\,k_L\,d\tau \tag{4.23}$$

and after integration:

$$\ln\frac{c_{A2}}{c_{A1}} = -ak_L\tau \tag{4.24}$$

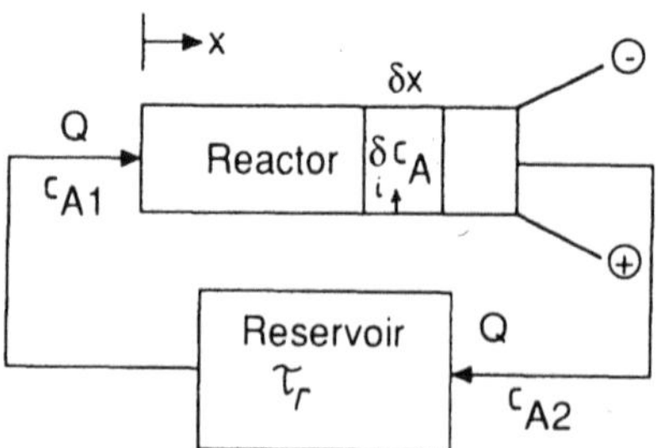


**FIGURE 4.2.** Recirculation through a plug-flow batch reactor.

$c_{A1}$ and $c_{A2}$ being the reactor inlet and outlet concentrations respectively. The analysis applies to a steady state situation where concentrations do not vary with time. This is not true in a recirculating batch reactor system where the concentrations change as the conversion proceeds; the material balance for the plug-flow reactor should reflect this time dependency. The material balance becomes:

$$\frac{\partial c_{Ax}}{\partial t} + \frac{\partial c_{Ax}}{\partial \tau} = -ak_L c_{Ax} \tag{4.25}$$

Solution of this partial differential equation requires long numerical routines which can be aided by Laplace transforms.[1] Fortunately, Eq. (4.25) need not be solved rigorously. Since in practice $V_r \gg V_c$, one can adopt a pseudo-steady state approximation where $\partial c_{Ax}/\partial t$ is assumed to be zero.

Turning to the electrolyte reservoir, an instantaneous material balance (input = output + accumulation) gives:

$$Qc_{A2} = Qc_{A1} + V_r \frac{dc_{A1}}{dt} \tag{4.26}$$

Eliminating $c_{A2}$ between Eqs. (4.24) and (4.26):

$$V_r \frac{dc_{A1}}{dt} + Qc_{A1} = Qc_{A1} \exp[-ak_L\tau] \tag{4.27}$$

Defining the residence time in the reservoir as $V_r/Q = \tau_r$, rearranging Eq. (4.27), separating variables, and integrating between the limits $c_{A1} = c_{A1}^0$, $t = 0$, and $c_{A1} = c_{A1}^f$, $t = t$, gives:

$$c_{A1}^f = c_{A1}^0 \exp\left[-\frac{t}{\tau_r}(1 - \exp[-ak_L\tau])\right] \tag{4.28}$$

Our analysis assumes the pipework to be of negligible volume but the latter can if necessary be allowed for by the magnitude of $V_r$.

With the simplification of a pseudo-steady state operation, use of Eq. (4.28) in a modeling procedure will be appropriate. It will, however, be helpful to see how far the behavior of the recirculating batch operation predicted by Eq. (4.28) departs from that of the simple batch reactor considered in Section 4.2.1. We will do this for the realistic condition that $\tau_r \gg \tau$. By introducing the approximation $\exp[-ak_L\tau] = 1 - ak_L\tau$,

Eq. (4.28) reduces to:

$$c_{A1}^{f} = c_{A1}^{0} \exp - \left[ \frac{tak_L \tau}{\tau_r} \right] \tag{4.29}$$

Redefining the specific electrode area in terms of the total volume of electrolyte, i.e., $a' = a\tau/(\tau_r + \tau)$ and with $\tau_r >> \tau$, Eq. (4.29) becomes:

$$c_{A1}^{f} = c_{A1}^{0} \exp[-a'k_L t] \tag{4.30}$$

which is identical with Eq. (4.17) derived for the batch reactor without recirculation. We must find out whether this simplification can be justified, by using typical conditions for turbulent flow in a parallel plate reactor as an example.

The group $k_L a\tau$ in Eq. (4.28) is equal to $k_L L/(ud)$ where $L$ is the electrode length, $d$ the electrolyte gap, and $u$ the superficial velocity. Using Eq. (2.18) for turbulent mass transfer in a rectangular channel, we have:

$$k_L d_e/D = 0.023(ud_e/\nu)^{0.8}(\nu/D)^{1/3}$$

For any industrial cell $d << W$ (the cell width), and hence $d_e = 2d$. Taking typical values $\nu = 10^{-6}\,\text{m}^2/\text{s}$ and $D = 3 \times 10^{-10}\,\text{m}^2\,\text{s}^{-1}$, the Schmidt number is:

$$Sc = \nu/D = 10^{-6}/3 \times 10^{-10} = 3300 \qquad \text{and} \qquad Sc^{1/3} = 14.9$$

Therefore:

$$\frac{k_L L}{ud} = \frac{0.023 \times (2 \times 10^6 ud)^{0.8} \times 14.9 \times L \times 3 \times 10^{-10}}{ud \times 2d} = \frac{5.6 \times 10^{-6} L}{u^{0.2} \times d^{1.2}} \tag{4.31}$$

For turbulent flow $Re$ must be at least 2000. We have the condition that $2 \times 10^6 ud = 2000$ or $u = 10^{-3}/d$. Substituting this equality in Eq. (4.30):

$$\frac{k_L L}{ud} = 2.2 \times 10^{-5} \frac{L}{d} \tag{4.32}$$

We also know that the simplifying condition $e^x = 1 + x$ is accurate to 0.5% if $x < 0.1$. This gives the condition that:

$$2.2 \times 10^{-5} L/d < 0.1 \qquad \text{or} \qquad L/d < 4.5 \times 10^3 \tag{4.33}$$

This aspect ratio is well within the design criteria for plate and frame cells and we conclude that the simplified Eq. (4.30) can be used for modeling.

#### 4.2.2.2. Recirculation Through a Stirred-Tank Batch Reactor

The system is illustrated in Fig. 4.3. Electrolyte flows into the reactor at a concentration of $c_{A1}$ and immediately mixes with the electrolyte in the cell at a concentration of $c_{A2}$ which is also the concentration of the exit liquor. A mass balance over the reactor gives:

$$Qc_{A1} = Qc_{A2} + V_c \frac{dc_{A2}}{dt} + \frac{i_A a V_c}{n\mathfrak{F}} \tag{4.34}$$

At steady state and assuming again limiting current operation, Eq. (4.34) becomes:

$$Qc_{A1} = Qc_{A2} + ak_L V c_{A2} \tag{4.35}$$

On rearrangement the reactor exit concentration is given by:

$$c_{A2} = c_{A1}\left[\frac{1}{1 + ak_L \tau}\right] \tag{4.36}$$

Adopting again the pseudo-steady state approximation for the recirculation system and eliminating $c_{A2}$ by combining Eq. (4.36) with Eq. (4.26), the material balance over the reservoir gives:

$$c_{A1}\left[1 - \frac{1}{1 + ak_L \tau}\right] = -\tau_r \frac{dc_{A1}}{dt} \tag{4.37}$$

Rearranging, separating variables, and integrating Eq. (4.37) over the same

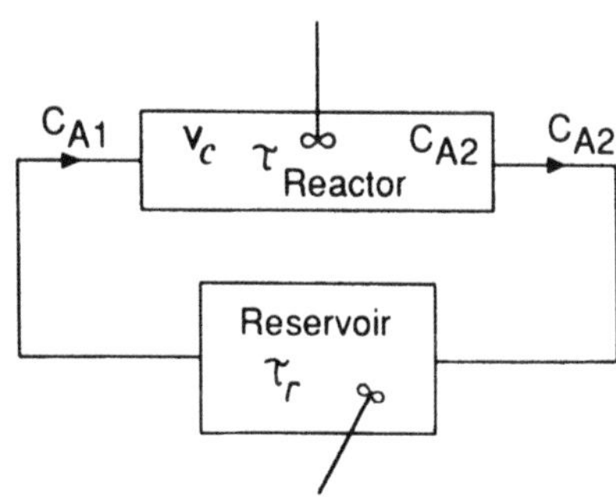


**FIGURE 4.3.** Recirculation through a stirred-tank batch reactor.

limits as in Section 4.2.2.1, we obtain:

$$c_{A1}^{f} = c_{A1}^{0} \exp\left\{\frac{-t}{\tau_r}\left[1 - \frac{1}{1 + ak_L\tau}\right]\right\} \tag{4.38}$$

Using the redefined specific area $a'$ and remembering that $\tau_r >> \tau$, Eq. (4.38) becomes:

$$c_{A1}^{f} = c_{A1}^{0} \exp\left[-\frac{a'k_L t}{1 + a'k_L\tau_r}\right] \tag{4.39}$$

Almost always $a'k_L\tau_r << 1$ and Eq. (4.39) reduces again to Eq. (4.30), the simple batch reactor equation.

---

EXAMPLE 4.2: Performance of Recirculating Batch Reactors at Limiting Current Density

THE PROBLEM: A batch recirculating system is used to recover metal from an aqueous electrolyte under limiting current conditions. Compare the conversions achieved after $2 \times 10^4$ seconds with a simple batch reactor, a recirculating plug-flow reactor and a recirculating stirred-tank reactor, assuming that the total electrolyte volume remains the same. The data in Table 4.3 can be used.

THE SOLUTION: The residence time in the reactor $\tau$ is:

$$0.005/0.00025 = 20\,\text{s}$$

The residence time in the reservoir $\tau_r$ is:

$$0.1/0.00025 = 400\,\text{s}$$

The specific electrode area $a'$, based on total volume, is:

$$100 \times 20/420 = 4.76\ \text{m}^{-1}$$

TABLE 4.3. Recirculating Batch Reactor

| | |
|---|---|
| Specific electrode area of reactor | $100\ \text{m}^{-1}$ |
| Mass transfer coefficient | $3 \times 10^{-5}$ m/s |
| Volume of reservoir | $0.1\ \text{m}^3$ |
| Reactor volume | $0.005\ \text{m}^3$ |
| Electrolyte recirculation rate | $0.00025\ \text{m}^3/\text{s}$ |

From the batch reactor Eq. (4.17), we have:

$$X_A = 1 - \exp[-ak_L t] = 1 - \exp[-4.76 \times 3 \times 10^{-5} \times 2 \times 10^4] = 0.943$$

From the plug-flow reactor with recirculation, Eq. (4.28):

$$X_A = 1 - \exp\left[-\frac{t}{\tau_r}(1 - \exp[-ak_L\tau])\right]$$

$$= 1 - \exp\left[-\frac{2 \times 10^4}{400}(1 - \exp[-100 \times 3 \times 10^{-5} \times 20]\right] = 0.946$$

From the stirred-tank reactor with recirculation, Eq. (4.39):

$$X_A = 1 - \exp\left[-\frac{a'k_L t}{1 + a'k_L \tau_r}\right] = 1 - \exp\left(-\frac{4.76 \times 3 \times 10^{-5} \times 2 \times 10^4}{1 + 4.76 \times 3 \times 10^{-5} \times 400}\right)$$

$$= 0.933$$

---

## 4.3. MODELING BATCH REACTORS

The agreement between the high conversion figures in Example 4.2 is much closer than the accuracy of the data, e.g., that of the mass transfer coefficient, and the results substantiate the conclusion in Section 4.2.2 that the simple batch reactor equation is satisfactory for modeling reactors with recirculation. So far our examples have been confined to limiting current operation. Is our conclusion also applicable to the general case of combined mass and charge transfer resistance when working below the limiting current density?

Basically, the simple batch equation predicts the behavior when recirculation occurs because of low conversion per pass through the reactor. If we use the data in Example 4.2, the conversion per pass will be $1 - \exp[-a'k_L\tau]$, less than 0.3%. Working below the limiting current density it will be even smaller. It is immaterial in practice whether in a batch recirculation system the reactor is operating in a plug-flow or stirred-tank mode. Therefore, we can use the model equations derived for the batch reactor unless there are special circumstances such as, for example, the use of a three-dimensional electrode cell with a high active specific electrode area.

Obtaining a reactor model for a batch system involves the following steps:

1. Decide on the key reactions in your process and their kinetics.

2. Select or derive an appropriate reaction model as discussed in Chapter 3.
3. Decide which quantities you would like to calculate, e.g., process time, electrode area, and chemical yield, as a function of which parameters, e.g., conversion and current density.
4. Derive simple difference equations appropriate to your model.
5. Set up a programing flow sheet, construct a program, and calculate the parameters you have decided on.

---

EXAMPLE 4.3: Reactor Model of a Pilot Plant Cell

THE PROBLEM: Construct a reactor model that will predict the effect of conversion, current density, and electrolyte recirculation rate on the chemical yield and run-time of a pilot plant cell producing *p*-anisidine.

The cell is a narrow-gap filter press type with recirculation. The electrode area is 0.1 $m^2$ and the volume of the reservoir is 5 liters.

THE SOLUTION: The reaction model for the production of *p*-anisidine has already been developed in Chapter 3, in Eqs. (3.190), (3.180), and (3.188) (shown again here):

$$i = \frac{c_A}{\dfrac{1}{4\mathfrak{F}k_L} + \dfrac{1}{k_A \exp(-b_A E')}} \frac{4\mathfrak{F}k_L + 3k_B \exp(-b_B E')}{4\mathfrak{F}k_L + 2k_B \exp(-b_B E')} + k_{\mathrm{H}} \exp(-b_{\mathrm{H}} E')$$

$$i_A = \frac{c_A}{1/4\mathfrak{F}k_L + 1/(k_A \exp[-b_A E'])}$$

$$i_B = \frac{i_A k_B \exp[-b_B e']}{4\mathfrak{F}k_L + 2k_B \exp[-b_B E']}$$

The numerical values of the kinetic constants, also determined in Section 3.2.3.2, are:

$$k_A = 0.22 \text{ Am/kmol},\ b_A = 0.034 \text{ mV}^{-1},\ k_B = 1.03 \text{ Am/kmol},$$
$$b_B = 0.019 \text{ mV}^{-1},\ k_{\mathrm{H}} = 4.9 \times 10^{-10} \text{ A/m}^2, \text{ and } b_H = 0.037 \text{ mV}^{-1}$$

The reaction scheme can be represented by:

$$A \xrightarrow[i_A]{4e} B \xrightarrow[i_B]{2e} D \qquad 2\mathrm{H}^+ \xrightarrow{2e} \mathrm{H}_2$$
$$\hphantom{A \xrightarrow[i_A]{4e}} \downarrow$$
$$\hphantom{A \xrightarrow[i_A]{4e}} C$$

A charge balance for the conversion of $A$ gives:

$$i_A S = -n\mathfrak{F}V\frac{dc_A}{dt}$$

Thus:

$$dt = -\frac{n\mathfrak{F}V}{Si_A}dc_A$$

$$t = \sum \delta t$$

$$c_A = c_A^0 - \sum \delta c_A$$

Production of by-product $R_D$:

$$R_D = \frac{S\sum i_B \delta t}{2\mathfrak{F}}$$

Production of product $R_C$:

$$R_C = \frac{S\sum i_A \delta t}{4\mathfrak{F}} - \frac{S\sum i_B \delta t}{2\mathfrak{F}}$$

Cumulative current efficiency of product C.E.:

$$\text{C.E.} = \frac{4\mathfrak{F}R_C}{Sit}$$

Chemical yield of product $Y_C$:

$$Y_C = \frac{R_C}{V(c_A^0 - c_A)}$$

These relationships are used to calculate the parameters as in the programing flow sheet in Fig. 4.4. The results are shown in Figs. 4.5 and 4.6. The former shows the chemical yield as a function of conversion, current density, and flow rate. The plot shows that reducing the flow rate by a factor of 4 and at the same time doubling the current density has a disastrous effect on the chemical yield once the conversion exceeds 60%. The latter figure provides the run-time again as a function of conversion, current density and flow rate. Run-time of a pilot plant is important since it enables one to optimize the use of an expensive facility.

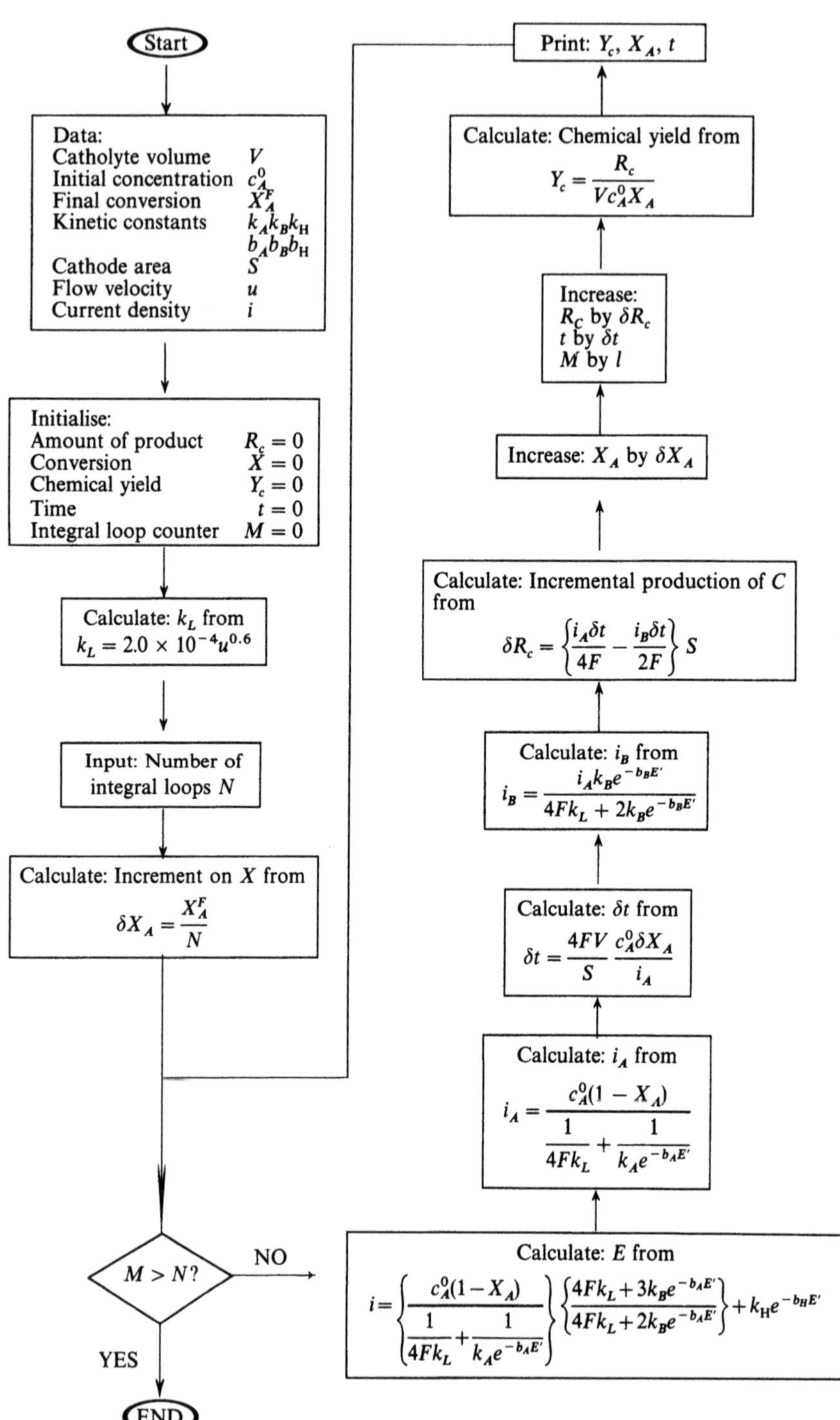


**FIGURE 4.4.** Programing flow sheet.

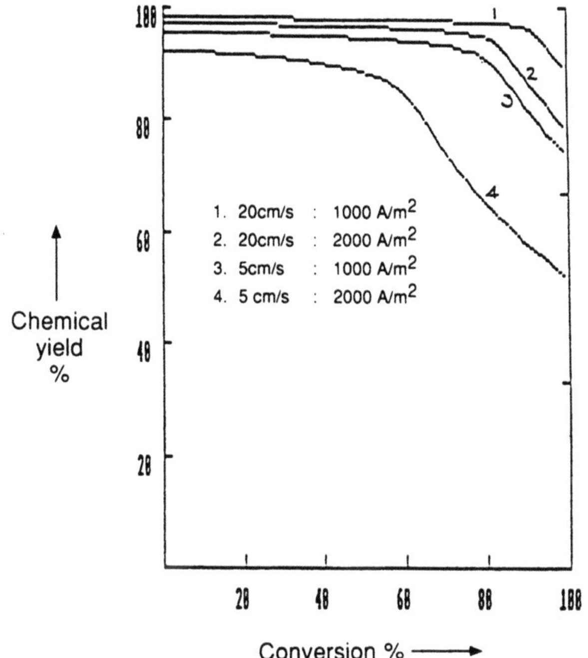


**FIGURE 4.5.** Chemical yield as a function of conversion, current density, and flow rate.

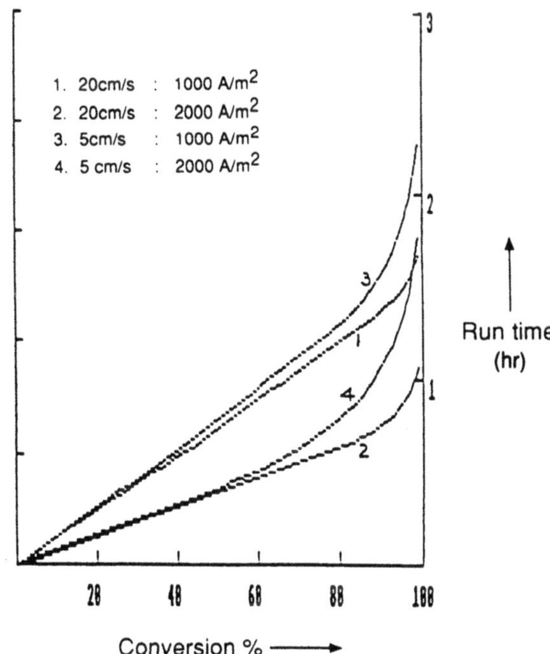


**FIGURE 4.6.** Run time as a function of conversion, current density, and flow rate

## 4.4. THE CONTINUOUS REACTOR

Continuous reactors are confined to processes with large throughput. In the electrochemical field this means a small number of organic syntheses, continuous operation being largely confined to the inorganic field and effluent treatment. The problem of low conversion per mass is as evident in the continuous reactor as it was in the batch one. On the basis of a charge balance such as in Eq. (4.3), for a specific electrode area of $100\,m^{-1}$, a current density of $2000\,A/m^2$, and $n = 2$, a residence time of 1000 seconds to convert $1\,kmol/m^3$ is required. Even at the relatively low velocity of 5 cm/s this requires an electrode length of 100 m. Lowering the velocity enough to obtain a reasonable residence time would make the mass transfer rate unacceptable since the electrode length of a single cell unit rarely exceeds 2 meters.

To meet sensible design requirements, a number of reactors have to be connected in series. This means that conversion per reactor unit will be low and the simplifications made for the batch reactor (Section 4.2.3) will also apply to the continuous reactor.

Another way to get sufficiently high residence time is to operate the reactor with a recycle of the product stream (Fig. 4.7). The overall flow rate through the reactor $Q$ is small compared to the recycle rate $q$, giving a high enough mass transfer rate. Again, conversion per pass is low and the same simplifications as for the case without recycle apply.

Models for a continuous reactor without recycle will be identical with those developed for the batch reactor. We will therefore confine ourselves to a continuous reactor with recycle. Although the low conversion per pass make the hydrodynamic classification of the reactor of little significance, it will still be convenient to develop the continuous reactor model on the basis of a plug-flow reactor.

(i) Continuous plug-flow reactor with electrolyte recycle. For simplicity we shall again use Eqs. (4.2) and (4.10) for the primary and parallel secondary reaction, respectively. The schematic arrangement of the reactor and recycle

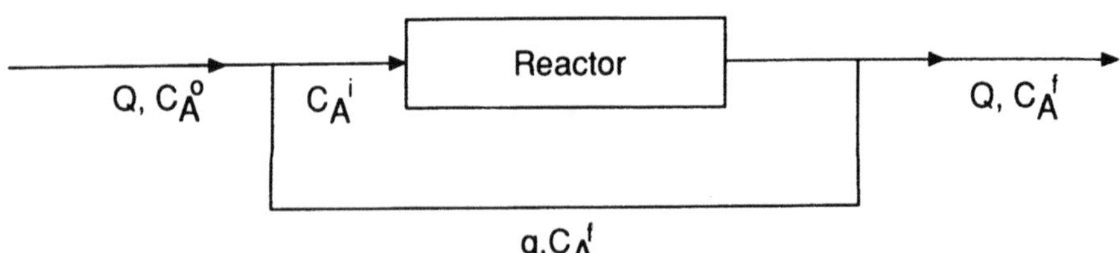


**FIGURE 4.7.** Continuous reactor with recycle.

has already been shown in Fig. 4.7 where some of the parameters are defined. Feed and product streams are supplied and removed at a volumetric flow rate of $Q$ and the recycle has a volumetric flow rate of $q$.

At the inlet the flow rate will be $Q + q$, and the inlet concentration $c_A^i$ will be:

$$c_A^i = \frac{Qc_A^0 + qc_A^f}{Q + q} \tag{4.40}$$

The reactor is analyzed as a plug-flow reactor beginning with a charge balance as the electrolyte flows past an elemental area $dS$ of the total cathode surface (Fig. 4.8). The charge balance is given by:

$$i_A\, dS = -n\mathfrak{F}(Q + q)dc_A \tag{4.41}$$

Tafel behavior (see Chapter 3) is represented by:

$$i_A = \frac{c_A}{\dfrac{1}{n\mathfrak{F}k_L} + \dfrac{1}{k_A \exp[-b_A E']}} \tag{4.42}$$

Eliminating $i_A$ between Eqs. (4.41) and (4.42) and integrating:

$$S = n\mathfrak{F}(Q + q) \int_{c_A^f}^{c_A^0} \left[\frac{1}{n\mathfrak{F}k_L} + \frac{1}{k_A \exp[-b_A E']}\right] \frac{dc_A}{c_A} \tag{4.43}$$

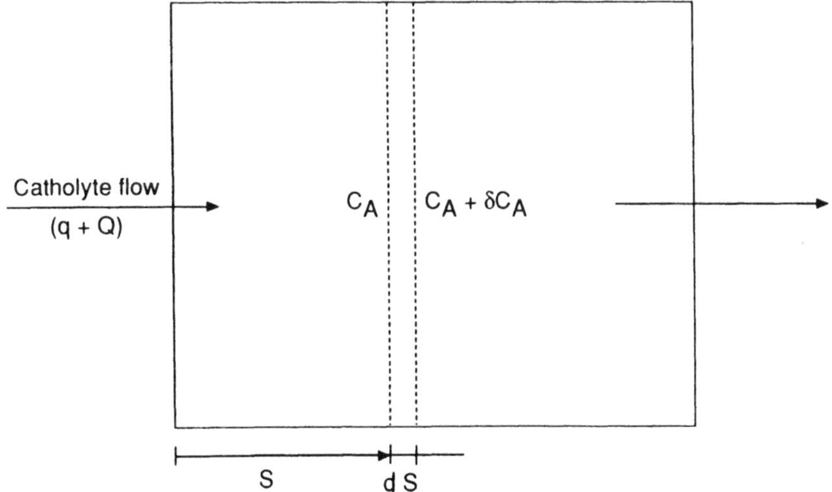


**FIGURE 4.8.** Elemental electrode area shown in charge balance.

The reaction model will be as in Eq. (4.11):

$$i = \frac{c_A}{\dfrac{1}{n\mathfrak{F}k_L} + \dfrac{1}{k_A \exp[-b_A E']}} + k_H \exp[-b_H E']$$

Equations (4.43) and (4.11) are again the reactor model.

To express the model in terms of conversion, proceed as follows. By definition:

$$X_A = \frac{\text{amount of } A \text{ converted}}{\text{amount of } A \text{ in the feed}}$$

Thus:

$$X_A = \frac{(Q+q)(c_A^i - c_A)}{Qc_A^0} = \frac{Qc_A^0 + qc_A^f - Qc_A - qc_A}{Qc_A^0} \quad (4.44)$$

Hence:

$$c_A = \frac{Qc_A^0(1 - X_A) + qc_A^f}{Q+q} \quad (4.45)$$

and

$$dc_A = -\frac{Q}{Q+q} c_A^0 \, dX_A \quad (4.46)$$

Also:

$$X_A^0 = 0 \quad (4.47)$$

and

$$X_A^f = \frac{c_A^0 - c_A^f}{c_A^0} \quad (4.48)$$

Substituting Eqs. (4.45) and (4.46) into Eqs. (4.43) and (4.11):

$$S = n\mathfrak{F}(Q+q) \int_0^{X_A^f} \left[ \frac{1}{n\mathfrak{F}k_L} + \frac{1}{k_A \exp[-b_A E']} \right] \frac{Qc_A^0 \, dX_A}{Qc_A^0(1 - X_A) + qc_A^f} \quad (4.49)$$

and

$$i = \frac{Qc_A^0(1 - X_A) + qc_A^f}{(Q+q)\left(\dfrac{1}{n\mathfrak{F}k_L} + \dfrac{1}{k_A \exp[-b_A E']}\right)} + k_H \exp[-b_H E'] \qquad (4.50)$$

If recycle is absent, $q$ becomes zero and the model simplifies to Eqs. (4.8) and (4.12):

$$S = n\mathfrak{F}Q \int_0^{X_A^f} \left[\frac{1}{n\mathfrak{F}k_L} + \frac{1}{k_A \exp[-b_A E']}\right] \frac{dX_A}{(1 - X_A)}$$

and

$$i = \frac{c_A^0(1 - X_A)}{\dfrac{1}{n\mathfrak{F}k_L} + \dfrac{1}{k_A \exp[-b_A E']}} + k_H \exp[-b_H E']$$

and thus is identical with the batch reactor model.

It follows that steps (1) to (5) in Section 4.3 can also be used to obtain a reactor model for the continuous reactor.

## 4.5. CONCLUDING REMARKS

The reader may be struck by the brevity of Chapter 4, considering that it deals with an important part of electrochemical process engineering. The reason is that the simplifications we have found possible lead to a simple way to get reliable rector models. Uncertainty in predicting performance is not usually due to the model but to an error in the values of mass transfer coefficients or kinetic data. Values of mass transfer coefficients found in the literature can vary by as much as 50%; it was demonstrated in Fig. 3.28 how sensitive reactor performance can be to relatively small changes in $k_L$. The reader is advised to test this sensitivity in a model by assuming a value of $k_L$ either side of the "experimental" value and doing the same for the kinetic constants—as we have seen in Chapter 3, changes in the nature of the electrode surface can lead to large changes in $k_j$ or $b_j$, particularly when there are catalytic effects. Since the model is usually obtained by means of a computer, little extra time is required for these sensitivity tests.

Sometimes conversion per pass may be high enough to warrant a more sophisticated reactor model and indeed an allowance for hydrodynamic nonideality. For this the reader should use a standard book on reactor design such as Ref. 2. Another area not covered here is multiphase reactions, where mass transfer across a phase boundary is important. For treatment of this topic, please consult Refs. 3 and 4.

Complexities due to changes in the electrode surface with electrode potential have already been mentioned. As we shall see in Section 5.3.8 there is, however, another phenomenon, particularly in organic electrosynthesis, that can lead to a gradual deactivation of the electrode. Although it is possible to allow for this in a model,[5] it is much better[6] to avoid this difficulty altogether.

## REFERENCES


1. Walker, A. T. S. and Wragg, A. A., 1977, "The modelling of concentration-time relationships in recirculating electrochemical reactor systems," *Electrochem. Acta* **22**: 1129–1134.
2. Westerterp, K. R., van Swaaiy, W. P. M., and Beenackers, A. A. C. M., 1984, *Chemical Reactor Design and Operation*, John Wiley & Sons, New York.
3. Doraiswamy, L. K. and Sharma, M. M., 1984, *Heterogeneous Reactions: Analysis, Examples and Reactor Design*, Vol. 2, John Wiley & Sons, New York.
4. Feess, H. and Wendt, H., 1982, "Performance of two-phase-electrolyte electrolysis," in *Techniques of Electroorganic Synthesis III*, (N. L. Weinberg and B. V. Tilak, eds.), John Wiley & Sons, New York, pp. 81–177.
5. Scott, K., 1986, Electrolytic reduction of oxalic acid to glyoxylic acid: A problem of electrode deactivation, *Chem. Eng. Res. Des.* **64**: 266–271.
6. Michelet, D., 1974, Glyoxylic acid, *Ger. Offen.* **2**, 359, 863.

# Chapter 5

# Electrolytic Reactor Design, Selection, and Scale-up

## 5.1. ELECTROLYTIC REACTOR DESIGNS

We mentioned in Section 1.3 some important industrial applications of electrolysis—in the chloralkali industry, metal winning and refining, and organic electrosynthesis. As indicated in Section 1.2, we do not intend to describe electrochemical processes in detail, since there are many books on electrochemical technology.[1] We will discuss the design of individual reactors, with emphasis on modularized, general purpose flow electrolyzers. We will classify reactors by their mode of operation.

### 5.1.1. Classifications of Reactors

There are many ways to classify cell designs. One is based on reactor engineering principles. Another is specific to electrolytic reactors, based on the nature of the working electrode, the method of current supply to individual cells, and the arrangement of the flow of electrolyte through the cell.

#### 5.1.1.1. Scheme Based on Reactor Engineering Principles

In Chapter 4 we classified reactors as batch or continuous, as well mixed or in plug flow, and with or without recirculation. Figure 5.1 summarizes the classifications.

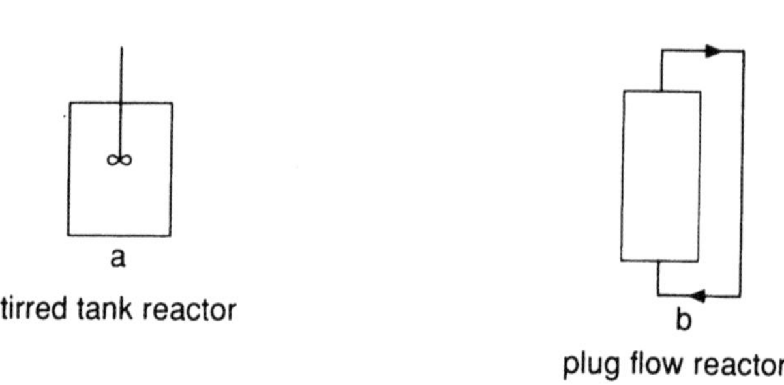


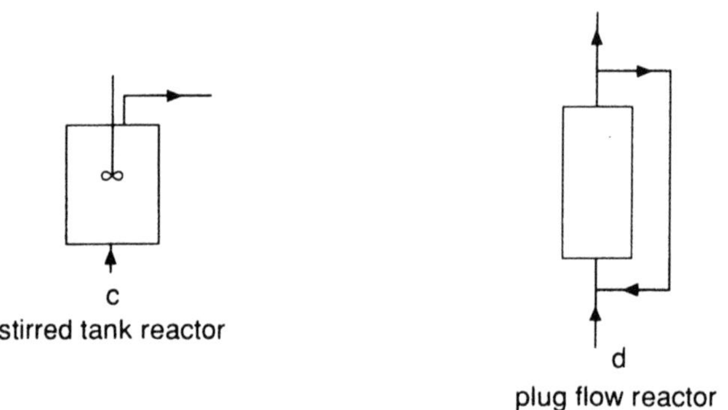


**FIGURE 5.1.** Classification scheme based on reactor engineering principles: (a) and (b), batch reactors; (c) and (d), continuous reactors; (a) and (c), stirred-tank reactors; (b) and (d), plug-flow reactors.

#### 5.1.1.2. Scheme Based on Electrochemical Modes of Operation

A plate electrode, for example, exhibits current distribution along the dimensions $y$ and $z$ (Fig. 5.2) and is called two-dimensional. But for a packed bed acting as an electrode, current distribution along a third dimension $x$ has to be considered. Electrolyte can penetrate the three-dimensional structure of the electrode by diffusion or convection. Reactor designs use electrodes that are composed of a porous matrix or particulate beds. The latter may be in the form of a packed bed, a fluidized (Fig. 5.3), or a circulating or spouting bed. A successful industrial application of a packed-bed electrode occurs in the production of tetra-alkyl lead.[2] The design is unusual in that the lead pellets in the packed bed react during the process and are continuously replaced. This is a good example of an

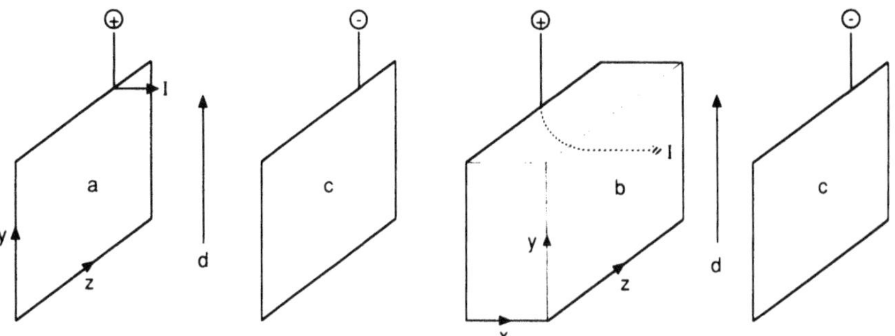


**FIGURE 5.2.** Classification scheme based on current distribution. (a) Two-dimensional electrode; (b) three-dimensional electrode; (c) counter electrode; (d) electrolyte flow.

electrolytic system where cell design must be tailored to match reaction characteristics. An interesting variation on the packed-bed scheme is the use by Fleet and Gupta[3] of packed carbon fiber electrodes for effluent treatment. Fluidized bed electrodes first attracted attention in the sixties,[4] initially for metal winning and later for effluent treatment, but so far have not fulfilled their early promise of revolutionizing industrial practice. A discussion of the detailed behavior of three-dimensional electrodes is beyond the scope of this book (see Reference 5).

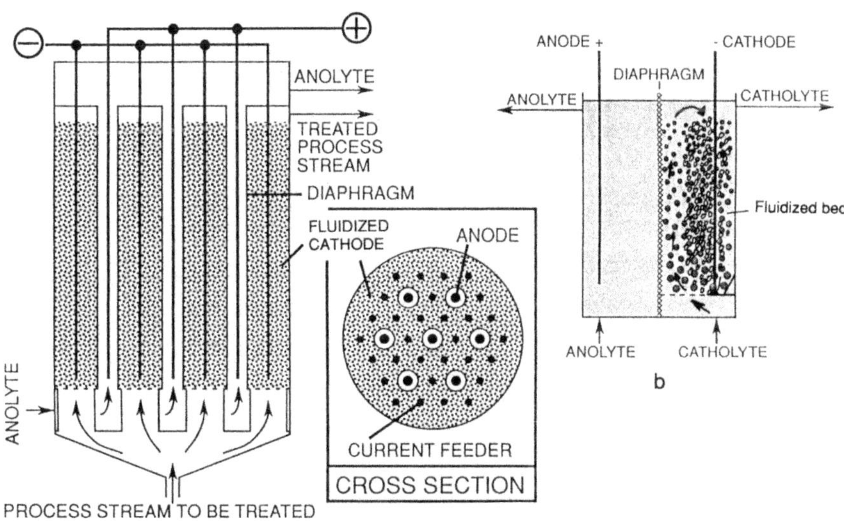


**FIGURE 5.3.** Two versions of fluidized bed cells. (a) Akzo cell; (b) circulating bed cell.

### 5.1.2. Electrical Connections to Cells

Current can be fed to cells in a monopolar or bipolar manner (Fig. 5.4). In a monopolar manner each unit cell is connected via bus bar to the electricity supply; in a bipolar manner only the end electrodes of a stack of cells are connected to the supply. The monopolar connection calls for a high current and low voltage, but a bipolar connection allows current and voltage to be of comparable magnitude, resulting in cheaper power supplies. The advantages and disadvantages of bipolar current connections are highlighted in the following list.

*Advantages:*

- Improved potential and current distribution
- Current and voltage extremes (can be avoided)
- Potentially lower resistance losses due to fewer electrical connections (see next advantage) and lower current transmission
- Avoiding expensive bus bars to each frame
- Avoiding the difficulty of individual connections to narrow gap cells

*Disadvantages:*

- Failure of the stack when one frame fails
- Greater problems in minimizing bypass currents

### 5.1.3. Hydraulic Connections

Since industrial cells are almost always flow cells, another important parameter is the arrangement of electrolyte flow through the cell stack. Electrolyte can flow through the cell stack in series or in parallel (Fig. 5.5).

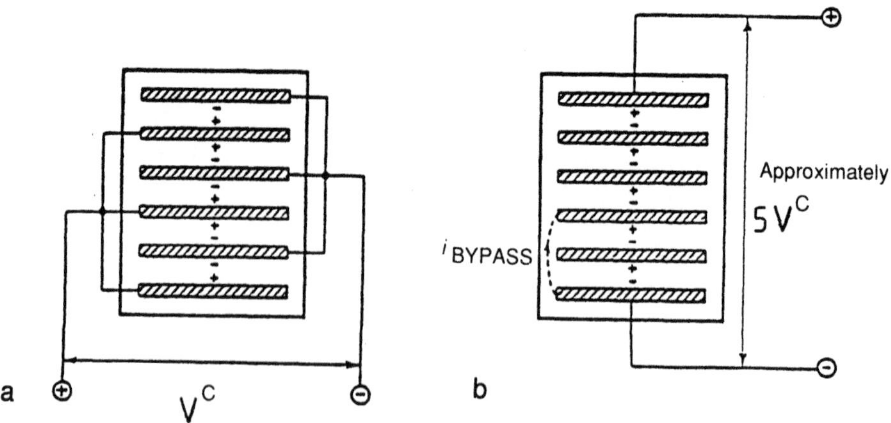


**FIGURE 5.4.** Ways of introducing current to electrodes. (a) monopolar; (b) bipolar.

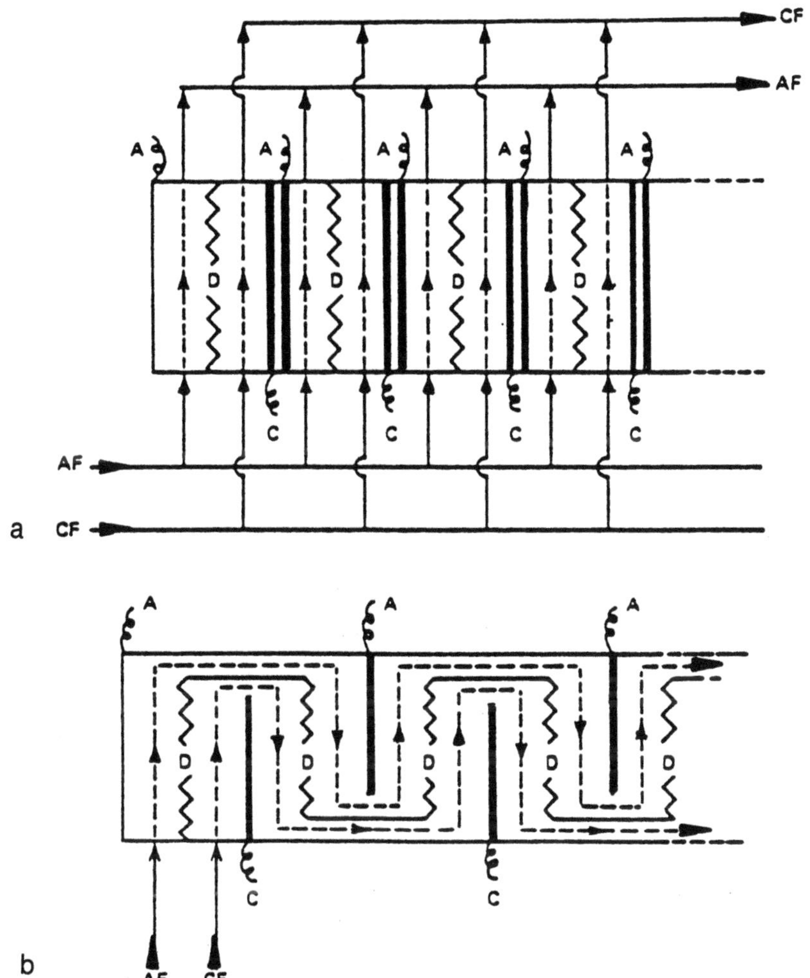


**FIGURE 5.5.** Electrolyte flow arrangements. (a) parallel flow; (b) series flow: $A$ = anode; $C$ = cathode; $D$ = diaphragm; $AF$ = anolyte flow; $CF$ = catholyte flow.

Series electrolyte flow allows greater conversion per pass of electrolyzer unit but pressure losses due to pumping are greater and problems associated with gas evolution may arise. This is why industrial practice favors parallel flow, the required conversion being achieved by connecting successive stacks hydraulically in series. Finally, hydraulic connections in cell stacks can be classified (Fig. 5.6) by internal or external electrolyte distribution. The advantages and disadvantages of the arrangements are shown in Table 5.1.

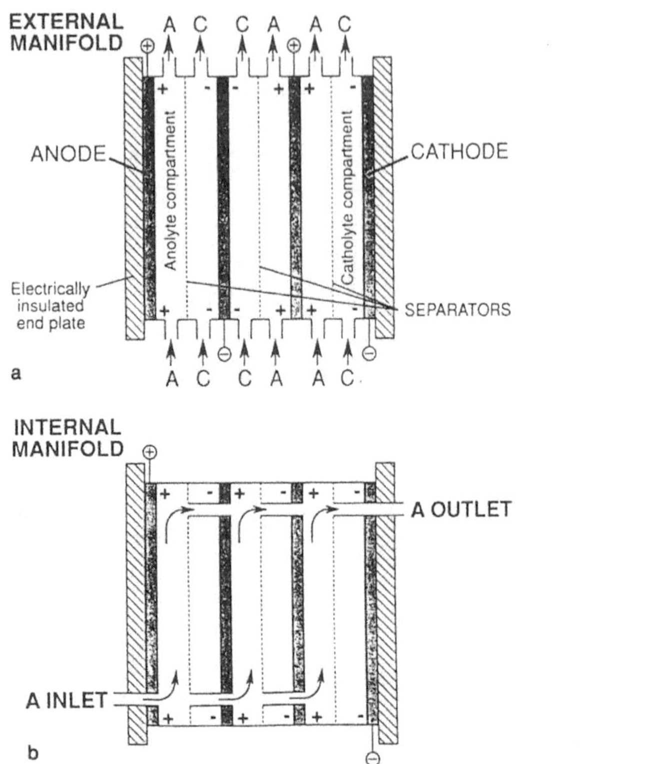


**FIGURE 5.6.** Electrolyte flow arrangements. (a) Internal; (b) external: $A$ = anolyte; $C$ = catholyte—catholyte flow not shown for (b).

TABLE 5.1. Characteristics of Internal and External Electrolyte Distribution

| Internal distribution | External distribution |
|---|---|
| *Advantages:* | |
| Easier frame design for narrow gap cells | If monopolar, a single frame can be isolated |
| More compact design specific area $>25\,m^2/m^3$ | |
| *Disadvantages:* | |
| Greater problems with current bypass | Separate inlet and outlet for each frame |
| Can lead to corrosion problems for bipolar electrodes | More complex design for narrow gap cells |
| Satisfactory electrolyte distribution more difficult | Less compact design specific area $<25\ m^2/m^3$ |

### 5.1.4. General Purpose Flow Electrolyzers

A problem that used to face industrial researchers was that cells for developing an industrial process were not commercially available. This has changed. Commercially available cell designs are based on modular units, usually with a parallel plate configuration. An advantage of such a concept is relative ease of scale-up from a pilot plant size to plant scale. As we shall see, however, scale-up from laboratory to pilot plant requires care due to the different mass transfer characteristics of the two cell sizes. Among the commercially available cells are the DEM cell[6] made by Electrocatalytic (Fig. 5.7a), the SU cell[7] made by Electrosyn (Fig. 5.7b), and the FM21 cell[8] made by ICI (Fig. 5.7c). See also Table 5.2. The DEM cell has external electrolyte flow and achieves a narrow electrolyte gap by using dished electrodes. SU and FM21 cells have internal electrolyte distribution, the latter starting life as a chlorine cell and being adapted later for organic electrosynthesis.

Design features these modular general purpose cells ought to have include:

- Narrow gaps between electrodes, to give a high specific area and a low cell voltage
- Good mass transfer characteristics
- Relatively easy replacement of electrodes
- Ability to withstand electrolytes of a wide range of *p*H and common organic solvents such as D.M.F., the lower aliphatic alcohols, acetonitrile, and methylene dichloride
- Ability to withstand operation at a temperature of 100°C and at modestly elevated pressure
- Being leakproof to liquids and gases
- Relatively low pressure drops and good gas release
- No contact between electrodes and diaphragms
- Ability to operate at current densities up to 5 kA/m$^2$

TABLE 5.2. Characteristics of Cells

| | Electro MP Cell | ElectroSynCell | ElectroProdCell |
|---|---|---|---|
| Electrode area (min) | 0.01 m$^2$ | 0.04 m$^2$ | 0.4 m$^2$ |
| (max) | 0.2 m$^2$/module | 1.04 m$^2$/module | 16.0 m$^2$/module |
| Current density (max) | 4 kA/m$^2$ | 4 kA/m$^2$ | 4 kA/m$^2$ |
| Interelectrode gap | 6–12 mm | 5 mm | 0.5–4 mm |
| Electrolyte flow, per cell | 1–5 l/min | 5–15 l/min | 10–30 l/min |
| Flow rate in each cell | 0.03–0.3 m/s | 0.2–0.6 m/s | 0.15–0.45 m/sek |

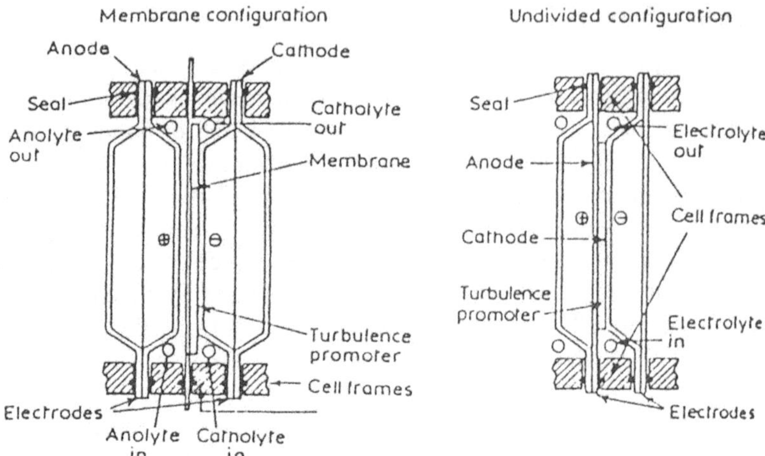


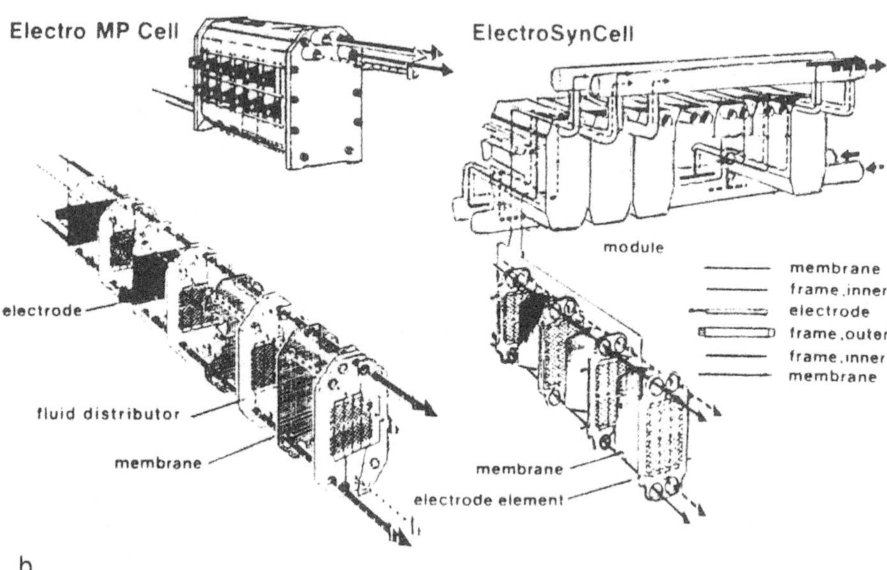


**FIGURE 5.7.** General purpose flow electrolyzers: (a) DEM cell: section of divided cell; section of undivided cell. (b) SU cell. (c) FM21 cell. Top: FM01 laboratory reactor; bottom: FM21 electrolyzer. Reproduced with permission.

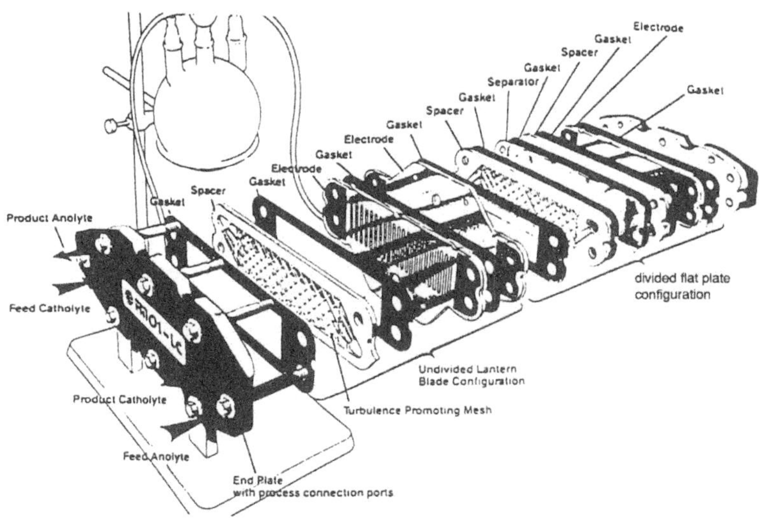


(each electrode has a projected area of approximately 64 $cm^2$)

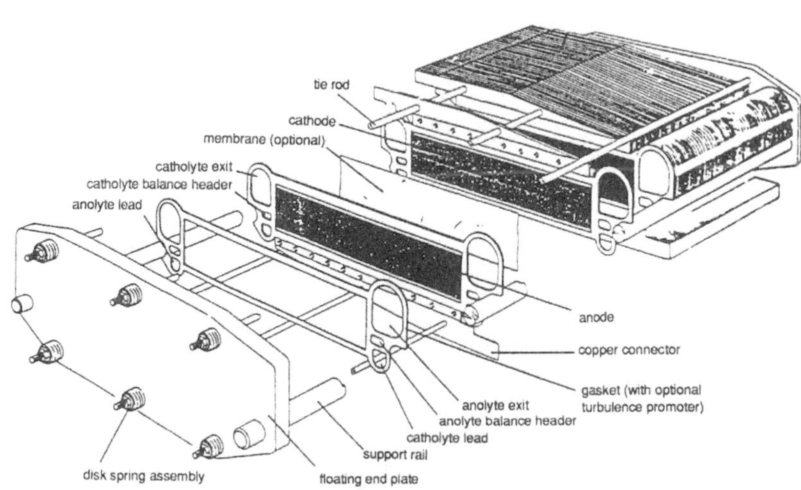


C (each electrode has a projected area of approximately 21 $dm^2$)

FIGURE 5.7. (*Continued*).

The electrodes and frames of general purpose cells have a rectangular geometry. However, electrodes with other geometries have been used sometimes.

### 5.1.5. Other Cell Designs

An alternative to using a rectangular plate as an electrode is to use a disk as in the BASF capillary gap cell, later called a disk-stack cell[9] (Fig. 5.8a). It is a stack of horizontally mounted plates in a cylindrical outer shell. Thin

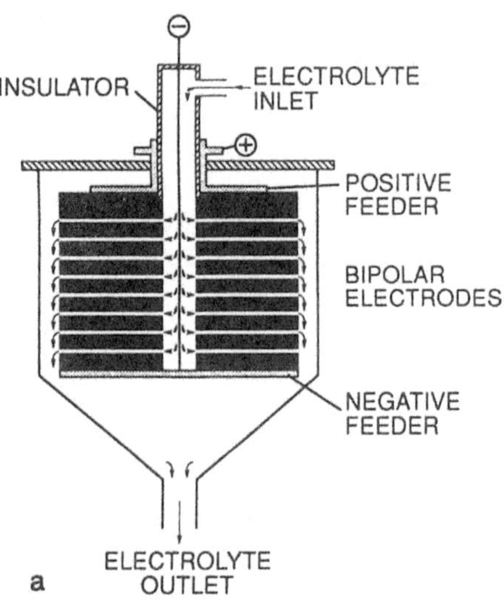


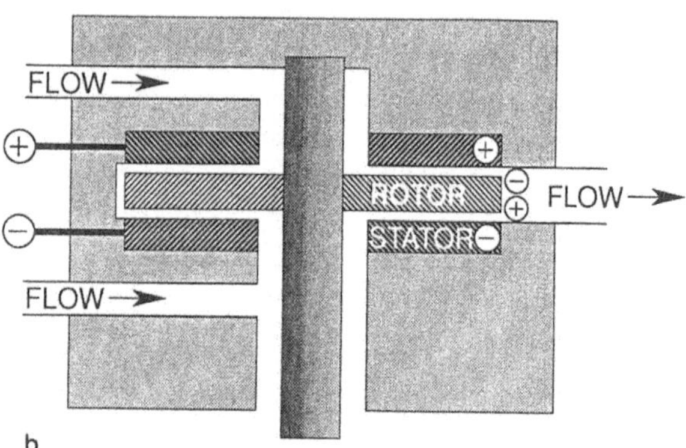


**FIGURE 5.8.** Narrow gap cells. (a) BASF capillary gap cell; (b) pump cell.

plastic spacers allow narrow electrode gaps to be used and electrolyte is pumped axially into the unit and flows radially outward between the disks. BASF has used the unit in a number of electro-organic synthetic processes.

Instead of pumping the electrolyte through the cell by means of an external pump, the cell itself can be made to perform this function by rotating the disk on a central spindle as in the pump cell (Fig. 5.8b).[10] Higher mass transfer rates and current densities can be achieved than in the disk-stack cell but have to be paid for in higher capital and maintenance costs, which limit the application of rotating devices in industry.

Rotation has also been applied to electrodes of cylindrical geometry (Fig. 5.9) where typically the electrode is rotated axially in a concentric cylinder unit. This has been applied to small-scale electroorganic synthesis[11] and to metal recovery from effluents[12] where the rotation assists in recovering the metal in powder form.

In an attempt to increase the specific electrode area of the reactor, electrodes, spacers, and separators are wound in a spiral form (Fig. 5.10) in the Swiss Roll cell.[13] Electrolyte flows axially in the narrow channels between electrodes and separator. In this compact design, the main problem is the need for flexible foil electrodes, spacers, and separators and the cost and difficulty of replacing damaged components. Nevertheless, the design has been used for relatively small-scale organic applications.[14]

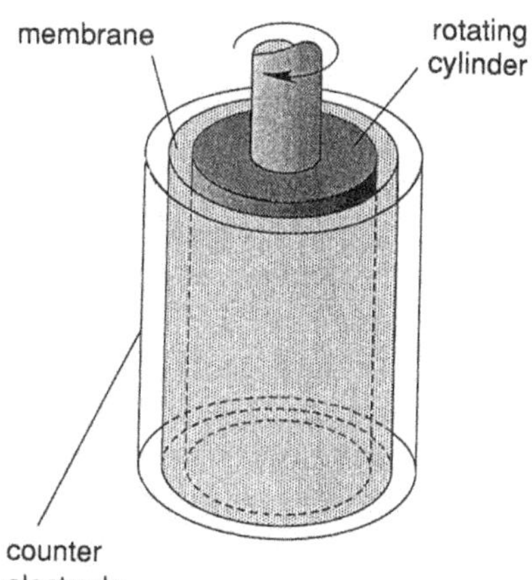


**FIGURE 5.9.** Rotating cylinder cell.

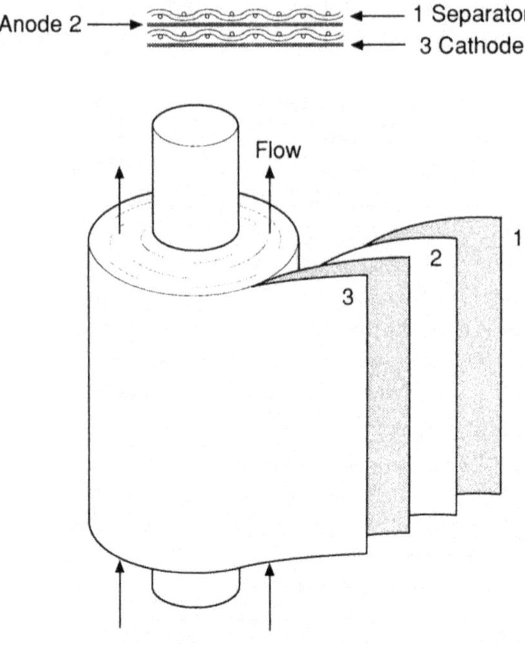


FIGURE 5.10. Swiss roll cell.

So far our reactor designs have concentrated on processes in a single phase (except when electrolytic gases or other substances are by-products). Special cell designs are needed when one reactant is a gas and has to be absorbed before charge transfer can take place. The reacting system can be an emulsion or can involve a two-phase liquid/liquid system.

Two designs have been employed for gas/liquid systems. One is the trickle tower cell (Fig. 5.11a),[15] which consists of layers of carbon Raschig rings, each layer separated from the next by a nonconducting net. The gaseous reactant is introduced at the base of the packed bed and meets the electrolyte, which trickles down the Raschig rings in a thin film. The second design is the sieve plate cell (Fig. 5.11b),[16] where the electrolyte flows horizontally in channels formed by bipolar electrodes, the gas bubbling through the liquid.

In recent years interest has focused on liquid/liquid systems, particularly on emulsions and phase transfer agents. Although large numbers of examples of reactions can be found in the literature,[17] there are few reactor designs that can be used in an industrial context. One such electrolytic cell is a pulse column reactor (Fig. 5.12),[18] which was originally designed for the separation of plutonium from uranium.

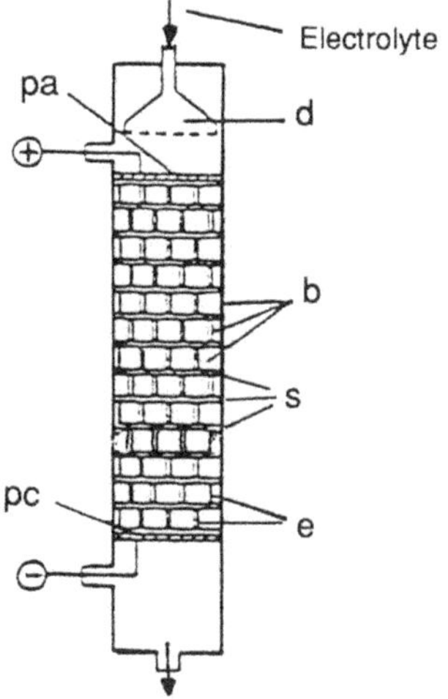


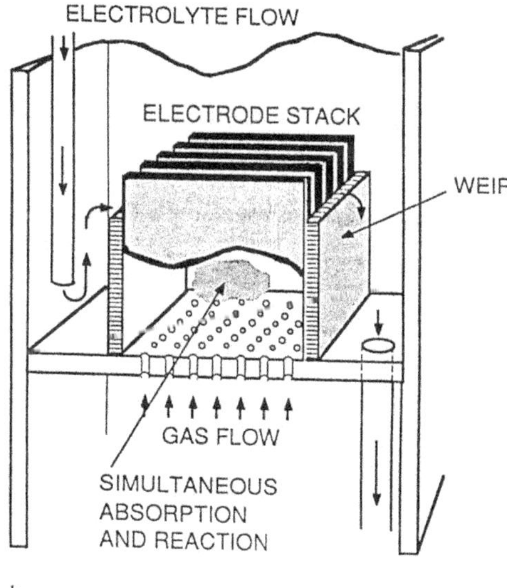


**FIGURE 5.11.** Cells for gas/liquid systems. (a) trickle tower cell, p.a. = Porous anode; p.c. = porous cathode, $d$ = distributor, $b$ = bipolar electrodes, $e$ = electrolyte thin film flow, $s$ = separator; (b) sieve plate cell.

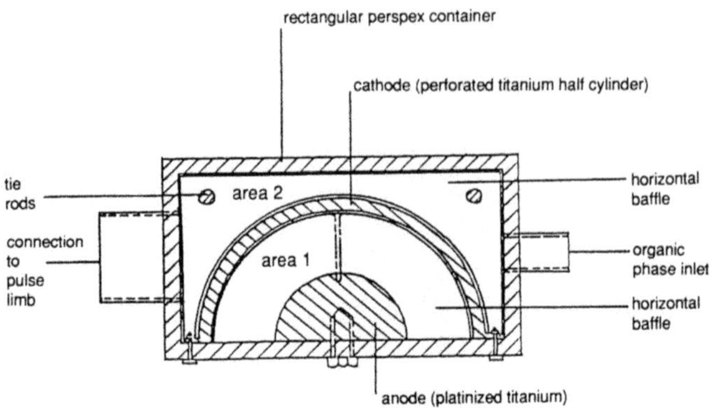


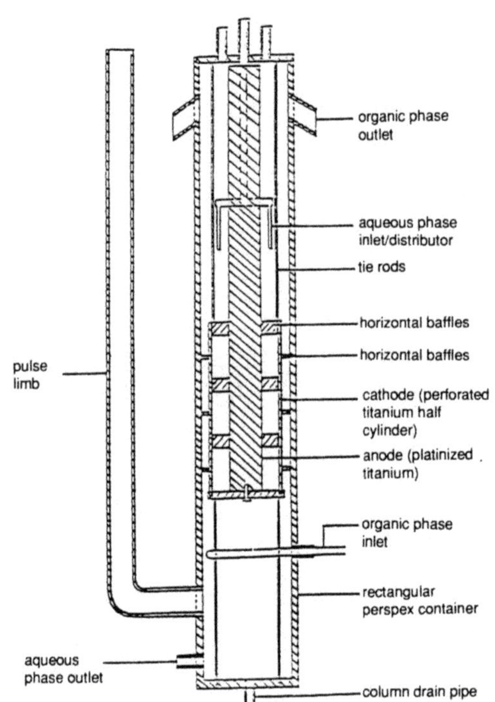


**FIGURE 5.12.** Electropulse column cell for liquid/liquid systems.

## 5.2. ELECTROLYTIC REACTOR SELECTION

Section 1.1 mentioned that in an electrolytic reactor or cell design it is necessary to start with the notional selection of a design, determine its performance, and then decide whether it will do or whether an alternative design has to be considered. Now we shall briefly consider classifications of cell designs, discuss the factors that have to be taken into consideration when selecting a particular cell geometry, and then give an example of selection. The reader will find some overlap between Chapters 4, 5, and 6, but we consider this not a bad thing in a book like this one.

(i) Notional reactor selection: Key cell parameters. In Section 1.5 we considered a number of quantitative criteria for judging the performance of an electrochemical reactor. These will be important when the selection of a cell is undertaken. Table 5.3 reminds you of five criteria which bear a direct relationship to cell design.

First, *current efficiency*, depends on good mass transfer characteristics of the cell. Details of how to achieve this have been discussed in Section 2.3.5. Uniformity of electrode potential over the electrode system and the nature of the electrode material can also have considerable impact.

Second, *chemical yield*, is like C.E. dependent on good mass transfer, uniform electrode potential, and electrode material.

In the third, *space-time yield*, the size of the cell is directly proportional to $a$, the electrode area, Table 5.4 gives values of $a$ for four typical designs (see Section 5.1). The "effective" electrode area for three-dimensional electrodes is smaller than what is shown in Table 5.4; nevertheless, it will be at least an order of magnitude larger than that for two-dimensional electrodes. In practical terms of cell selection (particularly for small-tonnage chemicals), small differences in cell volume are largely academic because cells are often dwarfed by separation equipment such as distillation columns.

Returning to Table 5.3, the fourth and fifth parameters, *energy yield* and *energy consumption*, depend on current efficiency (the first criterion) and cell voltage. As discussed in Section 1.4.5, this depends on electrode materials, the electrolyte gap between anode and cathode, the conductivity of the electrolyte, and the nature of the diaphragm if any.

TABLE 5.3. Criteria for Reactor Performance

| Criterion | Definition | Units |
|---|---|---|
| Current efficiency (C.E.) | $i_j/i$ | Dimensionless |
| Chemical yield ($Y_C$) | $m_{ACT}/m_{MAX}$ | Dimensionless |
| Space-time yield ($Y_{ST}$) | $\dfrac{\text{C.E. } iaM}{n}$ | kg/m$^3$ s |
| Energy yield ($Y_E$) | $\dfrac{0.036\,\text{C.E.}}{nV^c}$ | kmol/kWh |
| Energy consumption ($E_C$) | $1/Y_E$ | kWh/kmol |

TABLE 5.4. Specific Electrode Areas of Typical Cell Types

| Cell type | Specific area $a$ ($m^2/m^3$) |
|---|---|
| *With two-dimensional electrodes:* | |
| Plate and frame | $>10$ |
| Disk-stack | $>25$ |
| *With three-dimensional electrodes:* | |
| Packed bed | $<7000$ |
| Fluidized bed | $>4000$ |

We can now consider selecting a cell on the basis of points raised in Table 5.3. Four parameters stand out:

1. Permissible current density in relation to annual throughput
2. Sensitivity of the chemical yield to variation in electrode potential
3. Sensitivity of the product to reaction at the counterelectrode
4. Conductivity of the electrolyte

We illustrate selection procedure with a worked example taken from actual industrial practice.

---

EXAMPLE 5.1: Selection of a Cell Design for an Industrial Duty

THE PROBLEM: A cell design has to be selected for an organic electrosynthesis. The required process conditions are shown in Table 5.5.

THE SOLUTION: First calculate the cathode area needed for throughput. Since total batch time is 96 h, number of batches/annum $= (360 \times 24)/96 = 90$. kmol of $B$/batch $= 13500/90 = 150$. For a two-electron change and a C.E. of 80%, the number of coulombs required per batch will be:

$$\frac{2 \times 1000 \times 96500 \times 150}{0.8} = 3.62 \times 10^{10}\ \text{C}$$

Amperes required per batch will be:

$$\frac{3.62 \times 10^{10}}{72 \times 3600} = 1.40 \times 10^{5}\ \text{A}$$

The cathode area required for a current density of $1000\ \text{A/m}^2$ will be:

$$\frac{1.40 \times 10^{5}}{1000} = 140\ \text{m}^2$$

TABLE 5.5. Organic Electrosynthesis

| | |
|---|---|
| "Tafel-type" reduction | $A + 2H^+ \rightarrow B$ |
| Electrolyte | Aqueous 1M sulfuric acid |
| Current density | 1000 $A/m^2$ |
| Required amount of $B$ | 13500 kmol p.a. |
| Current efficiency | 80% |
| Sensitivity of $B$ to oxidation | High |
| Effect on $B$ of variation in $E'$ | Deleterious |
| Mode of cell operation | 72 h batches, 24 h downtime |
| Operational time of factory | 360 days p.a. |

The first principle in cell selection is to choose a simple design that has a proven record in industrial practice. A plate and frame cell, PFC (see Section 5.1.4) seems an obvious choice.

The typical cross-sectional area of an industrial module or unit cell is about 1 $m^2$; we would need 140 cells. If we are going to use cell stacks made up from 19 unit cells, we will need $140/19 = 7.4$ cell stacks. Using 8 cell stacks will leave some electrode area in reserve. This is always sensible, particularly in an organic electrosynthesis where current efficiency may deteriorate during a batch. Assuming that our PFC has a specific electrode area of 8 $m^2/m^3$ (Table 5.4), one cell stack will be a reasonable 2.4 m in length.

Looking at other parameters (as just listed), a PFC will satisfy parameter 2. For parameter 3 we will need diaphragms separating anolyte and catholyte. Aqueous molar sulfuric acid will pose no problems for parameter 4.

What would happen if current density were lower. If, for example, the current density were 200 $A/m^2$, then 700 $m^2$ and 37 cell stacks would be needed. A PFC would be just possible. But with a current density of 1–10 $A/m^2$ (this could be true if the reactant solubility is $10^{-3}$ molar) a PFC would no longer be feasible and a cell with three-dimensional electrodes would have to be considered.

## 5.3. SCALE-UP OF ELECTROLYTIC REACTORS

Change in the scale of a reactor will change its performance. This calls for suitable allowances in process design: knowledge of the probable variation in performance of units on a change of scale is required. For this we need mathematical models supported by key experimental trials. This section describes the procedures necessary to scale up an electrolytic reactor.

Variations in mass transfer rates and current distribution due to the change in the scale of operation are emphasized.

### 5.3.1. General View of Scale-up Procedures

An electrolytic process will proceed through a number of stages (as mentioned at the beginning of Chapter 1), relying on feedback of information, data, or calculations for continuous reassessment and update of earlier judgments, decisions, assumptions, or estimations—not to say guesses.

The first stage of process design is based on company policy or management objectives: the task of assembling a preliminary design is assigned to qualified personnel. Process flow sheets are drawn up identifying appropriate feed preparations and product recovery, specifying preliminary electrochemical reactor design and associated power supply (Fig. 1.1). Material and energy balances are made, along with a preliminary economic appraisal of the plant. If the decision is go rather than no-go, laboratory studies may be needed to produce more accurate or missing (guessed) data. This is when the scale of operation is changed, initially down to a suitable experimental size at which parametric investigations can conveniently be made. Such scale-down procedures are common in any branch of process engineering, as is the general philosophy of a preliminary process design. In this section we address only the factor peculiar to electrochemical processes: the electrochemical reactor.

### 5.3.2. Design Scheme for the Scale-up of Electrochemical Reactors

A stepwise procedure for reactor design adapted from that of MacMullin[19] starts with a specified production rate (Fig. 5.13). Scale-up is defined by production capacity which is used to characterize the scale of the reactor. A modular approach to reactor design is common in electrochemical process engineering because electrodes have to occur in pairs (one anode and one cathode) and because there must be external electrical connections. Each electrode pair should be identical in performance. This (we shall see later) does not always happen, which leads to scale-up problems.

As mentioned in Section 5.2, commercial scale reactors contain modules of multielectrode units. Scaling down for tests is done by reducing a cell stack to one or two electrode pairs. Preliminary tests may use geometrically scaled-down versions. This brings out facets of reactor design associated with scale-up. Problems that may arise are current distribution and mass transfer rates: these are dealt with later in this chapter. The design procedure

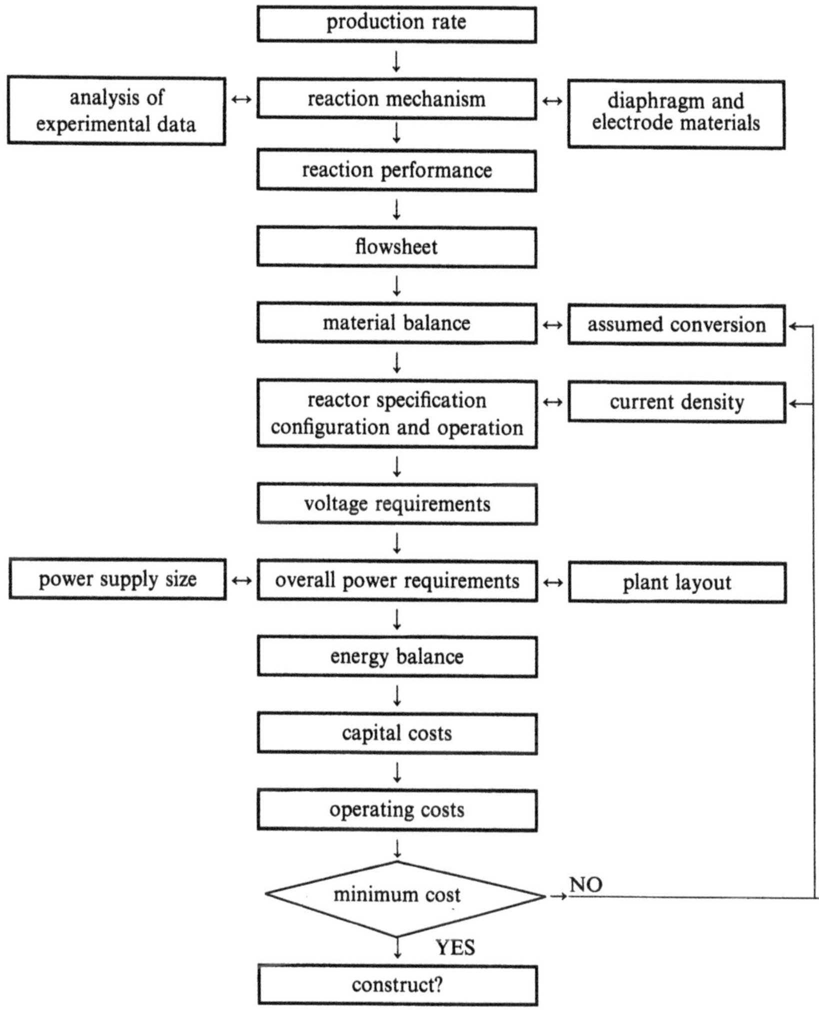


**FIGURE 5.13.** Design scheme for reactor scale-up.

outlined just below is based on MacMullin's 1963 paper.[19] Although the approach is sound in principle, the paper was written long before computers came into routine use in design, so MacMullin's approach must be supplemented by the mathematical modeling procedure in Fig. 1.1, in order to minimize expensive experimental runs. All decisions should be made in relation to the whole process plant and not the reactor alone. This latter point is considered in detail in Chapter 6.

Let us outline the design procedure for scaling up a reactor:

1. Stipulate all reactions and equilibria. Specify operating conditions of temperature, pressure, and pH for the calculation of physical and thermodynamic properties.
2. Specify a diaphragm if required:
   - Type required, i.e., permeable or semipermeable
   - Properties required, e.g., mechanical strength, chemical stability, porosity, resistivity
   - Cost: initial cost, operating life, frequency of replacement
3. Prepare a flow sheet, including the flow or transport of ions through the diaphragm.
4. Analyze experimental data on reaction kinetics, mass transport rates, current efficiencies, and chemical yields. Do laboratory experiments if gaps in knowledge are discovered.
5. Perform a material balance on the plant or pilot plant. Include assumed reactant conversions and estimated material yields. Incorporate possible ionic and water transport between catholyte and anolyte, vaporization losses and precipitation, etc.
6. Specify electrodes. The nature of the working electrode will be determined largely by data collected at the laboratory stage. Factors to be considered include:
   - Specificity of reaction
   - Overvoltages
   - Electrical conductivity and size
   - Corrosion, passivation, contamination, and wear
   - Durability and mechanical strength
   - Initial, replacement, and recoating costs
   - Thermal conductivity
   - Possible bipolar use
7. Specify reactor (see Sections 5.2 and 5.4):
   - Flow characteristics, volumetric throughput, and linear velocity
   - Estimation of mass transfer characteristics and conversion per pass
   - Recycle operation
   - Mode of current supply
   - Required electrode area and operating current densities for a specified production rate

8. Decide on detailed reactor geometry:
   - Configuration of electrodes, i.e., plates, cylinders, mesh or porous
   - Orientation, gas release, and polarization
   - Flow path, i.e., external or internal, parallel or series, gas disengagement, friction loss calculations
   - Current leads and connections
   - Features of thermal and electrical insulation
9. Produce an engineering drawing of the reactor.
10. Do a reactor voltage balance. This may be an integral part of (7) and includes electrode overvoltage; ohmic losses in catholyte, anolyte, diaphragm, and due to gas bubble corrections; bus bar; and contact resistance losses. Estimate values at selected current densities and reassess in the light of interelectrode spacing and items (7) and (8).
11. Determine power requirements:
    - Size of the power supply (current and voltage)
    - Preliminary plant layout, pipe sizes, and holding tanks
    - Estimation of pumping requirements based on pressure losses in pipes, fittings, and reactor. Size pumps
12. Perform an energy balance. Estimate cooling water, refrigeration, and fuel requirements. Size heat exchange equipment if needed for an assessed plant inventory.
13. Estimate cost of reactor and associated equipment.
14. Repeat steps as required for various reactor sizes, conversions, and production rates.

Other factors such as materials of construction, gas-tight seals, and safety are also vital. The order of procedure may vary, especially if there will be batch rather than continuous operation. Steps (7), (8), and (10)—in essence, the reactor design—can be drastically affected by changes in scale. So it is important to consider in detail the scale-up of the reactor. Scale-up has three stages: laboratory tests, pilot scale runs, and commercial or semicommercial operation. A critical point is the transition from laboratory to pilot plant scale operation, using a scaled-down version of the commercial reactor. Compatibility of performance of other plant units is an important feature.

### 5.3.3. Effect of Scale-up on Reactor Performance

To minimize or eliminate harmful effects of scale-up requires models. Modeling means dimensional analysis or physicochemical mathematical modeling. As we have seen in Chapters 2 and 4, the design of reactors uses both types with varying degrees of sophistication. Dimensional analysis is a standard technique in chemical engineering—if not familiar with the technique, the reader is referred to Ref. 20; it starts from the physical principles of the process and its important variables. By applying logical rules of dimensional consistency the functional relationship between variables is expressed in terms of dimensional groups. Mathematical modeling, on the other hand, brings together knowledge of the physical and chemical processes and conservation laws in the form of differential, algebraic, and nowadays most often numerical relationships.

**(i) Dimensional analysis in reactor scale-up.** We start with a simple worked example.

---

EXAMPLE 5.2: Illustration of the Use of Dimensional Analysis in Scale-up

THE PROBLEM: An electrolytic reaction takes place in a parallel plate cell of length $L$, width $w$, and interelectrode gap $d$, with $w >> d$. The reaction takes place under mass transport control with a mass transfer coefficient $k_L$. The flow regime is turbulent and it can be assumed that the electrolyte volume $V = w \times d \times L$. Use dimensional analysis to obtain a scale-up criterion that will keep conversion constant.

THE SOLUTION: Dimensional analysis leads to the conversion $X_A$ as a function:

$$X_A = f(Re, Sc, d/L, \tau)$$

where $Re$ and $Sc$ are the dimensionless Reynolds and Schmidt numbers (see chapter 2), and $\tau$, the residence time in the reactor, equals $L/u$.

A factorial experiment to evaluate this function would be most demanding. Fortunately, Eq. (4.17) leads to an expression for $X_A$:

$$X_A = 1 - \exp[-ak_L\tau]$$

where $a$ is the specific electrode area, which in this case reduces to $1/d$.

In Chapter 2 it was shown that for a parallel plate reactor under turbulent conditions, $k_L$ was given by the dimensionless correlation in Eq. (2.18):

$$\frac{k_L d_e}{D} = 0.023 \left[\frac{u d_e \rho}{\mu}\right]^{0.8} \left[\frac{\mu}{\rho D}\right]^{1/3}$$

Since $W >> d$, $d_e = 2d$.

The conversion is now expressed in terms of $ak_L\tau$. On rearrangement:

$$ak_L\tau = Ku^{-0.2}d^{-1.2}L$$

where the constant $K$ contains all properties of the electrolyte, namely density, viscosity, and diffusivity, that do not vary with scale-up.

Therefore the generalized group $(u^{-0.2}d^{-1.2}L)$ should be kept constant on scale-up to maintain conversion. It is often easy to determine these criteria but it may not be possible to maintain them for a cell of reasonable geometry. Indeed, the criteria as set often contain contradictory conditions.

---

### 5.3.4. Scale-up Methods and Similarity

The philosophy of scale-up is to ensure that corresponding units are similar as possible. To guide a design engineer, a number of so-called similarity criteria are defined. The criteria center around the use of a variety of dimensionless groups which describe the geometric, kinematic, thermal, and chemical characteristics of units. Clearly, for electrolytic systems we must also consider the very important electrical similarity. For complex systems such as electrochemical reactors, complete similarity on scale-up is impossible or uneconomical. Then design for scale-up should follow "dominant" criteria selected from the following list:

#### 5.3.4.1. Reactor Size and Geometric Similarity

Geometric similarity of two bodies of different scale is achieved by fixing corresponding dimensional ratios.

A heterogeneous system as complex as an electrochemical reactor arouses objections to this procedure. For example, an increase in $d$, the interelectrode gap, will dramatically increase cell voltage and energy costs (see Section 1.4.5). It is possible to reduce this effect by increasing the electrical conductivity of the electrolyte, but there is a limit to this since the "optimum" composition of the electrolyte will already be governed by considerations of maximizing chemical yield and minimizing polarization and I.R. losses. The specific electrode area of the reactor, which is inversely proportional to the interelectrode gap, decreases with any increase in $d$ and the production capacity of the cell $Y_{ST}$ (see Section 1.5) per unit volume of the cell will suffer; although this may not be a major part of total plant costs, it will lead to increased cell costs on scale-up. Also, the current distribution in the reactor (see Section 5.3.6) may be modified.

Geometric similarity, in practice applied only to the shape and area of the electrode, is ranked low among criteria. It is confined to specifying equivalent specific areas of electrodes. Scale-up is achieved by using multiple electrode pairs and reactor units.

#### 5.3.4.2. Fluid Mechanics and Kinematic Similarity

Kinematic similarity is concerned with the motion of phases within a system and the forces inducing that motion. For example, in the formation of boundary layers during flow past flat plates and during forced convection in regularly shaped channels, there are usually three dominant forces: pressure, inertia, and viscous forces. If corresponding points in two different-sized cells show at corresponding times identical ratios of fluid velocity, the two units are said to be kinematically similar and heat and mass transfer coefficients will bear a simple relation in the two cells. It can be shown[22] by means of dimensional analysis that for a closed system under forced convection the equation of motion for a fluid reduces to a function of *Re*, the Reynolds number, which we have met in Chapter 2. To preserve kinematic similarity under those circumstances, Reynolds numbers in the two cells must be identical.

As we shall see in Section 5.3.5, entrance and exit effects cause irregularities in the fluid flow and care has to be taken in predicting heat and mass transfer rates on scale-up.

#### 5.3.4.3. Concentration Distribution and Chemical Similarity

Chemical similarity calls for maintaining constant appropriate dimensionless holding times, residence times, or reaction times in batch or continuous reactors. Reactor mode of operation is important in defining these groups and must be the same in both units. In the context of dimensionless groups, conditions defining chemical similarity have been worked out by Damkohler,[23,24] the most important being expressed as rate of chemical formation divided by rate of bulk flow. This condition for a first-order reaction reduces to $kL/u$, where $k$ is the first-order reaction rate constant, $L$ a characteristic length, and $u$ the linear velocity of the electrolyte.

Overall chemical similarity is achieved not merely by the application of a few dimensionless numbers, but results from careful application of rigorous mathematical models.

#### 5.3.4.4. Current Distribution and Electrical Similarity

Electrochemical reactors, unlike their catalytic counterparts, require electrical similarity. This is often the most important consideration in the scale-up of electrochemical reactors. Considerable attention will be devoted to this subject in Section 5.3.6, dealing with current distribution.

Electrical similarity exists between two units when corresponding potential and current density differences bear a constant ratio. The two sizes of cells will require the same cell voltage and current distribution. The criterion often necessitates a constant interelectrode gap on scale-up. A variety of dimensionless groups have been proposed as similarity criteria. One for example is the so-called polarization parameter $P$, defined as:

$$P = \frac{\kappa}{L'} \frac{dE'}{di}$$

where $\kappa$ is the electrolyte conductivity, $L'$ a characteristic length of the cell, $E'$ the electrode potential, and $i$ the current density.

Dimensionless groups to be used have also been determined from models of current distribution.

#### 5.3.4.5. Heat Transfer and Thermal Similarity

Thermal similarity exists when temperatures at corresponding points of two cells have a constant ratio.

The heat exchange between an electrolytic reactor and its surroundings changes on scale-up. The electrode area of a reactor is usually held constant and thus on scale-up the increase in electrode area is proportional to the increase in volume. However, the surface area for heat transfer does not increase in proportion to the volume in most configurations. It is harder to remove heat from a larger than a smaller reactor: higher temperatures are therefore likely to be established.

On scale-up the heat exchange from the larger reactor to the surroundings should be evaluated to check that the rise in temperature is acceptable. Thermal similarity is assured by maintaining the required operating temperature for all reactor sizes. This may call for adjusting operating conditions or modifying cell design, e.g., by changing the thickness of thermal insulation.

### 5.3.5. The Effect of Scale-up on Mass Transfer

The effect of reactor size on behavior was identified in the discussion of fluid dynamics and mass transfer in Chapter 2. Variation of velocity with position and the formation of boundary layers cause changes in mass transfer coefficients with electrode coordinates. We saw that for developing laminar flow, mass transfer coefficients and hence limiting currents decrease with increasing electrode length. Reactor operation should be at current densities based on mass transfer rates associated with the scaled-up unit, not the smaller unit.

Mass transfer as discussed in Chapter 2 provided means for predicting mass transfer rates, but were restricted to reactors devoid of entrance and exit effects (see Section 2.3.2). Real reactors, however, can be affected significantly by such effects. This section will discuss the problem by looking at the limited data about electrolytic cells in general and parallel plate geometry in particular.

Workers investigating the distribution of mass transfer coefficients in channels have obtained distribution data by the use of so-called microelectrodes or segmented electrodes. Wragg *et al.*[25] have studied in detail the distribution of mass transfer rates for turbulent flow in tubular and rectangular channels when subject to entrance effects. The geometrical configuration they used is relevant to electrochemical reactor design when flow from inlet feed pipes suddenly expands into larger channels (Fig. 5.14). Then recirculating flow occurs in a region near the reactor entry with unusually large mass transfer rates at the reactor wall. The detachment of flow and reattachment to the wall farther downstream where the flow begins to develop in the usual manner cause a maximum value of mass transfer coefficients (Fig. 5.15a). The smaller the entry pipe in relation to the channel, the greater the value of the peak mass transfer coefficient. A nozzle diameter to channel diameter ratio of only 1:2 produces a peak mass transfer coefficient more than twice that of the fully developed turbulent flow downstream.

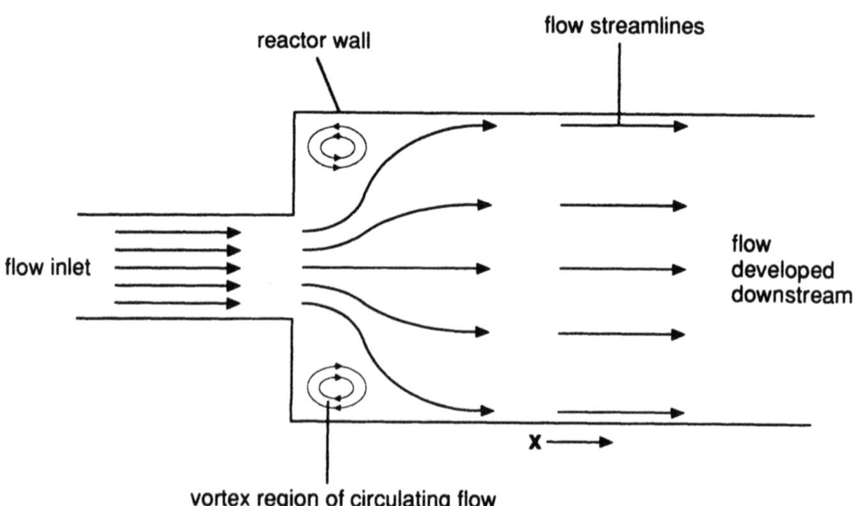


**FIGURE 5.14.** Effect of sudden expansion of flow into a square channel.

A reactor channel with a square cross section produces longitudinal variation of mass transfer and a spanwise distribution of the mass transfer coefficient (Fig. 5.15b). In the recirculating flow region mass transfer coefficients are highest at the center of the electrode and decrease toward

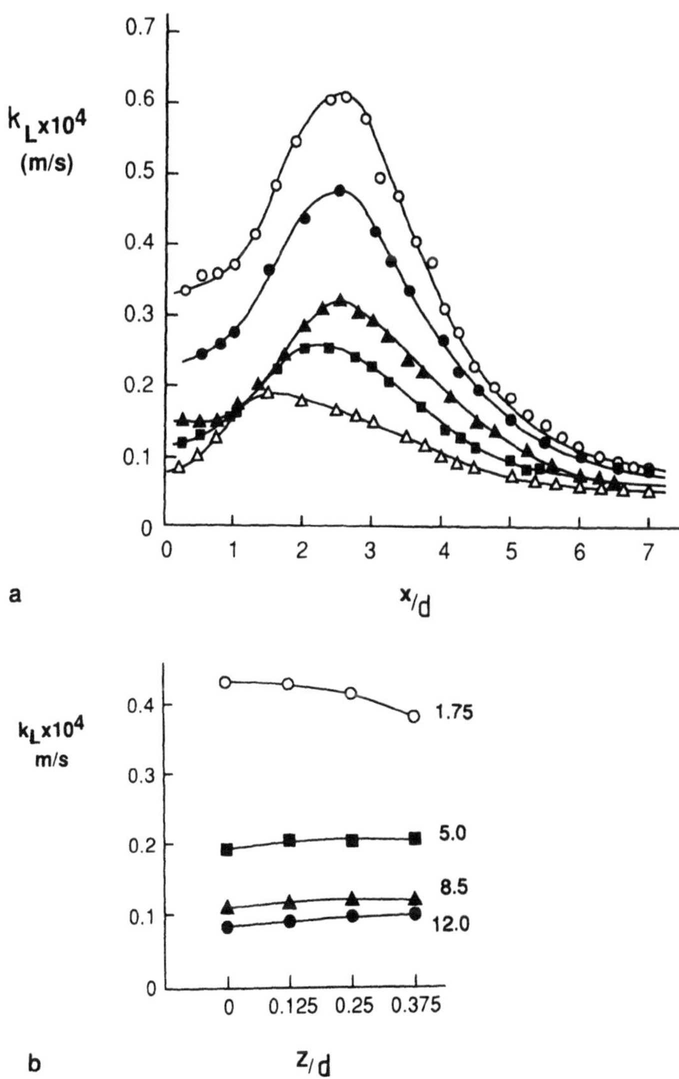


**FIGURE 5.15.** Values of local wall mass transfer coefficients due to sudden expansion from a circular entry into a square channel. (a) Variation with downstream distance. Re = 5000, expansion: ○ 1:10, ● 1:6, ▲ 1:4, ■ 1:3, △ 1:2; (b) spanwise variation. Re = 15670, 1:3 expansion parameters $x/d$.

the edges. This is somewhat reversed downstream—a phenomenon associated with the velocity distribution in rectangular and square ducts. According to results reported by Wragg *et al.*[25] for turbulent flow it takes approximately 10 equivalent reactor diameters to achieve a near uniform distribution of mass transfer rates. This, combined with the analogous although not as severe maldistribution of mass transfer rates near channel exits subject to sudden contraction means that short electrochemical reactors will not exhibit a uniform mass transfer distribution. It follows that when two reactors with different lengths are operated with the same mean velocity, the performance of the larger one may not match that of the smaller one, due to enhancement of mass transfer in the latter. This was confirmed by the work of Goodridge *et al.*,[26] who measured mass transfer distribution in two baffled-cell sizes of rectangular geometry in which the baffles are used as "turbulence promoters." The baffles divide the reactor into a "long" (Fig. 5.16) narrow serpentine flow path, subjected to a series of flow reversals through 180° and sudden expansions. A typical distribution of mass transfer rates is shown in Fig. 5.17 for a small-scale (electrode area $0.0225\,m^2$) and a larger-scale ($0.2025\,m^2$) reactor. Cycling behavior in the mass transfer rate is exhibited; values are highest at channel entry and lowest at channel exit. As predicted by the work of Wragg *et al.*,[25] the mass transfer rates in the smaller cell (channel length 0.12 m) are dictated by entrance effects and no even mass transfer distribution is seen. But the larger reactor of near industrial scale exhibits a large region of near uniform mass

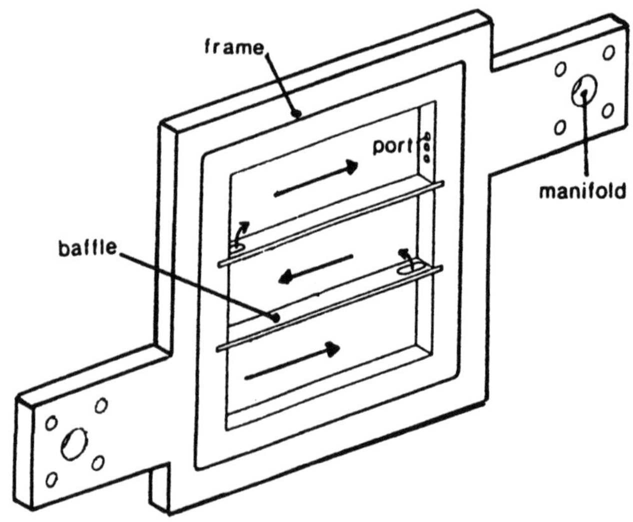


**FIGURE 5.16.** Frame of small baffled cell (cathode area = $0.0225\,m^2$).

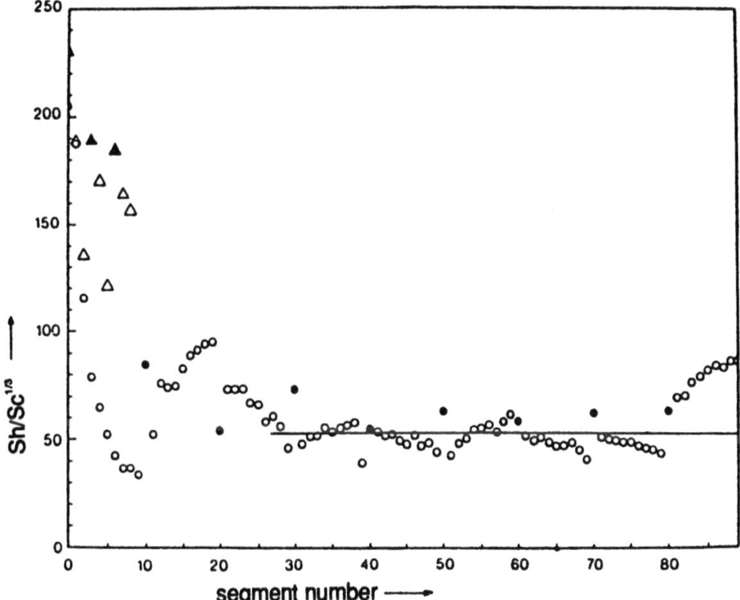


**FIGURE 5.17.** Distribution of mass transfer rates for small (cathode area = $0.0225\,m^2$) and large baffled cell (cathode area = $0.2025\,m^2$). $Re = 3900$, $\triangle$ = smaller cell, $\bigcirc$ = larger cell; the full symbols correspond to segments located at cell entry and where flow reversal occurs.

transfer rates (Fig. 5.17); in consequence, it can be used to predict the performance of a full-sized ($1\,m^2$) industrial reactor. Mass transfer rates in the uniform region are correlated by:

$$Sh = 0.36\,Re^{0.55}Sc^{0.33} \tag{5.1}$$

This equation, although predicting mass transfer rates some 10% or so below average for the reactor, is recommended for estimating mass transfer coefficients of baffled parallel plate reactors.

There is much research to be done in this area before "mass transfer design procedures" can be well documented.

### 5.3.6. The Effect of Scale-up on Current Distribution

Electrochemical reactors show nonuniform distributions of electrode potential and hence current density. This is because electrode dimensions, however large, are finite; even in the simplest situation the highest nonuniformity will occur at electrode edges and discontinuities. Current distribution problems can be divided into three categories of increasing complexity.

1. *Primary current distribution* occurs in the absence of electrode reaction, when the surface overpotential is neglected (i.e., the electrode surface is considered to be nonpolarized) and concentration gradients do not exist.
2. When an electrochemical reaction occurs, the electrode surface can no longer be considered an equipotential boundary and activation overvoltage has to be overcome to cause further reaction. The shift in electrode potential—the slope of the electrode potential/current density curve becomes significant—required to induce increased current flow gives rise to *secondary current distribution.*
3. When a shift in potential causes significant concentration changes so that mass transport limits to some extent the current flow, *tertiary current distribution* occurs. Tertiary distribution exists whenever current flows but at relatively low polarization the effect can be ignored.

Design methods try to include potential variations in the analysis. Local variations in product distribution and current efficiency are deduced and are part of the overall design. These approaches tend to assume an ideal reactor configuration in which the primary current distribution is uniform. This ideal is sometimes approached in practice but some reactor geometries exhibit severe nonuniform distributions, particularly at the higher current densities demanded in industry. It is important to analyze such systems and determine conditions which will minimize these effects. In the following sections we will consider the problem of current distribution and illustrate some calculations with worked examples.

#### 5.3.6.1. Primary Current Distribution

With no polarization of the electrode and an electrolyte of uniform concentration, we deal only with primary current distribution, which is given by the solution of the Laplacian equation:

$$\nabla^2 E_s = 0 \tag{5.2}$$

where $E_s$ is the potential in the solution.

Mathematical solution depends on the boundary conditions governed by the geometry of the reactor system. We will confine our attention to plate electrodes. Electrochemical reactors can be divided into two categories: tank electrolyzers and channel flow electrolyzers.

(i) Tank electrolyzers. In simple tank electrolyzers with a pair of flat electrodes the current distribution is influenced by the four containing walls. The potential field is such that a fraction of the current flows to the back of the electrode. This fraction decreases as the length of the electrodes increases in relation to the back walls. An analysis of this system, applying conformal mapping to the solution of the Laplace equation, can be found in Hine's book.[27] The problem is simplified if one or two approximations are adopted in the solution; these approximations either ignore the influence of the rear walls or of the side walls, one set of walls being considered to be remote from the electrodes (Fig. 5.18). Three dimensions dictate the distribution of current: the electrode length ($2L$), the interelectrode gap ($2d$), and the distance separating the insulating bounding walls ($2s$).

First, the effect of the side walls: The solution[27] for the primary current distribution for the reactor shown in Fig. 5.18a is presented in Fig. 5.19. The plot allows a rapid, practical, reasonably accurate determination of the primary current distribution. On the abscissa of Fig. 5.19 is plotted from left to right the percentage ratio of the length of the electrode to the width of the cell ($L/S \times 100\%$), so that the top scale shows from right to left the percentage gap between the edge of the electrode and the insulating wall ($[S - L]/S \times 100\%$). The upward ordinate scale indicates current at the inner electrode surface as a percent of total current the ratio for the outer electrode surface is given on the downward scale. Clearly, the inner surface carries the total current if the electrode is extended to the wall. The current distribution is shown by the sigmoid-shaped bold lines, which represent values of $S/d$ from 1 to 100. The finer lines (numbered from 10 to 90) are lines of equipercentage of the total current referred to the current when the electrodes touch the side walls. This percentage shows a characteristic decrease as the gap between electrode and wall grows larger.

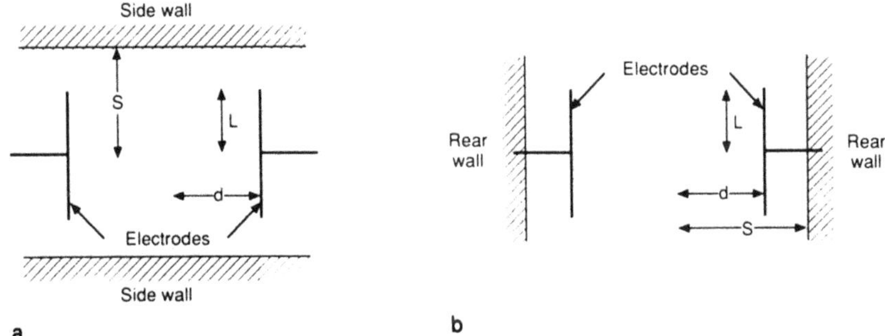


**FIGURE 5.18.** Approximations during the calculation of primary current distribution in a tank electrolyzer. (a) effect of the side walls; (b) effect of the rear walls.

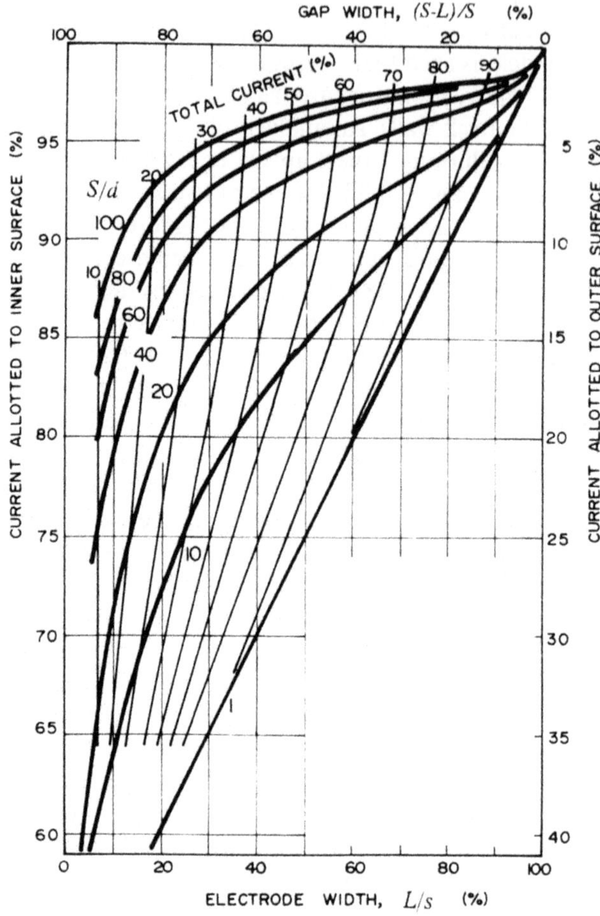


**FIGURE 5.19.** Primary current distribution for cell in Fig. 5.18a. (Reproduced from Hine, F., *Electrode Processes and Electrochemical Engineering*, Plenum, 1985.)

---

EXAMPLE 5.3: Primary Current Distribution in a Tank Electrolyzer: Rear Walls Remote

THE PROBLEM: A tank electrolyzer of $0.3\,m^3$ capacity is designed to contain ten pairs of electrodes. The interelectrode gap is 4 cm; the electrodes are 3 cm from the side walls. Initial design will try to achieve a good primary current distribution by ensuring that the current at the inner electrode surface is at least 95% of the total current and that the latter is at least 95% of the total current for the widest possible electrodes. If the tank length is 80 cm, estimate appropriate dimensions for the electrodes.

THE SOLUTION: Referring to Fig. 5.20, the height $x$ and width $y$ are to be sized.

The two ratios of cell parameters are defined in terms of $L$:

$$\frac{S}{d} = \frac{L+3}{2} \quad \text{and} \quad \frac{L}{S} = \frac{L}{L+3}$$

By selecting a number of values of $L$ a suitable electrode width can by interpolation be determined from Fig. 5.19.

Table 5.6, shows that a value for $L$ of 27 cm or more is satisfactory. Using a value of 30 makes the cell width $y$ equal to 60 cm. The cell height $x$ will be:

$$x = \frac{0.3}{0.6 \times 0.8} = 0.625 \text{ m}$$

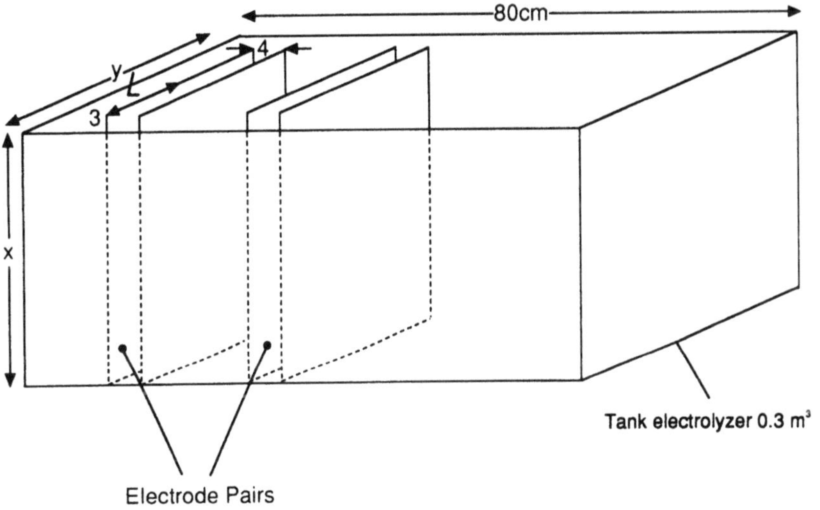

**FIGURE 5.20.** Tank electrolyzer for Example 5.3.

TABLE 5.6. Primary Current Distribution in a Tank Electrolyzer

| $L$ (cm) | $S/d$ | $L/s$ (%) | % total current | % current at inner electrode surface |
|---|---|---|---|---|
| 7 | 5 | 70 | ≈87 | ≈87 |
| 17 | 10 | 85 | ≈95 | ≈93.5 |
| 21 | 12 | 87.5 | 94 | 95.5 |
| 27 | 15 | 90 | ≈96 | ≈96 |

In this problem as well as in the derivation of Fig. 5.19 it has been assumed that the electrodes touch the bottom of the tank and protrude some distance above the surface of the electrolyte. If this were not so, the current would also be distributed along the length of the electrode.

---

Second, the effect of the back walls: Hine[27] has made a simplified analysis of the effect of the back walls on the primary current distribution. The side walls are considered remote (Fig. 5.18b). Results (Fig. 5.21) are conveniently presented as a chart giving the percentage of current sent to the inner electrode surface and the ratio of total current to that of a standard cell (infinite electrode) in terms of the three cell dimensions *L*, *d*, and *S*.

Primary current distributions are similar to those in the preceding section. The current at the edges is much higher than at the center of the electrodes; the shorter the electrode, the more uneven the distribution. If the reactor is designed with electrodes of reasonable length ($L/d > 8$) which are not situated a long way from the rear walls ($S/d > 2.5$), current distribution should be uniform. The effect of the rear walls will be insignificant.

The effect of scaling up electrode size is to improve the uniformity of current distribution, as shown in a short worked example.

---

EXAMPLE 5.4: Primary Current Distribution in a Tank Electrolyzer: Side Walls Remote

THE PROBLEM: A tank electrolyzer is to be scaled up from square electrodes with an area of $10\,cm^2$ to electrodes with an area of $100\,cm^2$. The

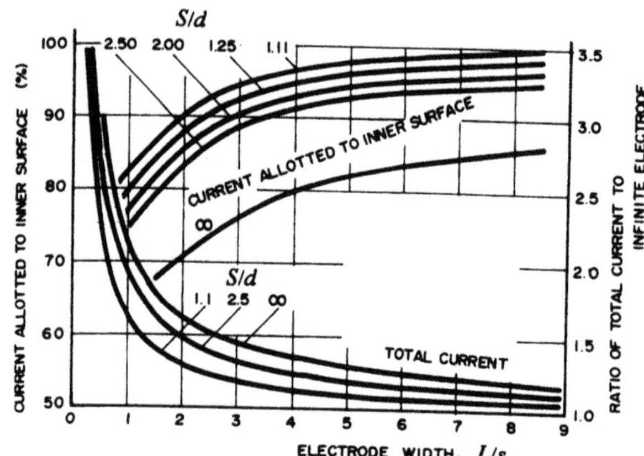


**FIGURE 5.21.** Primary current distribution for cell in Fig. 5.18b. Values of $S/d$ on figure; (% $i_{i.s}$ = current to inner surface; $i_{T/i_\infty}$ = ratio of total current to that to the infinite electrode; $S/d$: $a = 1.1$, $b = 1.25$, $c = 2.00$, $d = 2.50$) (Reproduced from Hine, F., *Electrode Processes and Electrochemical Engineering*, Plenum, 1985.)

interelectrode gap is to remain at 2 cm and the electrodes in both units are to be positioned 1 cm from the rear wall. Determine the improvement in the uniformity of current distribution on scale-up.

THE SOLUTION: For electrodes positioned 1 cm from the rear walls $S/d$ will be 2 for both units, but $L/d$ will be 1.58 and 5 for the smaller and larger unit respectively. For these parameters we obtain from Fig. 5.21 the results in Table 5.7.

---

In summary, Fig. 5.19 shows that electrodes with exposed front and rear faces will exhibit significant maldistribution of current. With the small-scale electrolyzers commonly encountered in laboratory studies the effect may be significant and have considerable influence on scale-up. If we look at a typical current distribution on one electrode of an identical pair (Fig. 5.20), this distribution, in what is essentially a poor reactor design, exhibits a virtually constant current over about two-thirds of the front face, but at the edges currents rise to over three times that value. At the rear of the electrode, apart from the edges, there is only a small amount of current flowing. A limiting case of this analysis is when the electrodes are embedded into the rear wall, i.e., $S = d$.

(ii) Flow electrolyzers. A flow electrolyzer is one where the electrodes are embedded in and flush with the insulating walls of the reactor. The primary current distribution for this[28,29] is:

$$\frac{i}{i_{av}} = \frac{(E^* \cosh E^*)/K[\tanh^2 E^*]}{\{\sinh^2 E^* - \sinh^2 [2x - L]E^*/L\}^{1/2}} \tag{5.3}$$

where $i$ and $i_{av}$ are the local and average current density, $K[m]$ is the complete elliptical integral of the first kind, $E^* = \pi L/2d$, and $x$ (Fig. 5.22) is measured from the center of the electrode. The current distribution is one-dimensional as the electrodes fill the channel in the $z$ direction and the lines of current flow in this plane are uniform. The current distribution is uneven since the current can flow into the space available at either side of the electrodes.

TABLE 5.7. Side Walls Remote

| Electrode area ($cm^2$) | % of current to inner electrode surface | Ratio of total current to that for infinite electrode |
|---|---|---|
| 10 | 85 | 1.25 |
| 100 | 94 | <1.1 |

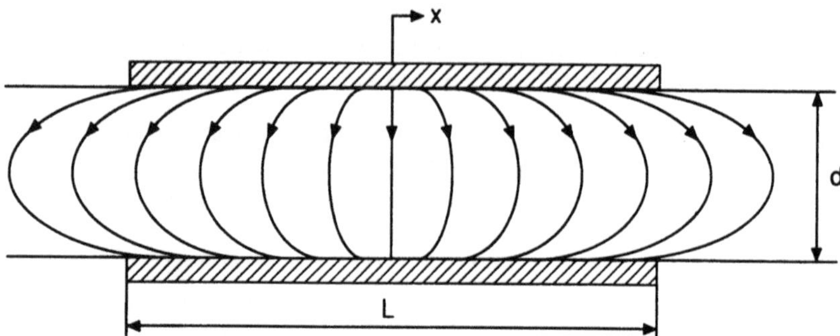


**FIGURE 5.22.** Primary current distribution for a flow electrolyzer.

Figure 5.23 shows the primary current distribution for the system, ranging from very short electrodes ($d/L \rightarrow \infty$) to somewhat longer electrodes ($d/L = 0.5$). The general rule for the distribution of primary current is that the more accessible parts of electrodes receive higher current densities. The local current density increases as we approach the electrode edges and indeed tends to infinity for $x/L = 0$ and 1 (Fig. 5.23). The larger the electrodes and the smaller the electrode gap, the more uniform will be the primary distribution.

Picket[30] argues that although Eq. (5.3) looks formidable, for practical systems where $L >> d$, it is not too important. If you evaluate $i/i_{av}$ at the midpoint of the electrode, then on putting $x = L/2$ Eq. (5.3) simplifies to:

$$\frac{i_{x=L/2}}{i_{av}} = \frac{E^*}{\tanh\ E^* \times K[\tanh^2 E^*]} \tag{5.4}$$

In practical reactors $E^*$ is large and $\tanh E^* \approx 1$. The limiting form of the

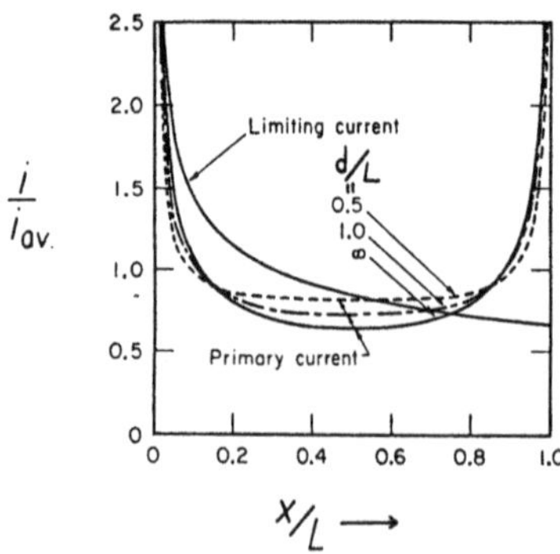


**FIGURE 5.23.** Primary and limiting current distribution at parallel plate electrodes in a flow channel with various $d/L$ ratios.

elliptic integral for an argument approaching unity can be written as[30]:

$$K(\tanh^2 E^*) = 1/2 \ln \frac{1}{1 - \tanh^2 E^*} = 1/2 \ln [16 \cosh^2 E^*]$$
$$= \ln 4 + \ln [\cosh E^*] \qquad (5.5)$$

Substituting Eq. (5.5) into Eq. (5.4) with tanh $E^* = 1$ and noting that as $E^* \rightarrow \infty$

$$\cosh E^* \approx 1/2 \exp E^*$$

it follows that

$$\frac{i_{x=L/2}}{i_{av}} = \frac{E^*}{\ln 2 + E^*} \qquad (5.6)$$

According to Eq. (5.6), if $L/d = 10$ (a comparatively small electrode), $i/i_{av} = 0.958$ at $x = L/2$. This is not far from the uniform current density ($i/i_{av} = 1$) for very large electrodes. Thus variations in primary current distribution have little significance in practical parallel plate electrode systems.

Although in theory only parallel plate flow electrolyzers with electrodes whose lengths tend toward infinity, or that fill the channel completely have the ideal geometry that leads to uniform primary current distribution, in practice these effects are not important. It is different, however, for laboratory cells with very small electrodes.

### 5.3.6.2. Secondary Current Distribution

We will consider only the influence of activation overpotential or overvoltage on secondary current distribution. It is useful[31] to regard the slope of the polarization curve $dE'/di$ (if any effect of concentration overpotential can be ignored) as a polarization resistance $R_a$. This represents the slowness of charge transfer across the interface and is based on the electrode kinetics of the reaction. If $R_a$ acts in series with $R_e$, the resistance of the electrolyte, we can distinguish between two situations. If $R_a >> R_e$, then the kinetics of charge transfer and not electrolyte resistance determine the current distribution, i.e., secondary current distribution dominates. Conversely, if $R_e >> R_a$, primary current distribution dominates. Secondary current distributions tend to smooth out the severe nonlinear variations of current associated with primary distributions; and they eliminate infinite currents associated with electrode edges.

Ibl[31] carries the argument on equalizing action of overvoltage a step further. As we have seen, current distribution is more uniform when ratio $R_a/R_e$ is greater. Now $R_e$ is the resistance of the solution per $m^2$ of cross-sectional area, and is given by the specific resistance times the distance between the two electrodes ($d$). Expressed in terms of the specific conductivity $\kappa$, $R_e = d/\kappa$. Remembering the definition of $R_a$, we obtain:

$$Wa = \frac{\kappa}{L'}\frac{dE'}{di} = \frac{R_a}{R_e} \tag{5.7}$$

This dimensionless group, the polarization parameter $P$ (Section 5.3.4.4), is sometimes called the Wagner number $Wa$. In Fig. 5.22, the characteristic length $L'$ is $d$.

Mathematical analysis of secondary current distribution requires solution of the Laplace equation with boundary conditions at the electrode surface described in terms of electrode kinetics. Because of the apparent complexity of the problem, solutions have been obtained with a linear boundary condition applicable to two approximations:

- At low electrode potentials, when the Butler–Volmer equation truncates to a linear form
- Linearization of the potential-current relationship over a limited range

Secondary current distributions obtained for channel electrolyzers from the work of Wagner[32] are shown in Fig. 5.24. The distributions obtained for small electrode gaps ($d << L$) are characterized in terms of the dimensionless Wagner number which we have met as Eq. (5.7). Evidently when $Wa \rightarrow 0$, primary current distribution is obtained. For large Wagner numbers the current distribution becomes uniform. Uniform distributions are obtained with low characteristic lengths, high electrolyte conductivities, and large slopes of voltage/current density polarization curves.

The Wagner number can be used as a guide to uniform current distribution during scale-up. The criterion is a constant value of $Wa$. If the characteristic length is increased on scale-up, the solution conductivity and/or the slope of the polarization curve must also be increased. Both approaches are limited in practice. Usually composition changes of electrolyte are unfeasible, as are modifications of the slope of the polarization curve. Fortunately in practical scale-up the charcteristic length is not always increased. This is particularly true of the interelectrode gap $d$, since scale-up is usually achieved by increasing electrode size (length $L$) and numbers.

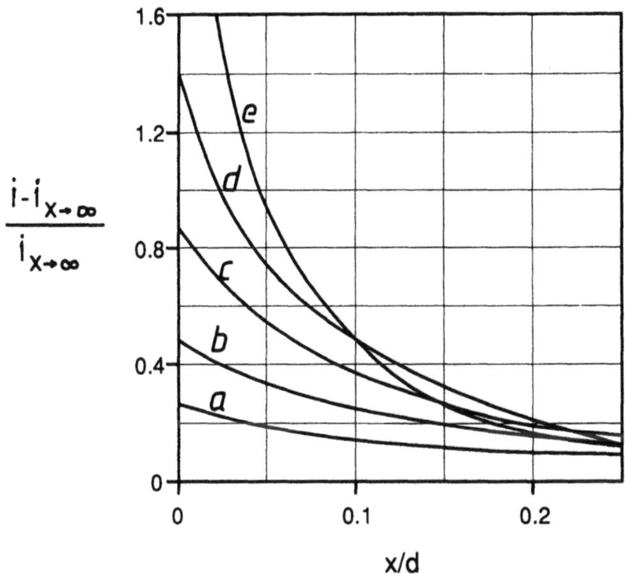


**FIGURE 5.24.** Secondary current distribution near the edge of an electrode in a channel. $d \ll L$ and $Wa$ has the values $a = 0.80$, $b = 0.40$, $c = 0.20$, $d = 0.10$, $e = 0$.

Thus for reactors with small interelectrode gaps uniformity of current distribution is maintained. Keeping current distribution uniform becomes difficult, however, when reactor dimensions are comparable or when no one dimension controls reactor behavior.

---

EXAMPLE 5.5: Secondary Current Distribution in a Parallel Plate Electrolyzer

THE PROBLEM: A parallel plate electrochemical reactor is used to generate a gaseous product in an aqueous solution. The reactor is 20 cm long with an interelectrode gap of 0.5 cm and operates at a current density of 1000 A/m$^2$. The reaction is not mass transfer–limited and kinetics are governed by the Tafel equation:

$$E' = -0.264 - 0.09 \log_{10} i \text{ volts} \qquad (i \text{ in A/m}^2)$$

The reactor operates with a uniform secondary current distribution such that the potential at the electrode edges is not greater than $-0.55$ volts. Determine the electrolyte conductivity required.

THE SOLUTION: For a potential of −0.55 volts the current density will be:

$$\log_{10} i = (0.55 - 0.264)/0.09 = 3.18 \qquad \text{and} \qquad i = 1505.8\ \text{A/m}^2$$

Assuming that over most of the electrode area the current density is $1000\ \text{A/m}^2$, the ratio $(i - i_{x\to\infty})/i_{x\to\infty}$ (Fig. 5.24) will be $(1505.8 - 1000)/1000 = 0.51$ and the Wagner number (also Fig. 5.24) is approximately 0.4. Applying a linear approximation to the Tafel equation, the slope is obtained by differentiation:

$$\frac{dE'}{di} = \frac{3.908 \times 10^{-2}}{i}$$

With $i = 1000\ \text{A/m}^2$ this gives:

$$\frac{dE'}{di} = 3.908 \times 10^{-5}$$

Now:

$$Wa = \frac{\kappa}{L'}\frac{dE'}{di} = \frac{\kappa \times 3.908 \times 10^{-5}}{5 \times 10^{-3}}$$

$Wa = 0.4$, which results in:

$$\kappa = \frac{0.4 \times 5 \times 10^2}{3.908} = 52.2\ \text{mho/m}$$

This conductivity assumes that the electrolyte is gas-free. In fact the evolution of gas will decrease the conductivity and alter the current distribution—a problem addressed in Section 5.3.6.4.

---

When current densities are too high for the polarization curve to be linearized, a more exact analysis of secondary current distribution should accommodate the typical exponential rise of current with electrode potential as described by the Butler–Volmer Eq. (3.83). If current density levels are pushed too high (to a level well within the normal range of industrial practice) then mass transfer effects can no longer be ignored and the problem of "tertiary current distribution" must be tackled.

### 5.3.6.3. Tertiary Current Distribution

At high polarization concentration changes can no longer be ignored and concentration overpotential has to be included with activation overpotential. For the latter, only $L'$ (a characteristic length) had to be considered. For concentration overpotential not only has $L'$ to be considered but $\delta_N$, the thickness of the diffusion boundary layer (see Section 2.3.1). Since we also have to consider concentration differences within the solution, the problem is formidable. More dimensionless parameters must be introduced to characterize behavior, but these serve only as a guide.

A complete integration of the relevant equations[31] is hardly possible, at least in the presence of convection where simplifying assumptions[31] have to be made. For laminar flow, Parrish and Newman[29] have determined tertiary current distribution on a flat plate electrode and Viswanathan and Chin[33] have done the same for continuous moving sheet electrodes. Using a number of simplifying assumptions Lapicque and Storck[34] have developed a model for a parallel plate plug-flow reactor in which the electrolyte is assumed to have three zones, a turbulent bulk where migration and convection take place and two thin boundary layers where migration and diffusion occur. The simplifying assumptions make evaluation of the current and concentration distribution simpler than for the laminar flow models above.

### 5.3.6.4. The Effect of Gas Evolution on Current Distribution

Evolution of gaseous products often occurs in electrochemical reactors, especially with aqueous electrolytes. Such gases may be the desired product, or by-product, at the working electrode or may be generated at the counterelectrode as an unavoidable supporting reaction in the cell. Formation of gas bubbles can affect reactor performance in several ways:

- Enhancement of mass transfer rates due to the disturbance of the boundary layer; there are a number of correlations now available to enable the prediction of this effect
- Decrease in the effective electrolyte conductivity as the current is forced to flow around the insulating bubbles

The effective conductivity $\kappa_e$ of an electrolytic solution containing dispersed gas bubbles and the volume fraction of the gas bubbles $f_g$ for stationary electrolytes are usually related thus:

$$\frac{\kappa_e}{\kappa} = F(f_g) \tag{5.8}$$

where $\kappa$ is conductivity in the absence of a bubble phase.

The function $F(f_g)$ takes several forms. One we have already met in the Bruggeman Eq. (2.76), where:

$$F(f_g) = (1 - f_g)^{1.5} \tag{5.9}$$

Another form is:

$$F(f_g) = 1 - 1.5f_g + 0.5f_g^2 \tag{5.10}$$

These equations give almost identical values. For example for a 30% bubble fraction the values are 0.585 and 0.595 respectively.

When gas bubbles are generated in electrochemical reactors their removal from the electrolyte is essential if conductivities are not to fall dramatically due to bubble accumulation. This is important with a narrow interelectrode gap and points toward minimum (or optimum) spacing. Removal of deleterious gas bubbles can be achieved by natural convection in which the buoyancy of the bubbles produces a gas lift. A gas-solution mixture rises through the electrode chamber, at the top of which the bubbles disengage and the liquid recirculates (Fig. 5.25b). (This is not possible in the arrangement shown in Fig. 5.25a.) Or gas bubbles can be removed by forced circulation of the electrolyte through the reactor (Fig. 5.25c), a situation most often met in industrial cells. This is the most effective way to reduce the gas fraction, and the only way if a plate-and-frame cell is used. The

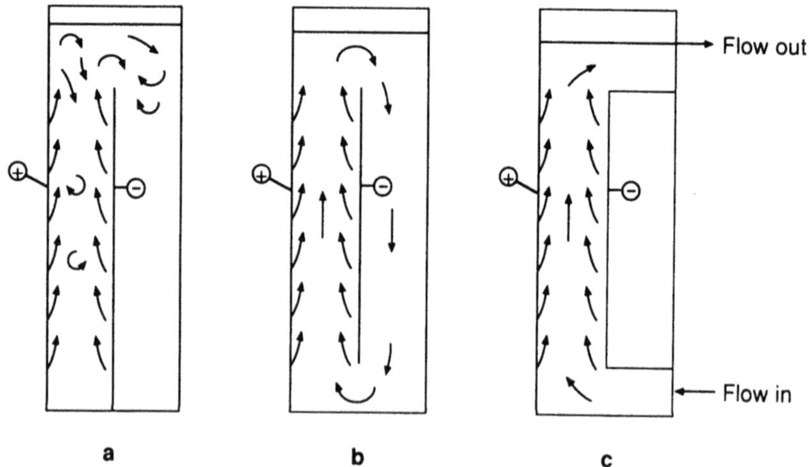


FIGURE 5.25. Flow patterns in vertical cell. (a) Blocked convection; (b) natural convection; (c) forced convection.

problem is made more severe by the modern practice of increasing the length of full-sized electrodes from 1 to 2 meters. The designer must describe adequately the two-phase flow in the reactor and the distribution of bubbles and gas fraction along the channel. The inevitable nonuniform distribution of conductivity associated with this effect will affect current distribution and reactor performance.

(i) Gas evolution assuming a stationary electrolyte. One of the first mathematical models describing the effect of the growth of gas fraction on the current distribution in a parallel plate electrochemical reactor is that of De La Rue and Tobias.[35] The model assumes a stationary electrolyte with a uniform distribution of gas bubbles between the electrodes. Bubbles are assumed to rise with a constant velocity $V_g$, given by Stokes's law, with no coalescence. For the system defined in Fig. 5.26, a local material balance on the gas bubble evolution gives the rate of increase of gas fraction $f_{gx}$ with

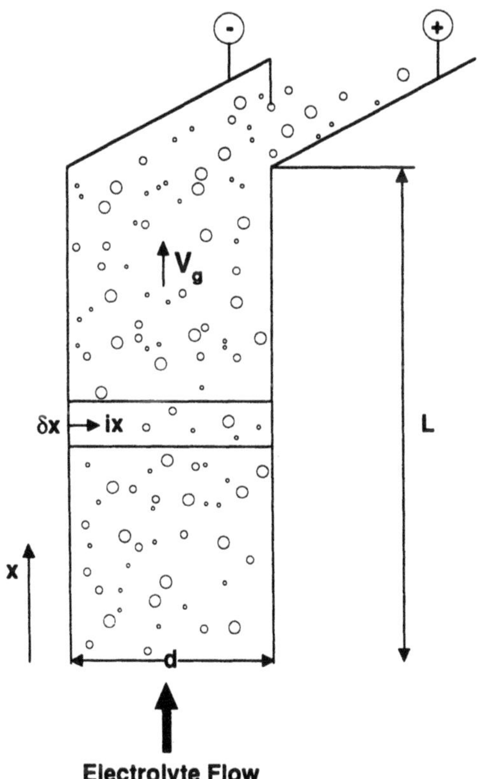


FIGURE 5.26. Cell with gas containing electrolyte.

distance $x$:

$$\frac{df_{gx}}{dx} = \frac{i_x RT}{nFPV_g d} \tag{5.11}$$

where $i_x$ is the current density associated with the gas evolution, $R$ the gas constant, $T$ the temperature, and $P$ the pressure. Equation (5.11) can be modified to account for simultaneous gas evolution from both electrodes.

By assuming that the voltage drop in the reactor is mainly due to the effective electrolyte resistance (in which the Bruggemans correlation is used) we obtain:

$$V^C = \frac{i_x d}{\kappa}(1 - f_{gx})^{-3/2} \tag{5.12}$$

The gas fraction position $x$, assuming a gas-free solution at entry, is given by integration of Eq. (5.11), using Eq. (5.12), as:

$$f_{gx} = 1 - (1 + Kx/2L)^{-2} \tag{5.13}$$

where $K = (RT\kappa LV_c)/(nFPd^2V_g)$ and is called a gas effect parameter.

From Eqs. (5.12) and (5.13) you obtain a distribution of current density along the electrode:

$$i_x = \frac{8V^C}{d(2 + Kx/L)^3} \tag{5.14}$$

The average current density $i_{av}$ is:

$$i_{av} = \frac{1}{L}\int_0^L i_x\, dx \tag{5.15}$$

Substituting Eq. (5.14) in Eq. (5.15) and integrating:

$$i_{av} = \frac{V^C(4 + K)}{d(2 + K)^2} \tag{5.16}$$

Expressing the current distribution along the electrode as $i_x/i_{av}$:

$$\frac{i_x}{i_{av}} = \frac{8(2 + K)^2}{(4 + K)(2 + Kx/L)^3} \tag{5.17}$$

This model is supported by some experimental data; within the limitations of the analysis it forms a sound basis for reactor design at relatively low current densities when bubble coalescence is minimal.

The gas effect parameter has a controlling influence on current distribution. Reactors with large electrode gaps, low cell voltages, and short electrodes operate with more uniform distributions. In scale-up this presents a problem in maintaining similarity of operation. An increase in length of reactor will force an alteration in cell voltage, electrolyte conductivity, or interelectrode gap to achieve comparable average current densities or current distributions in different scale units.

---

EXAMPLE 5.6: The Effect of Gas Evolution on Scale-up

THE PROBLEM: An electrochemical reactor with vertical parallel plates is used for the generation of a gaseous product involving a two-electron change at temperature 300 K and pressure $10^5$ $Nm^{-2}$. The electrolyte, which can be considered stationary, has a conductivity of 50 mho $m^{-1}$. The reaction is to be tested initially on a unit with electrodes 0.25 m long before the electrodes are scaled up to 0.75 m. Average current density on both units is 5000 $Am^{-2}$. You can assume the bubbles to rise with a mean velocity of $1.55 \times 10^{-2}$ $ms^{-1}$.

If the units have the same interelectrode gap of 0.04 m, determine their cell voltages and compare their current distributions.

THE SOLUTION: The gas parameter is:

$$K = \frac{8.31 \times 300 \times 50 \times V^C \times L}{2 \times 96500 \times 10^5 \times (4 \times 10^{-2})^2 \times 1.55 \times 10^{-2}} = 0.26 V^C \times L$$

Average current density is:

$$i_{av} = \frac{V^C \times 50(4 + 0.26 V^C L)}{4 \times 10^{-2}(2 + 0.26 V^C L)^2}$$

For $L = 0.25$ m and $i_{av} = 5000$ $Am^{-2}$:

$$5000 = \frac{V^C \times 5000(4 + 0.065 V^C)}{4(2 + 0.065 V^C)^2}$$

This is a quadratic equation in $V^C$. Its solution is:

$$V^C = 5 \text{ volts}$$

In comparison, the voltage of a gas-free cell is:

$$V^C = \frac{i_{av}d}{\kappa} = \frac{5000 \times 4 \times 10^{-2}}{50} = 4 \text{ volts}$$

In the scaled-up unit, $L = 0.75\,\text{m}$ for the same average current density and $V^C$ is obtained from:

$$5000 = \frac{V^C \times 5000(4 + 0.195V^C)}{4(2 + 0.195V^C)^2}$$

So $V^C = 11.1$ volts. Increase in scale by a factor of 3 severely affects the voltage requirements which are more than doubled. This could lessen the commercial viability of the process.

Current distributions can be conveniently compared by considering the values of current densities at the electrode edges, using Eq. (5.14). For the smaller unit:

$$i_x = \frac{8 \times 5 \times 50}{4 \times 10^{-2}(2 + 1.30x)^3}$$

Therefore when $x = 0$, $i_x = 6250\ \text{Am}^{-2}$, and when $x = L = 0.25\,\text{m}$, $i_x = 3978\ \text{Am}^2$.

For the larger unit:

$$i_x = \frac{8 \times 11.1 \times 50}{4 \times 10^{-2}(2 + 2.89x)^3}$$

When $x = 0$, $i_x = 13{,}875\,\text{Am}^{-2}$; when $x = L = 0.75$, $i_x = 1537\,\text{Am}^{-2}$.

The larger unit has a wider range of current density operation. This can make problems in, e.g., electrode corrosion and selectivity.

---

Electrode polarization was considered by Tobias[36] in terms of adopting a linear approximation to the Tafel equation. The influence of polarization, which is in essence a bubble-independent resistance, acts in series with the ohmic resistance, improving the uniformity of current distribution.

We have considered reactors operating at low current densities. At higher values when bubble motion is hindered and coalescence occurs, the rise velocity is likely to be affected by the local rate of gas generation at the electrode. The ability of the electrolyte to circulate between the electrodes will influence the distribution of gas voidage. Hine[37] has addressed this problem and has obtained expressions for the average gas void fraction.

(ii) Gas evolution under forced convection of the electrolyte. Forced convection is often used in electrosynthesis when high mass transfer rates are required, along with keeping gas void fractions low in order to minimize cell voltages. Nishiki *et al.*[38] have discussed a straightforward model of a vertical plate reactor supplied externally with electrolyte. A vertical configuration of electrodes is usually preferable, but a horizontal orientation may have to be used. Intuitively, the analysis for the vertical configuration can be applied to the horizontal case, as bubble characteristics or flow is controlled by volumetric electrolyte throughput in the cell. Caution is needed if electrolyte velocities are low: there is a tendency for bubbles to stay in the top section of the flow channel, causing additional nonuniformities in electrolyte conductivity or a partial blockage of the electrode surface. Such problems can usually be avoided by high electrolyte velocities. Hine[37] has studied the effect of gas bubbles in horizontal cells and expressed the ratio of conductivities with and without gas bubbles as a function of the ratio of gas to liquid volume, the electrode gap, and the Reynolds number.

A general approach is described by Pickett[39]: a material balance for bubble formation similar to Eq. (5.11) is combined with the Bruggeman relationship to give the conductivity of gas-bearing electrolytes. Conductivity, which varies along the reactor, is combined with a voltage balance and the resulting systems of equation solved numerically in order to evaluate current distribution.

Forced convection is used to prevent excessive buildup of gaseous products in the interelectrode space. Another approach is to allow release of gas behind the electrodes by using meshed or perforated electrodes. These devices have been used successfully in modern chloralkali cells and have the advantage of allowing narrow gaps between the electrodes. They can have, however, some disadvantages, since contours on the electrodes can cause nonuniformities in current distribution. Also, the open areas for gas release may cause higher current densities in operation, with a possible adverse effect on electrode life.

#### 5.3.6.5. The Effect of Finite Electrode Conductivity on Current Distribution

Electrode materials employed in industrial reactors often have relatively poor electrical conductivities. If cost requirements require thin electrodes, appreciable ohmic voltage loss may occur across the electrode, especially if large electrodes are used. Potential and current distribution may suffer. This results from having to make a mechanical connection between the bus bars and the electrode, usually at one point (Fig. 5.27) or, preferably, distributed along one edge. As current flows through the electrode structure a variation

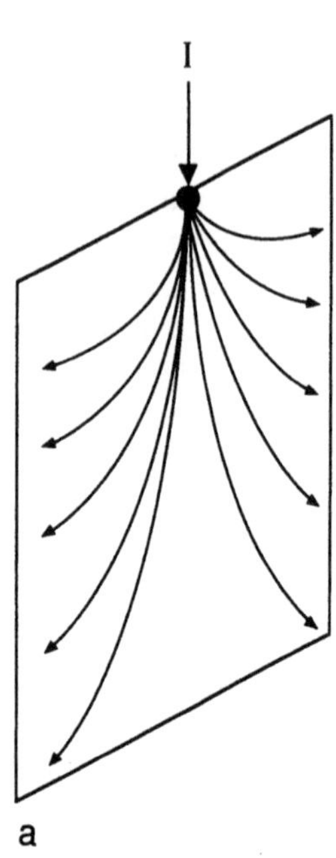


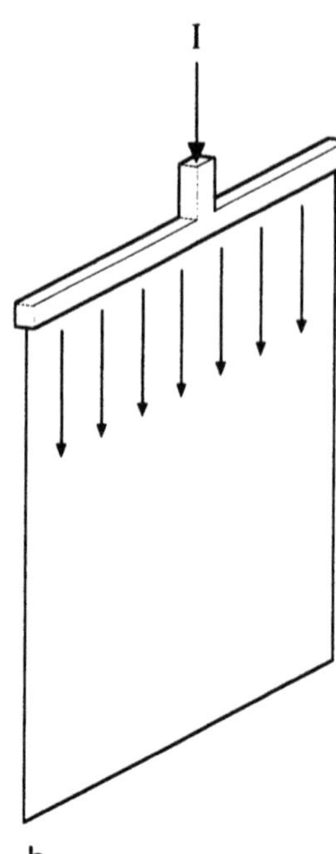


**FIGURE 5.27.** Bus bar connection to an electrode. (a) Single-point current feed; (b) uniform current feed.

of substrate potential occurs, which modifies the primary current distribution for a constant potential at the electrode-solution interface, etc.

Consider parallel plate electrodes with the current uniformly distributed along one edge, i.e., unidirectional current flow. For reactors with narrow electrode gaps, primary and secondary current density distributions for linear polarizations are given by:

$$\frac{i_x}{i_{av}} = \phi \frac{\cosh\ [\phi(1 - x/L)]}{\sinh\ \phi} \tag{5.18}$$

where

$$\phi^2 = \kappa_e L^2 \frac{\dfrac{1}{\sigma_a l_a} + \dfrac{1}{\sigma_c l_c}}{d + \kappa_e b_a + \kappa_e b_c}$$

with $\sigma_a$ and $\sigma_c$ the specific conductivities of anode and cathode, $l_a$ and $l_c$ the thicknesses of anode and cathode, $d$ the interelectrode gap, and $b_a$ and $b_c$

the linear polarization coefficients ($dE'/di$) for the anode and cathode reaction. Typical current density distributions[40] are shown in Fig. 5.28, where not surprisingly larger electrodes, higher ratios of electrolyte to electrode conductivity, and thinner electrodes all result in poorer current distributions.

If the mode of current supply to the electrode is replaced by a single point (Fig. 5.27a), the nonuniformity of current distribution is accentuated.[41] It is advisable, therefore, to adopt at least a multiple-point connection along one electrode edge to approximate a uniform current feed. Because of problems associated with limitations in electrode conductivity, industrial reactors often operate with banks of bipolar electrodes (see Section 5.1.2).

As we have seen in Section 5.1.2, bipolar connections lead to greater uniformity of current distribution on the electrodes in a cell stack. Analysis of current distribution on bipolar connected electrodes assumes that the electrochemical characteristics of the cell in the stack of electrodes can be represented by pure resistance. Results are conveniently expressed by an effectiveness factor $E_f$, denoting the ratio of the total current fed to the electrode to the maximum possible if the electrode is at a uniform (maximum) potential. Figure 5.29 shows the typical variation of $E_f$ with

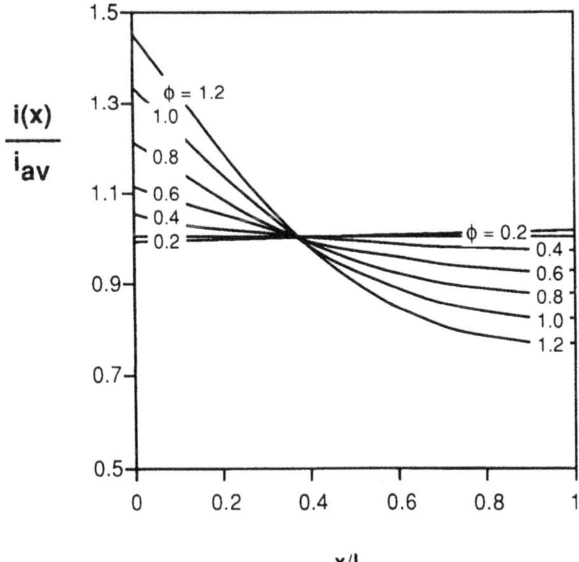


**FIGURE 5.28.** Current density distribution for parallel plate electrodes as a function of $\phi$ for the situation in Fig. 5.27b.

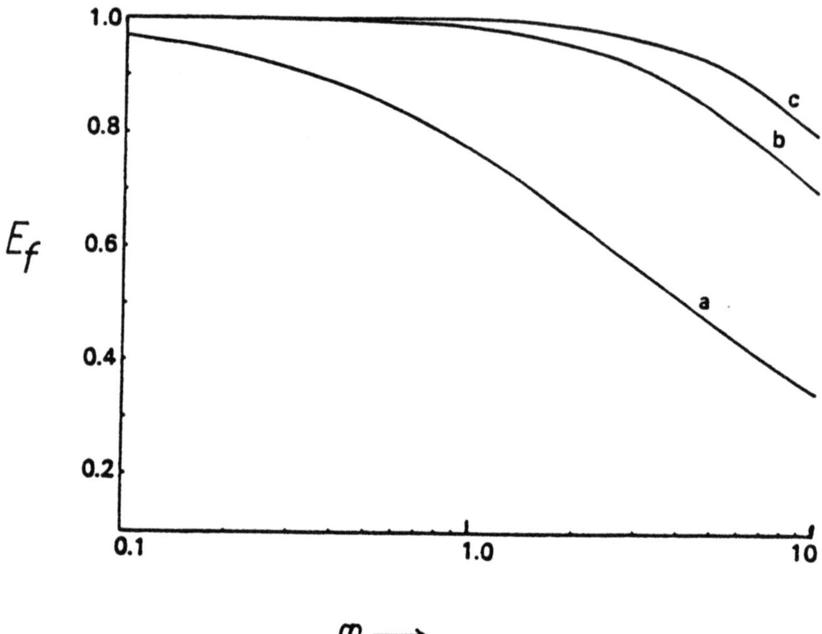


**FIGURE 5.29.** Effectiveness factor for a six-electrode cell bank. (a) Outer cells; (b) inner cells; (c) center cells.

$m = (\kappa_e L^2)/(dl\sigma)$ for a six-electrode stack, all electrodes with thickness $l$ and specific conductivity $\sigma$. When in addition polarization is negligible ($b_a$ and $b_c \to 0$), $\phi^2 = m/2$. The increase in uniformity of current density from the outer to the inner electrodes is clear. The central cell exhibits a high degree of uniformity over a large range of parameter $m$. Curve a is the best (limiting) distribution for a cell stack of six or more electrodes. On connecting bipolar electrodes, uniform current distribution is obtained in all but the two outermost cell pairs at either end of the stack.

In scale-up the problem of current distribution associated with a change in electrode length is accompanied by the change in the number of electrodes incorporated in a cell stack. A cell stack of a small number of electrodes can give different product specifications from one housing a large number of electrodes. The uniformity of current distribution throughout the stack can be improved by increasing the thickness of the feeder electrodes or by supporting them on appropriate backing pieces. Generally a bipolar mode of operation is preferred when electrode conductivity is limited.

Current distribution in parallel plate cells can be improved by introducing the current at opposite ends of the feeder electrodes rather than the same ends. A comparison can be made by using expressions derived for electrode effectiveness $E_f$. First, for current feeders at the same end:

$$E_f = \frac{\tanh(2m)^{1/2}}{(2m)^{1/2}} \tag{5.19}$$

where again $m = (\kappa_e L^2)/(dl\sigma)$.

Second, for current feeders at opposite ends:

$$E_f = (2/m)^{1/2} \tanh(m/2)^{1/2} \tag{5.20}$$

In Fig. 5.30 the opposite-end current feed results in more uniform current distribution. This arrangement gives effectiveness values close to those for a peripheral current feed, which requires relatively complex and expensive

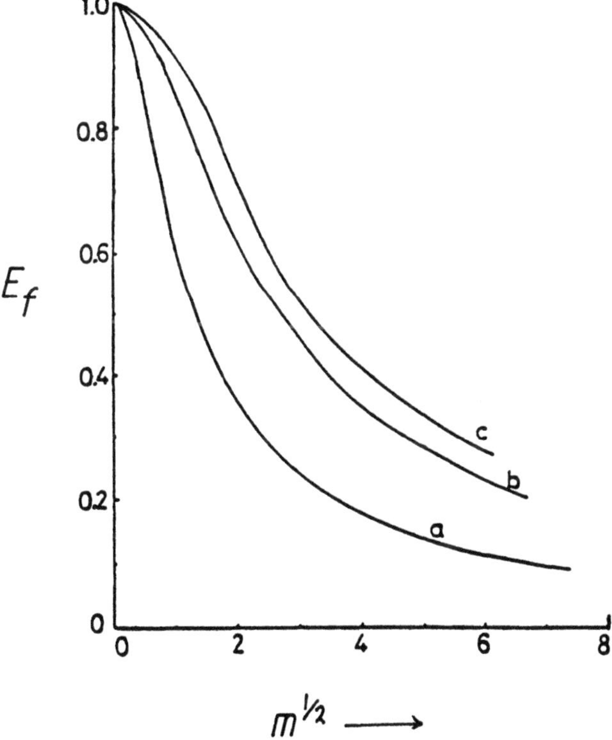


**FIGURE 5.30.** Comparison of the effectiveness of square electrodes. (a) Same-end current feed; (b) opposite-end current feed; (c) peripheral current feed.

electrical connections. Adoption of an opposite-end feed arrangement does not always result in improved current distribution. For low values of $m$—which are generally desirable—the same-end current feed is the better of the two.

---

EXAMPLE 5.7: Current Distribution in Electrodes of Finite Conductivity

THE PROBLEM: A parallel plate flow reactor with an interelectrode gap of $10^{-2}$ m is used in the oxidation of an aliphatic alcohol. Both electrodes used in the reactor are $2 \times 10^{-3}$ m thick and made of titanium substrate with a resistivity of $4.2 \times 10^{-7}$ ohm. Effective electrolyte conductivity is 40 mho $m^{-1}$. The kinetics of the anodic and cathodic reactions can be represented by:

$$E'_A = 0.51 + 0.09 \ln [i] \qquad \text{and} \qquad -E'_C = 0.23 + 0.05 \ln [i]$$

where $i$ is in units of $Am^{-2}$.

The decomposition voltage $E_D = 1.09$ volts.

The reactor operates with a terminal voltage not greater than 3 volts and with a current of 1000 amps.

Find an electrode size that produces uniform potential distribution such that the average current density is at least 90% of the maximum.

THE SOLUTION: The criterion is that electrode effectiveness $E_f = 0.9$. We compare electrode size requirements for two current end feed arrangements. A linearized Taylor approximation to the two Tafel equations will suffice. Total cell resistance is composed of two kinetic resistances and the ohmic resistance. The parameter $\phi$ in Eq. (5.18) is analogous to the parameter $m^{1/2}$ in Eqs. (5.19) and (5.20). The terminal voltage is used to calculate the current density:

$$V^C = E_D + E'_A + (-E'_C) + i\frac{d}{\kappa_e}$$

Therefore:

$$3 = 1.09 + 0.51 + 0.09 \ln i + 0.23 + 0.05 \ln i + i\frac{10^{-2}}{40}$$

giving:

$$1.17 = 0.14 \ln i + 2.5 \times 10^{-4} i$$

This is solved by iteration to give a current density of $880\,\mathrm{A\,m^{-2}}$. The gradients of the Tafel equations are:

$$\left|\frac{dE'_A}{di}\right| = \frac{0.09}{i} = \frac{0.09}{880} = 1.02 \times 10^{-4} \text{ V m}^2 \text{ A}^{-1}$$

and

$$\left|\frac{dE'_C}{di}\right| = \frac{0.05}{i} = \frac{0.05}{880} = 5.68 \times 10^{-5} \text{ V m}^2 \text{ A}^{-1}$$

Therefore the parameter $m$ is:

$$m = \frac{L^2}{l\sigma} \frac{1}{d/\kappa_e + b_a + b_c}$$

$$= \frac{L^2}{(2 \times 10^{-3})/(4.2 \times 10^{-7})} \frac{1}{2.5 \times 10^{-4} + 1.02 \times 10^{-4} + 5.68 \times 10^{-5}}$$

giving $m = 0.513L^2$.

For a value of $E = 0.9$ the value of $m$ can be obtained by solving Eqs. (5.19) and (5.20), as follows.

(i) Current feeders at the same end

$$0.9(2m)^{1/2} = \tanh (2m)^{1/2}$$

The solution is $m^{1/2} = 0.41$. Hence $L = 0.41/(0.513)^{1/2} = 0.57$ m.

880 Am$^{-2}$ for the current density is the maximum value, the average being $0.9 \times 880 = 790$ Am$^{-2}$.

For a current of 1000 amps the required electrode area is 1000/790 = 1.27 m$^2$. The electrode width is 2.22 m.

(ii) Current feeders at opposite ends

$$\frac{0.9m^{1/2}}{2^{1/2}} = \tanh (m/2)^{1/2}$$

The solution is $m^{1/2} = 0.81$. Hence $L = 0.81/(0.513)^{1/2} = 1.13$ m, giving an electrode width of 1.12 m.

The second electrode size seems the more attractive since the first with a width of over 2 meters would present problems of electrolyte flow distribution and of making satisfactory electrical connections.

### 5.3.6.6. Current Distribution in Three-Dimensional Electrodes

Three-dimensional electrodes were mentioned in Section 5.1.1.2. Table 5.4 indicates a potential advantage, namely a high space-time yield. Such electrodes differ from their two-dimensional counterparts in the distribution of potential and current density in the matrix of the electrodes. Rigorous analysis would require evaluation of three-dimensional potential distributions. Fortunately this is often unnecessary: one-dimensional approximate models or simplified two-dimensional models are sufficient. A comprehensive treatment of three-dimensional electrodes is beyond the scope of the present text (the reader has already been referred to the review in Ref. 5, particularly for information on fluidized bed electrodes. Further information can be found in Refs. 42–44.) We will concentrate on two limiting types of operation of packed-bed electrodes to illustrate the current distributions encountered and their relationship to scale-up.

(i) For kinetically controlled reactions. The concept behind the one-dimensional potential distribution of packed-bed electrodes (Fig. 5.31) is that the electrolyte and the electrode phase are continuous and Ohm's law applies to both.

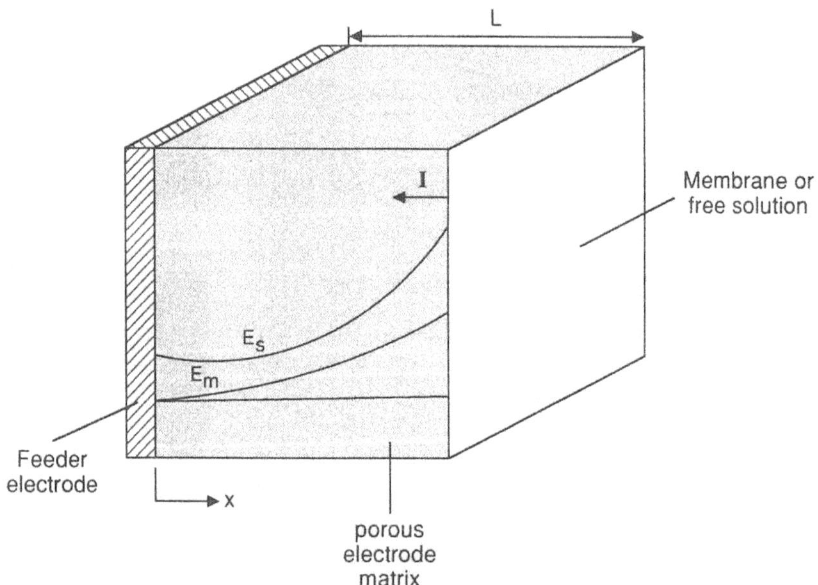


**FIGURE 5.31.** Schematic diagram of packed bed electrode. $E_s$ = potential in the liquid phase, $E_m$ = potential in the particulate phase.

The total current density $i$ at any cross-section is equal to the sum of current density in the metal phase $i_m$ and current density in the electrolyte phase $i_s$:

$$i = i_m + i_s \tag{5.21}$$

or

$$\frac{di_m}{dx} + \frac{di_s}{dx} = 0 \tag{5.22}$$

Boundary conditions are:

$$i_m = 0, \; E_s = 0 \qquad \text{at } x = 0 \tag{5.23}$$

and

$$i_m = i \qquad \text{at } x = L \tag{5.24}$$

The local reaction rate in the electrode is:

$$\frac{di_m}{dx} = af(E_m - E_s, c) \tag{5.25}$$

where $a$ is the specific area of the electrode and $c$ is a concentration term.

The complexity of Eq. (5.25) depends on the order of the electrochemical reaction and the spatial dependence of the concentration terms. For a first-order reaction you can use a form of the Butler–Volmer Eq. (3.83). We shall use the Tafel approximation, assuming that the reaction is cathodic; see Eq. (3.84):

$$\frac{di_m}{dx} = ai_o \exp\left[-\frac{\alpha nF}{RT}(E_m - E_s)\right] \tag{5.26}$$

Applying Ohm's law to the solid and liquid phases:

$$i_m = -\sigma \frac{dE_m}{dx} \tag{5.27}$$

and

$$i_s = -\kappa \frac{dE_s}{dx} \tag{5.28}$$

Differentiating Eq. (5.26), substituting Eqs. (5.27) and (5.28) in the resulting expression, and eliminating $i_s$ by means of Eq. (5.21):

$$\frac{d^2 i_m}{dx^2} = \beta \frac{di_m}{dx} \left\{ i_m \left[ \frac{1}{\kappa} + \frac{1}{\sigma} \right] - \frac{i}{\kappa} \right\} \tag{5.29}$$

where $\beta = \alpha\sigma nF/RT$. Transforming Eq. (5.29) into dimensionless form:

$$\frac{d^2 i_d}{dy'^2} = \frac{di_d}{dy'} (\delta' i_d - \varepsilon') \tag{5.30}$$

where

$$i_d = i_m / i \tag{5.31}$$

$$y' = x/L \tag{5.32}$$

$$\delta' = Li\beta \left[ \frac{1}{\kappa} + \frac{1}{\sigma} \right] \tag{5.33}$$

and

$$\varepsilon' = Li\beta/\kappa \tag{5.34}$$

Boundary conditions from Eqs. (5.23) and (5.24) transform to:

$$i_d = 0 \qquad \text{at } y' = 0 \tag{5.35}$$

and

$$i_d = 1 \qquad \text{at } y' = 1 \tag{5.36}$$

For an anodic reaction Eq. (5.30) remains unchanged but $|i_d|$ replaces $i_d$ and $\beta = (1 - \alpha)nF/RT$ in Eqs. (5.29), (5.33), and (5.34).

The solution of Eq. (5.30) is:

$$i_d = \frac{2\theta}{\delta'} \tan\,(\theta y' - \psi) + \frac{\varepsilon'}{\delta'} \tag{5.37}$$

$\theta$ and $\psi$ are integration constants found by trial and error from:

$$\tan\,\theta = \frac{2\delta'\theta}{4\theta^2 - \varepsilon'(\delta' - \varepsilon')} \tag{5.38}$$

and

$$\tan\psi = \frac{\varepsilon'}{2\theta} \qquad (0 < \theta < \pi) \tag{5.39}$$

Differentiating Eq. (5.37) gives us an expression for the current distribution through the electrode:

$$\frac{di_d}{dy'} = \frac{2\theta^2}{\delta'} \sec^2 (\theta y' - \psi) \tag{5.40}$$

The parameters $\delta'$ and $\varepsilon'$ control the current distribution in the electrode structure. $\delta'$ represents the competition between the reaction current and the ohmic resistance, with low values (small $L$, high conductivities, and low current densities) tending to produce a more uniform current distribution (Fig. 5.32).[42] The second parameter $\varepsilon'$, viewed in conjunction with $\delta'$ as the ratio of electrolyte to electrode conductivities, has a significant effect on current distribution. In practical electrodes the conductivity of the metallic matrix will usually be very close to that for the pure electronic conductor; hence, $\sigma \gg \kappa$. This corresponds to the situation when $\varepsilon' = \delta'$. Most of the activity is near the counterelectrode (Fig. 5.32).

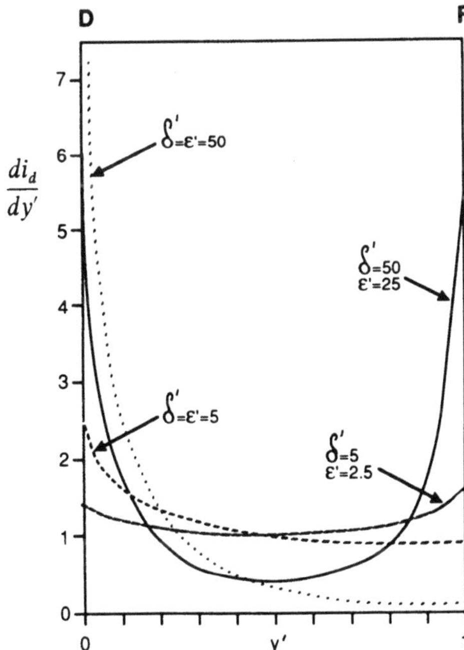


**FIGURE 5.32.** Current distribution in a packed bed electrode.

The uniformity of current distribution can be conveniently expressed in terms of the concept of effectiveness $\bar{\eta}$ given by[45]:

$$\bar{\eta} = \frac{\text{observed electrolytic current}}{\text{current obtained if the overpotential at all values of } y' \text{ is equal to the maximum observed}}$$

For Tafel-type kinetics this produces:

$$\bar{\eta} = \frac{\delta' \cos^2 \psi}{2\theta^2} \tag{5.41}$$

The variation of $\bar{\eta}$ with $\delta'$ is shown in Fig. 5.33 for values of $\varepsilon'$. An increase in $\sigma$ reduces effectiveness; a decrease in $\varepsilon'/\delta'$ has the opposite effect.

Using "effectiveness" allows the predicted maximum reaction rate to be corrected to give the actual reaction rate or can be used to decide on use of the electrode area (effective specific electrode area $a_e$), from the equation:

$$a_e = \bar{\eta} a \tag{5.42}$$

In the scale-up of three-dimensional electrodes, identical potential distribution is the criterion for maintaining similar performance. The parameter $\delta'$ must be held constant and any increase in length $L$ will require

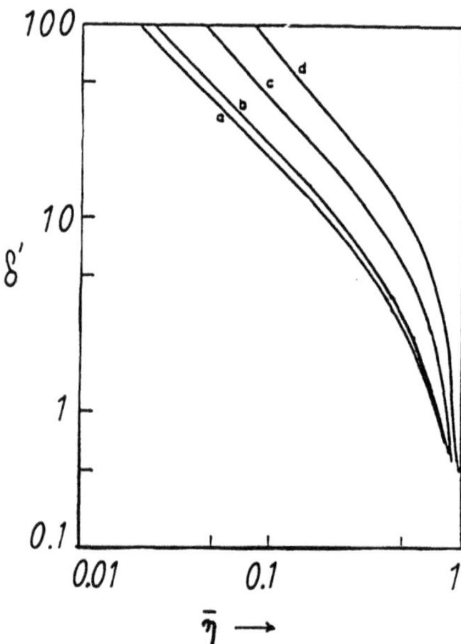


**FIGURE 5.33.** Effectiveness values for Tafel polarization as a function of $\delta'$, with $\varepsilon'$ as a parameter. (a) $\varepsilon' = \delta'$ ($1/\sigma = 0$); (b) $\varepsilon' = \delta'/1.1$ ($\kappa = \sigma/10$); (c) $\varepsilon' = 0.67\ \delta'$ ($\kappa = \sigma/2$); (d) $\varepsilon' = \delta'/2$ ($\kappa = \sigma$).

(1) reduction in operating current density or (2) increase in electrolyte conductivity. (1) Is likely to conflict with other operating criteria, notably space-time yield. (2) Calls for restrictions, especially if a low conductivity electrode material is used with the requirement of a fixed $\kappa/\delta'$.

The general method of scale-up is to increase the dimensions of the electrode perpendicular to the current flow between the electrodes and to use many thin electrodes, thereby maintaining similarity in potential distribution. Although our analysis is restricted to a few types of reaction(s) and ignores mass transport effects, the general concepts are a useful starting point in the design of industrial-sized units.

---

EXAMPLE 5.8: Sizing a Three-Dimensional Electrode

THE PROBLEM: A potential electrochemical synthesis has been studied experimentally in the laboratory and results indicate that good selectivity can be achieved if the electrode potential is kept to $-200\,\text{mV}$. Unfortunately, such an electrode potential results in very low current densities; a three-dimensional electrode must be used to achieve acceptable space-time yields. Using the following data, estimate the specific electrode area which will give an operating current density of $500\,\text{A/m}^2$.

kinetic rate equation $i = 4 \times 10^{-3} \exp[-40(E_m - E_s)]$

conductivity of electrolyte = 60 mho/m

bed width $L = 10^{-2}$ m

THE SOLUTION: We begin with a one-dimensional analysis of the potential distribution in the electrode, Eq. (5.26). Assuming a linear distribution of the local reaction rate, i.e., $di_m/dx = i/L$, we get:

$$(E_m - E_s) = \frac{1}{\beta} \ln \left[\frac{Lai_o\bar{\eta}}{i}\right]$$

which rearranges to:

$$\frac{i}{La\bar{\eta}} = i_o \exp\left[-(E_m - E_s)\right]$$

Since electrode potential is $-200\,\text{mV}$:

$$\frac{i}{La\bar{\eta}} = 4 \times 10^{-3} \exp\left[-40 \times 0.2\right] = 11.9\ \text{A/m}^2$$

This is the maximum current density for a planar electrode. We are looking for an increase of more than fortyfold. The effectiveness of a particle electrode is a function of $\delta'$ and may be obtained from Fig. 5.33 or from the empirical expression[45]:

$$\bar{\eta} = 0.739 - 0.2445 \ln \delta'$$

Now:

$$\delta' = \frac{Li\beta}{\kappa} = \frac{10^{-2} \times 500 \times 40}{60} = 3.33$$

giving:

$$\bar{\eta} = 0.739 - 0.2445 \ln 3.33 = 0.445$$

Hence the specific electrode area $a$ is:

$$a_e = \frac{i}{L \times \bar{\eta} \times 11.9} = \frac{500}{10^{-2} \times 0.445 \times 11.9} = 9442 \text{ m}^{-1}$$

If electrode material is available as approximately spherical particles with a voidage of say 0.42, the particle diameter $d_p$ will:

$$d_p = \frac{6}{a}(1 - 0.42) = 3.69 \times 10^{-4} \text{ m}$$

---

**(ii) For mass transfer controlled reactions.** It is often desirable to operate electrolytic reactions at or near limiting current conditions when low reactant concentrations limit available current density levels. The two basic reactor designs are shown in Fig. 5.34. In Fig. 5.34a current and electrolyte flow in parallel; this arrangement is called a flow-through electrode. Figure 5.34b shows a three-dimensional electrode in which current and electrolyte flow at right angles to each other. The advantage of this is that the length of the current and electrolyte paths can be varied independently. It follows that the dimension of the reactor in the direction of flow can be considerably greater than that in the flow-through arrangement. High conversions can be achieved without unwanted side reactions. Modeling these systems under sublimiting current conditions calls for complex numerical methods of solution. For reactions under complete mass transfer control the analysis is analogous to that for the flow-through situation if you assume a one-dimensional flow of current between the electrodes.

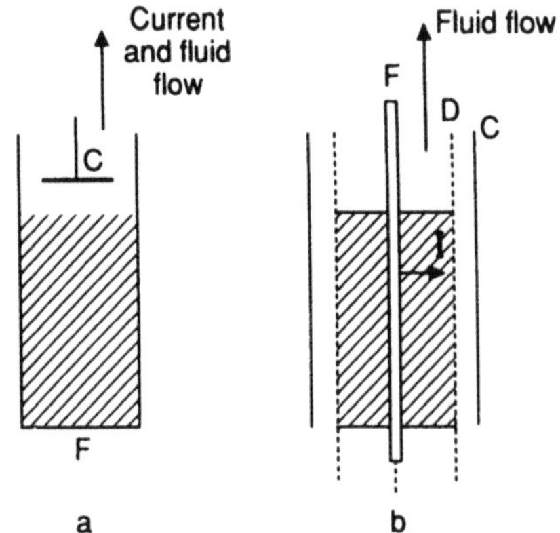


FIGURE 5.34. Basic characteristics of three-dimensional electrodes. (a) Flow-through cylindrical electrode; (b) flow-by cylindrical electrode. $F$ = current feeder, $C$ = counter electrode, $D$ = diaphragm.

Considering the reactor as a plug-flow device operating under mass transport control with a constant mass transfer coefficient, you can use Eq. (4.17), which for the flow-by situation is:

$$\frac{c_A}{c_A^o} = \exp\left[-\frac{ak_L y}{u}\right] \tag{5.43}$$

where $u$ is the flow velocity of electrolyte in the direction $y$.

For operation at limiting current we analyze the two-dimensional distribution of potential in the electrode, based on the Poisson equation:

$$\frac{\partial^2 E_s}{\partial x^2} + \frac{\partial^2 E_s}{\partial y^2} = \frac{ai_{lim}}{\kappa} \tag{5.44}$$

where the limiting current density $i_{lim}$ varies in the direction of electrolyte flow.

Figure 5.35 shows typical nonuniform potential distributions based on

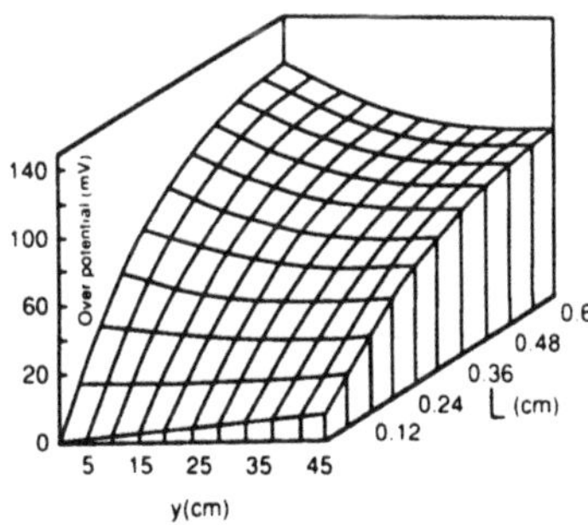


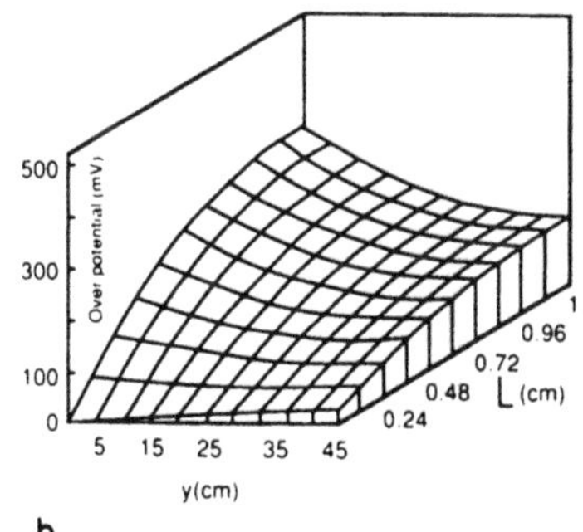


FIGURE 5.35. Calculated overpotential distributions[96] for different experimental conditions. (a) $u = 11.1$ cm s$^{-1}$, $c^o = 0.5$ kg m$^{-3}$, $L = 0.6$ cm; (b) $u = 35$ cm s$^{-1}$, $c^o = 0.5$ kg m$^{-3}$, $L = 1.2$ cm.

this analysis.[46] The general effect of an increase in velocity or electrode width $L$ is to increase the variation in electrode potential distribution. It is possible[47] to derive a simple approximate expression which predicts the maximum electrode width in the direction of current flow $L_{max}$ for a maximum allowable value of overpotential $\Delta\eta_{max}$:

$$L_{max} = \left(\frac{2\kappa\Delta\eta_{max}}{nFak_L c_A^o}\right)^{1/2} \tag{5.45}$$

The value of $\Delta\eta_{max}$ is given approximately by the difference $\eta(L,0) - \eta(0,0)$.

---

EXAMPLE 5.9: Operating Requirements of a Packed-Bed Electrode at Limiting Current

THE PROBLEM: A packed-bed electrode is used to process an effluent stream containing 1 mol/m$^3$ of heavy metal ions. The reactor operates under limiting current conditions in order to reduce the metal ion concentration to 0.1 mol/m$^3$. A bench scale unit is used to determine the feasibility of the procedure. The maximum range of overpotential at which limiting current conditions prevail is 0.28 volts.

Determine a suitable size of flow-by electrode to process $10^{-5}$ m$^3$/s of electrolyte if particles $5 \times 10^{-4}$ m in diameter are used and the electrolyte feeder plate has a width of 0.1 m. Available data are shown in Table 5.8.

THE SOLUTION: Using an appropriate correlation to calculate the mass transfer coefficient:

$$k_L = \frac{0.6(1-e)D}{e \times d_p}(Sc)^{0.33}\left[\frac{u \times d_p}{(1-e)v}\right]^{0.5}$$

$$= \frac{0.6 \times 0.55 \times 8.33 \times 10^{-10}}{0.45 \times 5 \times 10^{-4}}\left[\frac{10^{-6}}{8.33 \times 10^{-10}}\right]^{0.33}\left[\frac{5 \times 10^{-4}}{0.55 \times 10^{-6}}\right]^{0.5} u^{0.5}$$

$$= 3.82 \times 10^{-4}\, u^{0.5} \text{ m/s}$$

TABLE 5.8. Packed-Bed Electrode

| | |
|---|---|
| Conductivity of electrolyte | 33.11 mho/m |
| Kinematic viscosity | $10^{-6}$ m$^2$/s |
| Metal ion diffusivity | $8.33 \times 10^{-10}$ m$^2$/s |
| Number of electrons for reactant conversion | 2 |
| Voidage of the bed electrode | 0.45 |

Calculating the effective conductivity[35] from:

$$\kappa_e = \kappa e^{1.5} = 33.11 \times (0.45)^{1.5} = 10 \text{ mho/m}$$

The specific electrode area is:

$$\frac{6}{d_p}(1 - e) = \frac{6}{5 \times 10^{-4}}(1 - 0.45) = 6600 \text{ m}^{-1}$$

Using Eq. (5.45) to determine the width of bed in the direction of current flow gives:

$$L_{max} = \left[\frac{2 \times 10 \times 0.28}{2 \times 96540 \times 6600 \times 1 \times 3.82 \times 10^{-4} u^{0.5}}\right]^{0.5} = \frac{3.4 \times 10^{-3}}{u^{0.25}} \text{ m}$$

Equation (5.43) rearranges to:

$$-\ln \frac{c_A}{c_A^o} = \frac{k_L a y_o}{u}$$

where $y_o$ is the bed length in the direction of electrolyte flow. Therefore:

$$-\ln [0.1] = \frac{3.82 \times 10^{-4} u^{0.5} \times 6600 \times y_o}{u}$$

and

$$y_o/u^{0.5} = 0.913$$

The superficial velocity $u$ is:

$$u = \frac{10^{-5}}{0.1 \times L_{max}}$$

Substituting for $u$ in the expression for $L_{max}$:

$$L_{max}\left[\frac{10^{-4}}{L_{max}}\right]^{0.25} = 3.4 \times 10^{-3}$$

giving $L_{max} = 1.1 \times 10^{-2}$ m.

The velocity $u$ is:

$$u = \frac{10^{-5}}{0.1 \times 1.1 \times 10^{-2}} = 9.1 \times 10^{-3} \text{ m/s}$$

The bed length in the direction of flow $y_0$ is:

$$y_o = 0.913 \times (9.1 \times 10^{-3})^{0.5} = 8.7 \times 10^{-2} \text{ m}$$

The value of the mass transfer coefficient for the electrode, then, is:

$$k_L = 3.82 \times 10^{-4} \times (9.1 \times 10^{-3})^{0.5} = 3.64 \times 10^{-5} \text{ m/s}$$

#### 5.3.6.7. Concluding Remarks

Evaluating current distribution in reactor design and scale-up is not a separate exercise but an integral feature of design and scale-up strategy, the goal of which should be the determination of reactor performance by means of mathematical models which adequately describe reactor behavior. It does not really matter whether the model has a number of simplifying assumptions, is rigorously complex, or can only be solved numerically. Many problems can be avoided by using a geometric reactor configuration which yields "uniform" primary and secondary current distributions.

### 5.3.7. Multiple Electrode Modules

It is common in the scale-up of electrolytic reactors to increase the number of smaller modules rather than go for an increase in electrode size. This may appear more costly in capital investment, but it is often more than compensated for by more efficient maintenance and process operation. Design limitations often make this the only choice.

In building up multiple units (Sections 5.1.2 and 5.1.3), alternative systems of electrical and hydraulic connections are possible, each of which have advantages. The choice often depends on the scale of the process and whether batch or continuous operation is to be used. In batch processes with flow electrolysis, the conversions per cell pass are usually small and the type of hydraulic connection will depend on mechanical aspects of reactor design and operation. With parallel plate cells, parallel electrolyte flow is certainly the most common option. In continuous processes the type of hydraulic connections—series or parallel—may affect the production capacity of a unit. Further information can be found in Picket's book[48] on reactor design.

### 5.3.8. Time Factors

Scale-up procedures tend to focus on spatial increase in reactor and plant in order to accommodate the required production capacity. But another important factor is change in time scale of operation. Bench scale units tend to operate for hours, pilot plant scale units for weeks and perhaps months, and production plant runs continuously for many months or even years. Transferring a process from a bench scale to industrial production can bring problems, especially for heterogeneous processes that are sensitive to adsorption. Adsorption characteristics of electrode reactions can be modified by small amounts of impurities. Sources of impurities can be the feedstock, process piping and hardware, reactor vessels, or pumps. The degree of contamination may vary with time over long periods of operation before a steady state is approached. The level of impurity may decrease with a change in scale, because on increasing the capacity of a vessel the ratio of wall area to volume decreases. This presupposes that geometric similarity in the reactor is maintained, which is not always true.

One important source of impurities in electrochemical reactors is the anode, which invariably suffers from corrosion over the operating life of the plant. Corrosion is affected by the nature of the electrolyte, temperature, and current density. If any of these change on scale-up, the rate of contamination could change. Systems which use aggressive media such as molten salts are generally susceptible to contamination risk.

Recent examples of contamination of electrochemical reactors include the production of (1) adiponitrile and (2) glyoxylic acid.[49] In (1) the source of the problem is the anode and polymer formation; in (2) the anode, process hardware, and feedstock impurities contribute. Operation in both suffers from a gradual decrease in current efficiency due to increasing hydrogen evolution which accompanies the cathodic reaction.

In (2), again, trace amounts of additives[50] seem to provide a solution. The commonest use of additives is in the metal-winning and finishing industries.

There is no way to predict the effect of contaminants on reactor performance; prolonged pilot plant trials are necessary. Contaminants may cause modifications in the selectivity of a reaction and reduce overall production capacity. For example, in the reduction of acetophenone to ethylbenzene,[51] reaction selectivity was almost halved from its initial high value after only a few hours of operation, due to adsorption effects on the electrode surface.

Besides upsetting reaction specificity, variations in electrode structure may influence mass transfer characteristics. Estimates of mass transfer coefficients are usually made from correlations obtained with "clean" systems. Electrodes operating for weeks and months will suffer from roughening, erosion, pitting, and scaling, and so on, which will affect mass transport rates. The significance of this may be almost unknowable without prolonged tests on plant or under simulated conditions. There is a need for such data if electrochemical processes are to perform at their best.

## REFERENCES


1. Pletcher, D. and Walsh, F. C., 1990, *Industrial Electrochemistry"*, 2d ed., Chapman & Hall, New York.
2. Blackmar, G. E., U.S. Patent no. 3,573,178 (1971).
3. Fleet, B. and Gupta, S. D., 1976, Novel electrochemical reactor, *Nature (London)*, **263:** 5573, 122–123.
4. Backhurst, J. R., Coulson, J. M., Goodridge, F., Plimley, R. E., and Fleischmann, M., 1969, "A Preliminary investigation of fluidised bed electrodes", *J. Electrochem. Soc.*, **116:** 1600–1607 (1969).
5. Goodridge, F. and Wright, A. R., 1983, "Porous flow-through and fluidised bed electrodes," in *Comprehensive Treatise of Electrochemistry*, Vol. 6, (E. Yeager, J. O'M. Bockris, and S. Sarangapany, eds.) Plenum Press, pp. 393–443.
6. Hughes, D., 1988, "The dished electrode membrane cell facilitates wide range of syntheses," *Spec. Chem.*, **8:** 16, 17.
7. Carlsson, L., Holmberg, H., Johansson, B., and Nilsson, A., 1982, "Design of a multipurpose modularised electrochemical cell," in *Techniques of Electroorganic Synthesis*, III, (N. L. Weinberg and B. V. Tilak, eds.), John Wiley & Sons, New York, pp. 179–194.
8. Brooks, W. N. 1986, "The ICI (Mond) FM21 cell as a multipurpose electrolyser, *Instit. Chem. Eng. Symp. Ser., Electrochemical Engineering*, **98:** 1–12, 320–321.
9. Degner, D., 1982, "Scale-up of electroorganic processes: Some examples for a comparison of electrochemical syntheses with conventional syntheses," in Weinberg and Tilak, eds., p. 256.
10. Jansson, R. E. W., Marshall, R. J., and Rizzo, J. E., 1978, "The rotating electrolyser. I: The velocity field," *J. Appl. Electrochem.* **8:** 281–285; R. E. W. Jansson and R. J. Marshall, "The rotating electrolyser II: Transport properties and design equations," *J. Appl. Electrochem.*, 287–291.
11. Udupa, H. V. K. and Udupa, K. S., 1982, "Use of rotating electrodes for small-scale electroorganic processes," in Weinberg and Tilak, eds., pp. 385–422.
12. Holland, F. S., 1978, "The development of the Eco-Cell Process," *Chem. Ind. (London)* **7:** 453–458.
13. Robertson, P. M., Berg, P., Reimann, H., Schleich, K., and Seiler, P., 1983, "Application of the Swiss-Roll Cell in vitamin-C production," *J. Electrochem. Soc.* **130:** 591–596.
14. Robertson, P. M., Cettou, P., Matic, D., Schwager, F., Storck, A., and Ibl, N., 1979, "Electrosynthesis with the Swiss-Roll Cell. Properties of the cell components and their selection for electrosynthesis," *Am. Instit. Chem. Eng. Symp. Ser. Electroorganic Synthesis Technology*, **75:** 115–124.
15. Oloman, C., 1979, "Trickle bed electrochemical reactors," *J. Electrochem. Soc.*, **126:** 1885–1892.
16. Goodridge, F., Harrison, S., and Plimley, R. E., 1986, "The electrochemical production of propylene oxide," *J. Electroanal. Chem. Interfacial Electrochem.*, **214:** 283–293.
17. Feess, H. and Wendt, H., 1982, "Performance of two-phase-electrolyte electrolysis," in Weinberg and Tilak, eds., pp. 81–177.

18. Dafana, R., 1987, *The Design and Performance of a Novel Electropulse Column*, Ph.D. dissertation, University of Newcastle upon Tyne, U.K.
19. MacMullin, R. B., 1963, "The problem of scale-up in electrolytic processes," *Electrochem. Technol.*, **1:** 5–17.
20. Coulson, J. M. and Richardson, J. F., *Chemical Engineering*, Vol. 1, 4th ed., Pergamon Press, pp. 9–15.
21. Johnstone, R. E. and Thring, M. W., 1967, *Pilot Plants, Models and Scale-up Methods in Chemical Engineering*, McGraw-Hill, New York.
22. Johnstone and Thring, p. 80.
23. Damkohler, G., 1936, "The influence of flow, diffusion and heat transfer on the performance of reaction furnaces. I. General considerations of the transfer of chemical processes from small to large size equipment," *Z. Elektrochem.*, **42:** 846–862.
24. Johnstone and Thring, p. 90.
25. Wragg, A. A., Tagg, D. J., and Patrick, M. A., 1980, "Diffusion controlled current distributions near cell entries and corners," *J. Appl. Electrochem.*, **10:** 43–47.
26. Goodridge, F., Mamoor, G. M., and Plimley, R. E., 1986, "Mass transfer rates in baffled electrochemical cells," *Inst. Chem. Eng., Symp. Ser.*, **98:** pp. 61–71.
27. Hine, F., 1985, *Electrode Processes and Electrochemical Engineering*, Plenum Press, New York, p. 313.
28. Parrish, W. R. and Newman, J., 1969, "Current distribution on a plane electrode below the limiting current," *J. Electrochem. Soc.*, **116:** 169–172.
29. Parrish, W. R. and Newman, J., 1970, "Current distribution on plane parallel electrodes in channel flow," *J. Electrochem. Soc.*, **117:** 43–48.
30. Pickett, D. J., *Electrochemical Reactor Design*, 2d ed., Elsevier Scientific Publishing, New York, p. 114.
31. Ibl, N., 1983, "Current Distribution," in *Comprehensive Treatise of Electrochemistry*, Vol. 6, (E. Yeager, J. O'M. Bockris, and S. Sarangapany, eds.) Plenum Press, New York, pp. 239–315.
32. Wagner, C., 1951, "Theoretical analysis of the current distribution in electrolytic cells," *J. Electrochem. Soc.*, **98:** 116–128.
33. Viswanathan, K., and Chin, D. T., 1977, "Current distribution on a continuous moving sheet electrode," *J. Electrochem. Soc.*, **124:** 709–713.
34. Lapicque, F. and Storck, A., 1985, "Modelling of a continuous parallel plate plug flow electrochemical reactor: Electrowinning of copper," *J. Appl. Electrochem.*, **15:** 925–935.
35. De La Rue, R. E. and Tobias, C., 1959, "On the conductivity of dispersions," *J. Electrochem. Soc.*, **106:** 827–833.
36. Tobias, C. W., 1959, "Effect of gas evolution on current distribution and ohmic resistance in electrolysers," *J. Electrochem. Soc.*, **106:** 833–838.
37. Hine, p. 89.
38. Nishiki, Y., Aoki, K., Tokuda, K., and Matsuda, H., 1986, "Effect of gas evolution on current distribution and ohmic resistance in a vertical cell under forced convection conditions," *J. Appl. Electrochem.*, **16:** 615–625.
39. Pickett, p. 347.
40. Tobias, C. W. and Wijsman, R., 1953, "Theory of the effect of electrode resistance on current density distribution in electrolytic cells," *J. Electrochem. Soc.*, **100:** 459–467.
41. Hine, p. 329.
42. Goodridge, F. and Hamilton, M. A., 1980, "The behaviour of a fixed bed porous flow-through electrode during the production of p-amino phenol," *Electrochim. Acta*, **25:** 481–486.
43. Goodridge, F., Plimley, R. E., and Leetham, R. P., (1985) *Purifying Mixed-Cation Electrolyte*, Eur. Pat. Appl. EP 197,769.

44. Scott, K. and Lui, W. K., 1986, "The performance of a moving bed electrode during the electrowinning of cobalt," *Inst. Chem. Eng., Symp. Ser.*, **98:** 143–154.
45. Scott, K., 1982, "The effectiveness of particulate bed electrodes under activation control," *Electrochim. Acta,* **27:** 447–451.
46. Weise, L., Giron, M., Valentin, G., and Storck, A., "A chemical engineering approach to selectivity analysis in electrochemical reactors," *Inst. Chem. Eng., Symp. Ser.*, **98:** 49–59.
47. Storck, A., Enriques-Granados, M., and Roger, M., 1982, "The behaviour of porous electrodes in a flow-by regime. I. Theoretical study," *Electrochim. Acta,* **27:** 293–301.
48. Pickett, pp. 362–388.
49. Scott, K., 1986, "Electrolytic reduction of oxalic acid to glyoxylic acid: A problem of electrode deactivations," *Chem. Eng. Res. and Des.*, **64:** 266–271.
50. Michelet, D., 1974, "Glyoxylic acid," *Ger. Offen.*, **2**, 359, 863.
51. Pletcher, D., and Razaq, M., 1981, "The reduction of acetophenone to ethylbenzene at a platinised platinum electrode," *Electrochim. Acta,* **26:** 819–824.

Chapter 6

# Cost Estimation, Profit Appraisal, Process Modeling, and Optimization

## 6.1. COST ESTIMATION AND PROFIT APPRAISAL

The process modeling and optimization procedures of this chapter require a knowledge of capital investment and operating costs of chemical plant, as well as methods of appraisal of profitability. To find optimum design parameters of a plant, a process model has to be set up so that capital and running costs as a function of the variables under consideration such as conversion, current density, and product purity can be determined. The duty or throughput of the plant is based on a market survey and is usually fixed. Plotting process capital costs against process running costs leads to the "payback" time, how many years it will take for the production cost savings to recover the extra investment incurred for equipment. Profitability is then assessed, usually in terms of $i$, the discounted cash flow rate of return (DCFRR), which is related to payback time. The company decides whether $i$ is acceptable or if another process must be looked at. Unfortunately, an optimized plant design does not guarantee profitability.

It is routine to use software packages for costing and subsequent optimization. But it is vital to understand the criteria on which software is based.

We will first look at costing procedures; then process modeling, with emphasis on interaction between individual items of chemical plant; and finally optimization, by means of examples.

### 6.1.1. Costing Procedures

*Equipment* design is the province of general chemical engineering, and so are the techniques of cost estimation and profit appraisal. We will consider some techniques generally and illustrate them with worked examples. For further details on costing techniques and profit appraisal, the reader is referred to standard texts.[1-3]

#### 6.1.1.1. Capital and Capital-Related Costs

The nature of a cost estimate depends what it is used for. An order-of-magnitude or ratio estimate gets total capital cost from a single equation; this, needless to say, is hazardous. Or a detailed or final estimate can be prepared when design, drawing, and specification work is virtually complete; this will take a professional design team a considerable time to complete. Better is a predesign or study estimate where cost figures for installed individual plant items are obtained from a correlation. It is preferable to get a cost estimate of the delivered unit from the manufacturer and then to calculate installed costs with a factorial method, as in Section 6.1.2.

The reliability of estimates varies from $> \pm 30\%$ for the ratio estimate to $\pm 5\%$ for the final estimate. The cost of estimates ranges from less than 0.1% to as much as 5 or 10% of final project costs.[1]

#### 6.1.1.2. Production and Production-Related Costs

Estimating the cost of operating the plant and selling the product should be done on an annual basis. You can subdivide costs into production costs and general expenses.[2] The former includes costs of chemicals, energy, and labor, while the latter includes administrative expenses, distribution and marketing, research and development. General expenses vary from company to company and are not usually included in a preliminary or predesign estimate. An important cause of serious error is to ignore the cost of maintaining a fixed stock of raw materials (usually a three-month supply), which for a large-tonnage or high-value process can be considerable.

### 6.1.2. Example of a Predesign Estimate for Producing Glyoxylic Acid by the Electrolytic Reduction of Oxalic Acid

#### 6.1.2.1. Basis for Cost Estimation

The costs refer to the reduction step:

$$\begin{matrix} \text{COOH} \\ | \\ \text{COOH} \end{matrix} + 2e + 2\text{H}^+ \longrightarrow \begin{matrix} \text{CHO} \\ | \\ \text{COOH} \end{matrix} + \text{H}_2\text{O}$$

TABLE 6.1. Glyoxylic Acid Process Conditions

| | |
|---|---|
| Catholyte[a] | Saturated solution of oxalic acid |
| Anolyte | 0.1M sulfuric acid |
| Conversion | 50% of total oxalic acid added to the system |
| Chemical yield | 100%, discounting the small percentages of glycolic acid and glyoxal formed |
| Current efficiency | 60% |
| Current density | 1000 $A/m^2$ |
| Anode-to-cathode Voltage | 4 V |
| Temperature | 20 °C |
| Pressure | Atmospheric |

[a]Saturation is maintained by adding solid oxalic acid as the reduction proceeds.

Losses in current efficiency are attributed to hydrogen evolution. The companion reaction at the anode is the evolution of oxygen. For process conditions, see Table 6.1.

The cost has been based on the production of 500, 1000, 1800, and 4000 kg/annum of glyoxylic acid in the form of an aqueous 10% solution, by weight, saturated with oxalic acid. It does not, therefore, include the cost of oxalic acid recovery and product concentration. The costing of the process is in terms of added value: it does not contain the costs of raw materials. All costs are in pounds sterling (£).

Using the nomenclature of Chapter 5, a unit cell or module is made up of an electrode pair. A number of modules form a cell stack. A notional flow sheet is shown in Fig. 6.1.

#### 6.1.2.2. Process Conditions (Table 6.1)

#### 6.1.2.3. Capital Costs

(i) The electrolytic cell stack. Calculations (Table 6.2) are made on the assumption that an 8-hr batchtime is most convenient and that the cell stack will be operated one batch per day, 5 days per week, 45 weeks per year.

(ii) The power supply. Voltage drop across the stack is $2 \times 4 = 8$ V. Current supplied to the cell stack is $347/2 = 137.5$ A. A rectifier delivering 10 V, 200 A will give a margin of safety.

- Cost of power supply (including regulator, transformer, rectifier, and instrument panel): £800

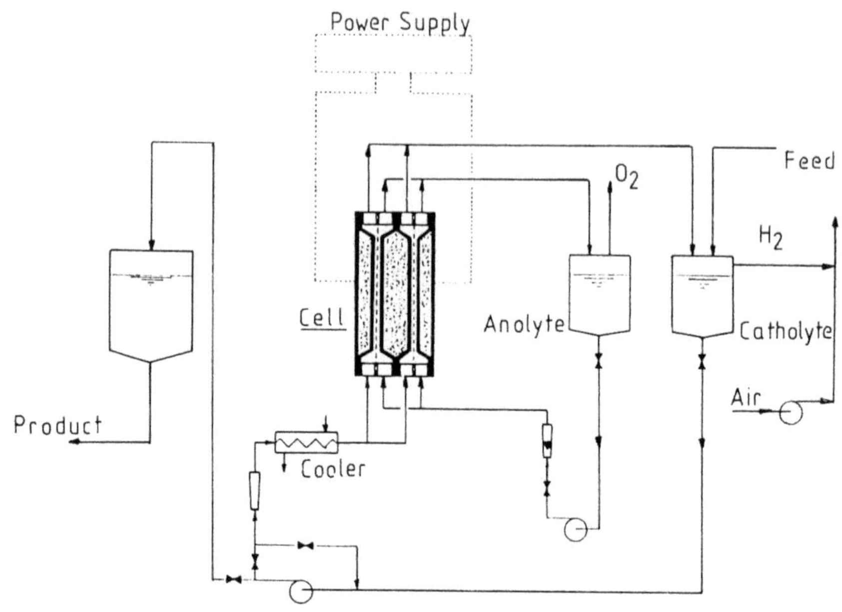


FIGURE 6.1. Flow sheet for the production of glyoxylic acid.

TABLE 6.2. Electrolytic Cell Stack

| | |
|---|---|
| Total kmols of product | 500/74 = 6.76 |
| Batches | 225 |
| kmols of product per batch | 0.03 |
| Coulombs per batch = gmols × number of electrons × coulombs/Faraday<br>= $0.03 \times 10^3 \times 2 \times 10^5 = 6 \times 10^6$ A s | |
| Current required $(6 \times 10^6)$/(current efficiency × batchtime) | $6 \times 10^6/(0.6 \times 8 \times 3600) = 347$A |
| Cathode area required when working at 1000 A/m$^2$ | 0.347 m$^2$ |
| Cathode area of commercial filter press cell module | 0.175 m$^2$ |
| Modules required | 2 |
| Costs (£): | |
| Cost of each module | 3000 |
| Cost of two modules | 6000 |
| Cost of press | 3500 |
| Cost of minimum order of sheets of cationic membrane material (3 sheets each 3 m × 1 m) | 585 |

(iii) The pumps. Assuming a maximum flow velocity of 30 cm/s and an electrode-membrane gap of 5 mm, the volumetric flow rate per gap will be 46 $dm^3$/min with an estimated pressure loss of 20 kN/$m^2$. The total flow rate past both cathodes will be 92 $dm^3$/min; allowing for further pressure losses in the flow circuit, a total pressure loss of 40 kN/$m^2$ will be assumed. Catholyte and anolyte pumps are required.

- Cost of a plastic-lined centrifugal pump (92 $dm^3$/min – 40 kN/$m^2$): £377

(iv) The heat exchanger. Remembering Example 2.14 in Section 2.5 (see Section 6.5.2), the total heat load in the exchanger $Q$ is estimated by an energy balance for an isothermal batch reactor system:

$$Q = W + E'' - \Delta H_{reaction}$$

where $W$ is the energy due to pumping, $E''$ the energy input from the power supply, and $\Delta H_{reaction}$ the uptake of enthalpy due to the reactions. All are treated as rates in kilowatt units.

$W_c$, the energy derived from pumping the catholyte, is:

$$\begin{aligned} W_c &= \text{volumetric flow rate } (\text{m}^3/\text{s}) \times \text{pressure loss } (\text{kN/m}^2) \\ &= [(92 \times 10^{-3})/60] \times 40 = 0.06 \text{ kW} \end{aligned}$$

Total energy $W$ derived from pumping catholyte and anolyte is $2W_c$:

$$W = 0.12 \text{ kW}$$

$E''$, the energy input due to the electrical supply to the cell, is:

$$E'' = \text{cell current} \times \text{cell voltage} = (173.5 \times 8)/1000 = 1.39 \text{ kW}$$

Overall reactions in the cell are:

$$(COOH)_2 \longrightarrow \underset{\displaystyle COOH}{\overset{}{CHO}} + 0.5\,O_2 \qquad 30 \times 10^3 \text{ kJ/kmol}$$

and

$$H_2O \longrightarrow H_2 + 0.5\,O_2 \qquad 287 \times 10^3 \text{ kJ/kmol}$$

0.03 kmol of the first reaction occurs per batch and, at a current efficiency of 60%, 0.02 kmol of the second reaction. For the entire batch the enthalpy change $\Delta H_R$ is:

$$\Delta H_R = 0.03 \times 30 \times 10^3 + 0.02 \times 287 \times 10^3 = 6.64 \times 10^3\,\text{kJ}$$

The average rate of change of enthalpy due to reaction over the batch $\Delta H_{reaction}$ will be:

$$\Delta H_{reaction} = (6.64 \times 10^3)/(8 \times 3600) = 0.23\,\text{kW}$$

Therefore:

$$Q = W + E'' - \Delta H_{reaction} = 0.12 + 1.39 - 0.23 = 1.28\,\text{kW}$$

The required heat transfer area $A$ is:

$$A = Q/(h \times \Delta T_m)$$

where $h$ is the overall heat transfer coefficient and $\Delta T_m$ the mean temperature difference.

For a glass heat exchanger handling an aqueous stream, a value of $0.6\,\text{kW/m}^2$ for $h$ is reasonable. On the assumption that water is available at a temperature of 10°C, with a temperature rise of 2°C, $\Delta T_m$ will be 9°C. Therefore:

$$A = 1.28/(0.6 \times 9) = 0.24\ \text{m}^2$$

- Cost of a $0.25\,\text{m}^2$ glass heat exchanger complete with reducers and coupling: £241

(v) The electrolyte vessels. A 10% solution by weight, containing 0.03 kmol of glyoxylic acid as the final product, requires that:

$$0.1 = \text{mass of product/mass of solution} = (0.03 \times 74)/\text{mass of solution}$$

Hence mass of solution = 2.22/0.1 = 22.2 kg, i.e., volume is about $22\,\text{dm}^3$. This is very small, so pipe runs will have to be kept to a minimum (e.g., 4.5 m of 4 cm diameter tubing for the catholyte flow) if a reasonable volume of liquid is to be kept in the catholyte reservoir.

- Cost of a $50\,\text{dm}^3$ cylindrical polypropylene tank with cover: £75

A similar tank will have to be used for the anolyte.

(vi) The product-receiving tank. The size of the product-receiving tank depends on the concentration steps that follow. If concentration will weight the arrival of 5 batches, the tank's volume must be $150\,dm^3$, which would be sufficient to store a product volume of $110\,dm^3$.

(vii) Summary of itemized purchase costs. (Table 6.3)

TABLE 6.3. Summary of Itemized Purchase Costs

| Item | Purchase Costs[a] (£) |
|---|---|
| Cell stack | |
| Modules (2 off) | 6000 |
| Membrane | 585 |
| Power supply | 800 |
| Pumps | |
| Catholyte | 377 |
| Anolyte | 377 |
| Heat Exchanger | |
| Catholyte | 241 |
| Vessels | |
| Catholyte | 75 |
| Anolyte | 75 |
| Product receiving | 100 |
| Total purchase costs | 8630[a] |

[a] Purchase costs in this example do not include the cost of the cell press (see Section 6.1.2.10.2) nor do they include local taxes, e.g., value-added tax. Some allowance has been made, however, for the latter in the installed costs (see Section 6.1.2.3.viii).

(viii) Installed capital costs. Installed costs are usually determined from the purchase costs by multiplying by a factor to allow for piping, instrumentation, safety equipment for hydrogen removal, labor costs, and taxation if any. Usually this factor is about 2.6. In view of the small scale of the plant the figure will be nearer 1.6 and it is also judged that inclusion of the cell press in the purchase costs will distort the costing figures (see Section 6.1.2.10.2).

$$\text{Installed capital costs} = 8630 \times 1.6 = £13{,}800$$

### 6.1.2.4. Production Costs

The major running costs, apart from labor—which is indeterminate in this case and has been considered to be covered by the installation factor—are power for the cell and pumps, and cooling water. From Section 6.1.2.1, the costing is of the added-value type and does not include, therefore, the cost of oxalic acid.

(i) Cost of electrical power. The power consumed by the cell is the product of the total voltage and the current passed, i.e., $(8 \times 173)/1000 = 1.39\,\text{kW}$. The electrical supply to the pumps, assuming them to be 80% efficient, is (see Section 6.1.2.3.iv) $0.12/0.8 = 0.15\,\text{kW}$. The total power required $= 1.39 + 0.15 = 1.54\,\text{kW}$.

For an 8-hr batch $1.54 \times 8 = 12.3\,\text{kWh}$ will be needed. Assuming the cost of a kWh to be £0.04, cost is £0.49 per batch.

(ii) Cost of cooling water. $F$, the flow of cooling water required, is:

$$F = \text{heat load } Q/(\text{heat capacity} \times \text{temperature rise})$$

In Section 6.1.2.3.iv, $Q$ was calculated to be 1.28 kW and a temperature rise of 2°C was assumed. Therefore $F = 1.28/(4 \times 2) = 0.16\,\text{kg/s}$. Taking cooling water obtained from a small-scale supply as costing £0.1/$m^3$, the cost per batch is:

$$(0.16 \times 8 \times 3600 \times 0.1)/10^3 = £0.46$$

(iii) Total running costs.

$$\text{Total running costs per batch} = 0.49 + 0.46 = £0.95.$$

### 6.1.2.5. Capital Depreciation

In Section 6.1.2.3.viii, installed capital cost was calculated to be £13,800. Writing the investment off over three years, the capital element of the cost per batch is:

$$13{,}000/(3 \times 225) = £20.44$$

### 6.1.2.6. Added Value per kg of Glyoxylic Acid

Added value per batch is the sum of the capital element and running costs as calculated in Sections 6.1.2.5 and 6.1.2.4.iii, i.e., $20.44 + 0.95 = £21.39$.

Each batch produces $0.03 \times 74 = 2.22$ kg of glyoxylic acid. The added value per kg of glyoxylic acid is £21.39/2.22 = £9.64.

### 6.1.2.7. Costing Figures for a Product Throughput of 1000 kg/annum

We assume the use of the same two-module cell, operating two 8-hour batches per day, 5 days per week, 45 weeks per year.

Calculating as in Section 6.1.2.3, we make 1000/74 = 13.51 kmols, i.e., double the amount of product. Since, however, the number of batches has also doubled to 450, the values of Sections 6.1.2.3 and 6.1.2.4 remain unchanged.

The capital element is now $13{,}800/(3 \times 450) =$ £10.22 and the added value per kg of glyoxylic acid is $(10.22 + 0.95)/2.22 =$ £5.03.

### 6.1.2.8. Maximum Amount of Glyoxylic Acid that can be Produced per Annum with the Two-Module Cell

It will be useful to consider the minimum capital element that can be achieved and the amount of product that can be produced by running the two-module cell as fully as possible.

With three 7-hour batches per day, 7 days per week, but still 45 weeks per annum to provide some slack in the system, there will be 945 batches per annum, each producing $(0.03 \times 7)/8 = 0.0263$ kmol. (Since catholyte volumes will be lower than before, minimization of holdup becomes more important.) Amount of product will be $0.0263 \times 74 \times 945 = 1800$ kg per annum. The capital element will be $13{,}800/(3 \times 945) =$ £4.87 and the running costs £0.83. The amount of glyoxylic acid produced per batch will be 1.95 kg and the added value per kg of product will be $(4.87 + 0.83)/1.95 =$ £2.92.

### 6.1.2.9. Costing Figures for a Product Throughput of 4000 kg/annum

Finally, let us see how the added value figures change for a significantly higher throughput of 4000 kg/annum. There are many ways to achieve such a throughput, but it will be instructive to retain the methods of Section 6.1.2.8.

The new capital costs will be:

(i) The electrolytic cell stack. The cell stack is operated by three 7-hour batches per day, 7 days per week, 45 weeks per year (Table 6.4).

(ii) The power supply. With five modules the cell current is 159 A. A *rectifier delivering* 25 V *and* 200 A is the choice.

TABLE 6.4. Electrolytic Cell Stack

| | |
|---|---|
| Total kmols of product | $4000/74 = 54.05$ |
| Batches | $3 \times 7 \times 45 = 945$ |
| kmols of product per batch | $54.05/945 = 0.06$ |
| kumber of Coulombs per batch | $0.06 \times 10^3 \times 2 \times 10^5 = 1.2 \times 10^7$ A s |
| Current required | |
| [$6 \times 10^6$/(current efficiency × batchtime)] | $1.2 \times 10^7/(0.6 \times 7 \times 3600) = 794$ A |
| Cathode area working at 1000 A/m$^2$ | 0.794 m$^2$ |
| Cathode area of commercial filter press cell module | 0.175 m$^2$ |
| Modules required | 5 |
| Costs (£): | |
| Cost of each module | 3000 |
| Cost of five modules | 15000 |
| Cost of minimum order of sheets of cationic membrane material (3 sheets, each 3 m × 1 m) | 585 |

- Cost of power supply (including regulator, transformer, rectifier, and instrument panel): £1050

(iii) The pumps. The total flow rate is 230 dm$^3$/min and the pressure drop remains at 40 kN/m$^2$.

- Cost of a plastic-lined centrifugal pump (230 dm$^3$/min, 40 kN/m$^2$): £500

(iv) The heat exchanger.

- Cost of a 1.2-m$^2$ glass heat exchanger complete with reducers and coupling: £550 (see Table 6.5).

(v) The electrolyte vessels.

Similar tank will be used for both anolyte and catholyte (see Table 6.6).

TABLE 6.5. Heat Exchanger

| | |
|---|---|
| $W_c$ (see section 6.1.2.3.iv) $= [(230 \times 10^{-3})/60] \times 40$ | 0.15 kW |
| Total energy $W$ | 0.30 kW |
| $E''$ $(159 \times 40)/1000$ | 6.36 kW |
| $\Delta H_R$ $(0.06 \times 30 \times 10^3 + 0.04 \times 287 \times 10^3)$ | $13.28 \times 10^3$ kJ |
| $\Delta H_{reaction}$ $(13.28 \times 10^3)/(7 \times 3600)$ | 0.53 kW |
| $Q = W + E'' - \Delta H_{reaction}$ $0.30 + 6.36 - 0.53$ | 6.13 kW |
| Heat exchanger area $A$ $[6.13/(0.6 \times 9)]$ | 1.2 m$^2$ |

TABLE 6.6. Electrolyte Vessels

| | |
|---|---|
| Mass of product | $0.06 \times 74 = 4.44$ kg |
| Mass of solution | $4.44/0.1 = 44.4$ kg |
| Cost of 100-dm$^3$ cylindrical polypropylene tank with cover | £82 |

(vi) The product-receiving tank. Assuming again an accumulation of five batches:

- Cost of 300-dm$^3$ cylindrical polypropylene tank with cover: £155

(vii) Summary of itemized purchase costs. (See Table 6.7.)

(viii) Installed production costs. Using the same factor as before:

$$18{,}504 \times 1.6 = £29{,}600$$

The new production costs are, first, electrical power cost:

$$\text{Power to cell} = (40 \times 159)/1000 = 6.36 \text{ kW}$$

$$\text{Power to pumps} = 0.3/0.8 = 0.38 \text{ kW}$$

$$\text{Total power required} = 6.36 + 0.38 = 6.74 \text{ kW}$$

TABLE 6.7. Summary of Itemized Purchase Costs

| Item | Purchase costs (£) |
|---|---|
| Cell stack | |
| Modules (5 off) | 15000 |
| Membrane | 585 |
| Power supply | 1050 |
| Pumps | |
| Catholyte | 500 |
| Anolyte | 500 |
| Heat exchanger | |
| Catholyte | 550 |
| Vessels | |
| Catholyte | 82 |
| Anolyte | 82 |
| Product receiving | 155 |
| Total | 18,504 |

TABLE 6.8. Summary of Cost Calculations

| Throughput (kg/annum) | Added value per kg of glyoxylic acid (pounds) | |
|---|---|---|
| | 3-year writeoff | 10-year writeoff |
| 500[a] | 9.69 | 3.19 |
| 1000[a] | 5.03 | 1.81 |
| 1800[a] | 2.92 | 1.17 |
| 4000[b] | 3.21 | 1.57 |

[a]Two-module cell.
[b]Five-module cell.

For a 7-hour batch, $6.74 \times 7$, i.e., 47.2 kW is required at a cost of $47.2 \times 0.04 = £1.89$.

Second, there is the cost of cooling water:

The required flow rate $F$ is:

$$6.13/(4 \times 2) = 0.77 \text{ kg/s}$$

cost per batch:

$$(0.77 \times 7 \times 3600 \times 0.1)/10^3 = £1.94$$

Third are the total running costs per batch $= 1.89 + 1.94 = £3.83$.

(ix) **Added-value calculation.** Writing off the plant again after three years, the capital element is $29{,}600/(3 \times 945) = £10.44$. The added value per batch is $10.44 + 3.83 = £14.27$. Each batch produces 4.44 kg of glyoxylic acid. Therefore, the added value per kg of glyoxylic acid is $14.27/4.44 = £3.21$. (See Table 6.8).

### 6.1.2.10. Comments on the Above Cost Calculations

1. Cost calculations have been done on the basis of 3-year and 10-year write-offs. The capital item is important, particularly at low throughputs. However, with a 10-year write-off electrode renewal would almost certainly be necessary.

2. To calculate installed capital costs a factor of 1.6 has been used in view of the small scale of the plant (see Section 6.1.2.3.viii). To test the

reliability of the factor, a detailed costing (not shown here) was undertaken taking into account the actual cost of instrumentation, hydrogen safety equipment (i.e., air blower to reduce possible hydrogen concentration below the explosion limit and flame trap), piping, valves, framework, cell press, labor cost, and value-added tax.

Installed capital cost from this detailed costing was £5071. This compares well with the figure of £5178 obtained from the factor 1.6.

3. As explained in Section 6.1.2.1, costing has been done on a value-added basis and does not include cost of the reactants or costs of the concentration stages.

Other items not included in the costing are working capital necessary for production for say three months (see Section 6.1.1.2), and the cost of an automatic feeding device which might be used to supply oxalic acid to the systems as the run proceeds.

4. Costs quoted in this chapter do not represent up-to-date figures but are only used illustratively. This is why no attempt has been made to present examples using U.S. dollars.

## 6.2. PROFITABILITY CRITERIA FOR OPTIMIZATION

Optimization methods for electrochemical processes are no different from those used for other chemical plants. The criterion applied is the same; before considering process modeling and optimization, we will discuss this further.

In any chemical process we are faced at some design stage with choices between alternative technologies, or modes of operation. The underlying choice objective is maximizing financial return.

The simplest measure of financial return is profit $Pr$, which is the difference between income $I$ and production or running costs $P$ (i.e., $Pr = I - P$). Given that income is usually fixed by the total sales value of the product, maximization of profit becomes minimization of production costs. But is profit alone a fair measure of financial return?

For example, take two processes A and B for manufacturing a proposed product, which have the economic characteristics shown in Fig. 6.2. If the only criterion is to minimize $P$ we will select process B at once, but let us take a closer look by considering two cases which satisfy the economic descriptions in Fig. 6.2, but which lead to different decisions. They are presented graphically in Fig. 6.2.

In case I it can be seen that in going from process A to process B, the slight increase in capital investment, $\Delta C$ produces a large saving $\Delta P$ in running costs. We spend a little to save a lot. In case II, however, although

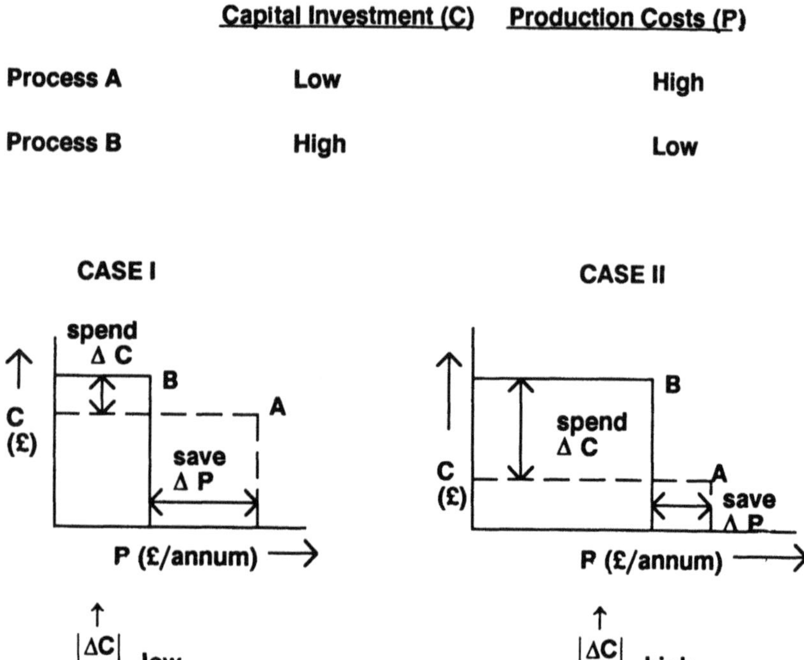


FIGURE 6.2. Choice between alternatives. (a) low capital investment, high production costs; (b) high capital investment, low production costs.

B offers lower production costs the improvement over A is too slight to justify the enormous jump $\Delta C$ in capital outlay. We would have to spend a lot to save a little.

More formally, $|\Delta C/\Delta P|$ is low in case I, and the extra capital is attractive; in case II $|\Delta C/\Delta P|$ is high and the extra capital outlay is unattractive. Our decision is dependent upon the ratio of the incremental or marginal costs $|\Delta C/\Delta P|$. The unit of this ratio is time, and it measures the payback time, i.e., the time it will take for savings in production costs to recover increased capital expenditure. The more expensive alternative is selected if the payback time is low, and the less expensive if the payback time is high. "High" and "low," however, only have meaning if the figure can be compared with some standard. This standard is referred to as the maximum acceptable payback time ($Y_m$) and will be set by company policy, although it will of course be in line with the minimum discounted cash flow rate of return (DCFRR) which the company demands.

It can be easily shown that $Y_m$ is related to the depreciation ($n$ years) and the minimum required DCFRR, $i$ according to the relationship:

$$Y_m = \frac{(1 + i)^n - 1}{i(1 + i)^n} \tag{6.1}$$

(i) What is optimization? Optimization is a special case of the choice between alternatives, one in which the choice lies between a large number and, in the extreme, between an infinite number of alternatives—which arises when a process variable can be varied continuously between an upper and lower value, as for example the operating current density of an electrochemical reactor.

The corresponding capital and production costs $C$ and $P$ are related to operating current density by the curve in Fig. 6.3, where increasing capital investment $C$ (to allow lower current densities) leads to decreasing production costs $P$ (due to lower power costs).

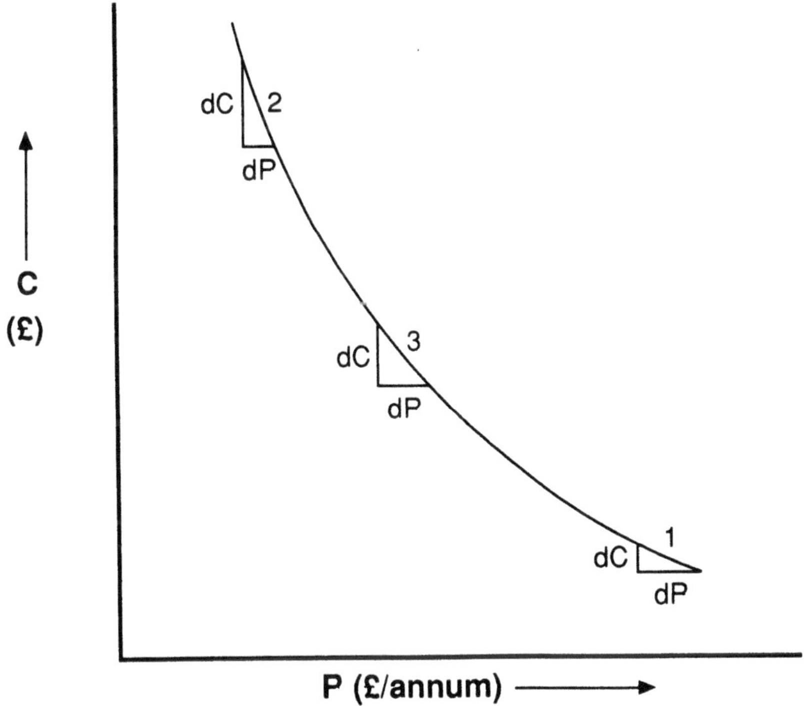


FIGURE 6.3. Plot of capital investment against production costs for the selection of an optimum investment.

Optimization identifies the most attractive level of capital investment. How does it? Let us consider two alternatives, those represented by the points of location 1 near the bottom of the curve in Fig. 6.3, separated by an increase in capital $dC$ and a reduction in running costs $dP$. The payback time in moving from the lower alternative at 1 to the upper alternative at 1 is by definition $|dC/dP|$, and the absolute value of the slope of the curve is also $|dC/dP|$. The slope and payback time can be equated. As we move up the curve in Fig. 6.3 it becomes steeper, i.e., the payback time increases.

Assume that the payback time has increased so much at the top of the curve that for the points located at 2 it exceeds $Y_m$, while at the bottom, for the points located at 1, it is less than $Y_m$. At location 1, therefore, we would choose to move up the curve, whereas at location 2 we would choose to move down. Where the slope exceeds $Y_m$ we move down the curve to regions of lower capital investment; where the slope is less than $Y_m$ we move up to regions of higher investment.

The limiting case, which corresponds to the optimum capital investment, occurs at the point on the curve where the slope equals $Y^m$, at location 3, say. At lower points on the curve the plant will be undercapitalized; at higher points, overcapitalized. The criterion which identifies the optimum is the equality:

$$-\frac{dC}{dP} = Y_m \tag{6.2}$$

This can be developed further. If the R.H.S. of Eq. (6.2) is multiplied by unity in the form of $dP/dP$, the terms can be combined into one differential:

$$\frac{d(C + Y_m P)}{dP} = 0 \tag{6.3}$$

Dividing by $Y_m$, our optimization criterion becomes:

$$\frac{d}{dP}\left[\frac{C}{Y_m} + P\right] = 0 \tag{6.4}$$

indicating that the optimum is where $C/Y_m + P$ is a minimum.

The search for the minimum can be made with respect to any convenient variable $Z$, and need not be confined to $P$, so that in its final form the optimization criterion becomes:

$$\frac{d}{dZ}\left[\frac{C}{Y_m} + P\right] = 0 \tag{6.5}$$

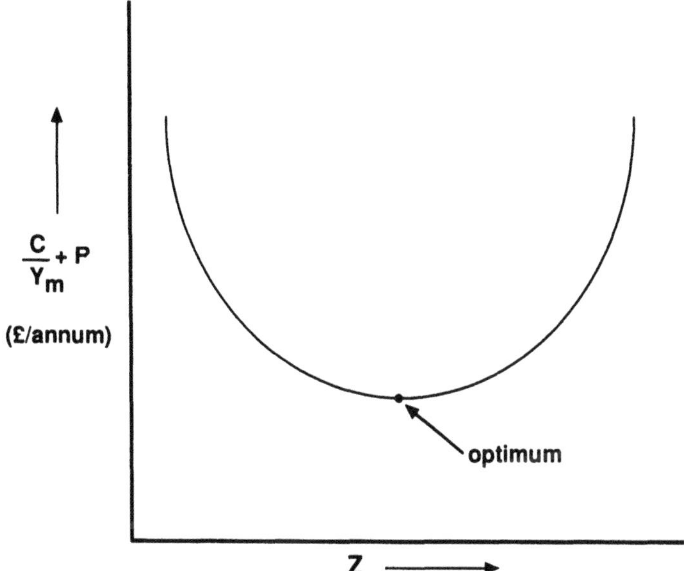


FIGURE 6.4. Plot of $C/Y_m + P$ against variable $Z$.

The optimum level of investment and the corresponding optimum value of a process variable are located at the minimum of a $(C/Y_m + P)$ versus $Z$ plot (Fig. 6.4).

## 6.3. PROCESS MODELING AND OPTIMIZATION

Process modeling applies mathematical modeling to a complete chemical process. A mathematical model is a family of equations that describes a system in terms appropriate to a specified objective. For process modeling, a system is broken down into subsystems. Each subsystem is modeled in turn; collectively, the models comprise the process model. The way a system is divided into subsystems is a matter of choice, a matter of convenience. One subsystem, for example, is an electrochemical reactor as modeled in Chapter 4. The others could be separation stages, heat exchangers, etc. The simple process shown in Fig. 6.5, which comprises an electrochemical reactor coupled to a distillation unit which recovers unconverted reactant and recycles it through a cooler, has three subsystems.

The design of individual units of a plant and the subsequent combination into a flow sheet is the province of chemical engineering and is dealt with in standard chemical engineering texts.[3–5] The optimization of complex integrated plant is done with commercial software that can solve many

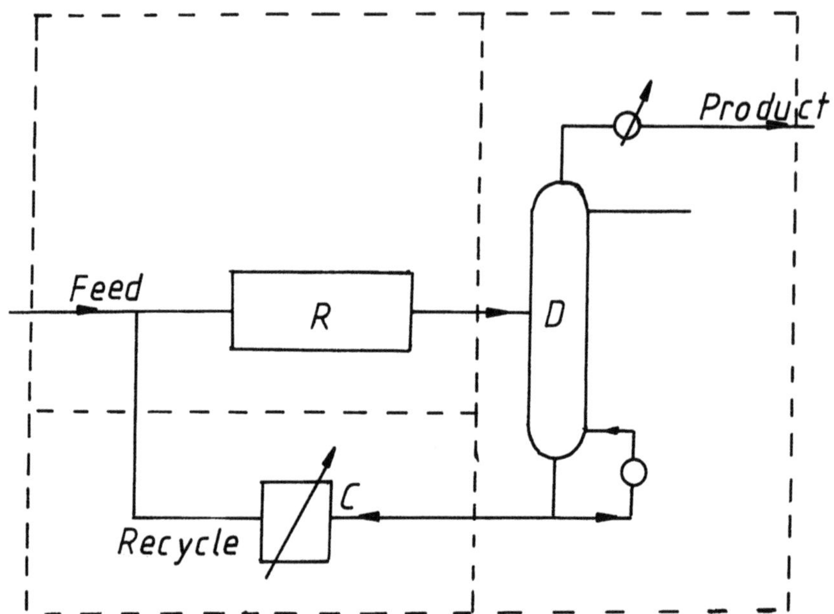


**FIGURE 6.5.** System divided into subsystems: $R$ = Reactor; $D$ = distillation column; $C$ = cooler.

nonlinear equations and perform rapid multivariable optimizations. In the remainder of the present chapter, we will concentrate on presenting relatively straightforward examples of optimization procedures which may readily be applied to uncomplicated electrochemical processes. Fundamental concepts and methods will be emphasized as a basis to the reader's understanding or optimization of more complex chemical plant and processes. The "first-order" optimizations discussed below are suited to the early conceptual stages of electrochemical process design, where comparison with nonelectrochemical processes is needed early, or when direction to areas of research or pilot plant operation is required.

Our examples start with the optimization of the performance of an electrolytic batch reactor with respect to its operating current density.

### 6.3.1. Example of the Optimization of Current Density in a Batch Reactor

Let us look at what kind of data we need to perform optimization. Table 6.9 is a sort of checklist for this task.

We need to know the major reactions occurring in the cell, and the desired production rate. We should have a good idea what type of cell will

TABLE 6.9. Data Employed in Example 6.3.1

*Reactions:*

| | | |
|---|---|---|
| "Tafel-type" reduction | $A + 2H^+ + 2e \rightarrow B$ | cathode reactions |
| Side reaction | $2H^+ + 2e \rightarrow H_2$ | cathode reactions |
| Anode reaction | $H_2O - 2e \rightarrow 2H^+ + 0.5\ O_2$ | |

*Production rate:*
11,880 kmol/annum of product $B$

*Reactor:*

| | |
|---|---|
| Plate and frame | 1 $m^2$ electrode area/module |
| Electrolyte/membrane resistance | 0.002 ohm/module |
| Batch time | 64 h |
| Downtime | 8 h |
| Availability | 360 days/annum |

*Process data:*

| | |
|---|---|
| Initial reactant concentration ($c_A^0$) | 1.0 kmol/$m^3$ |
| Required conversion ($X_A^f$) | 0.99 |
| Temperature | 313 K |

*Pumping:*
20 W per module per stream

*Pertinent reactor and reaction models:*

$$S = \frac{2\mathfrak{F}V}{t_B} \int_0^{X_A^f} \left[ \frac{1}{2\mathfrak{F}k_L} + \frac{1}{k_A \exp\left[-b_A E'\right]} \right] \frac{dX_A}{(1 - X_A)}$$

$$i = \frac{c_A^0(1 - X_A)}{\dfrac{1}{2\mathfrak{F}k_L} + \dfrac{1}{k_A \exp\left[-b_A E'\right]}} + k_H \exp\left[-b_H E'\right]$$

*Reaction data (313 K):*

$b_A = b_H = 19\ V^{-1}$
$k_A = 19.3$ Am/kmol
$k_H = 0.386\ A/m^2$
$k_L = 3.0 \times 10^{-5}$ m/s
$A + H_2O \rightarrow B + 0.5\ O_2 \quad \Delta H_R = 2.0 \times 10^5$ kJ/kmol A
$H_2O \rightarrow H_2 + 0.5\ O_2 \quad \Delta H_R = 3.0 \times 10^5$ kJ/mol $H_2$

*Cooling:*

| | |
|---|---|
| Cooling water supply temperature | 288 K |
| Maximum allowable temperature rise of c.w. | 10 K |
| Cooler overall heat transfer coefficient | 1.7 kW/$m^2$ K |

*Costs:*
(Capital costs for the reactor, power supply, and cooler(s), process costs for cooling water and electricity)

| | |
|---|---|
| Nominal plant life | 10 years |
| Maximum payback time ($Y_m$) | 3.6 years |
| Minimum required DCFRR | 25% |

be used and its characteristics (see Section 5.2), and the proposed run-time and downtime; these are subject to optimization themselves as an extension of this exercise.

We need basic process data on the composition of the feed, the final conversion, and the temperature of operation, important in sizing and costing the coolers.

Power consumption by the pumps must be known, and can usually be conveniently expressed as power consumption per module.

We need a pertinent reactor model (see Chapter 4), including a reaction model (see Section 3.2) for the chemistry at the working electrode. Strictly speaking, there should be a model for the anode reaction so that the potential drop at the anode can be calculated for each current density. Allowance for the drop has been made by simply doubling the cathode potential. This will suffice for an illustrative example but is not recommended in practice. As we have seen in Section 3.2, to use the reaction model we need values for the kinetic constants and mass transfer coefficient, and for the energy balance we need the enthalpies of reaction for the major reactions in the cell.

We need data to help us size the coolers, and finally we must have costing data for the capital equipment, plus information on company policy on return on investment.

Next we draw up the flow sheet (Fig. 6.6), which is not very taxing for a batch reactor. If the modules are divided by a membrane two circuits, catholyte and anolyte, are required. Only one cooler is included, on the

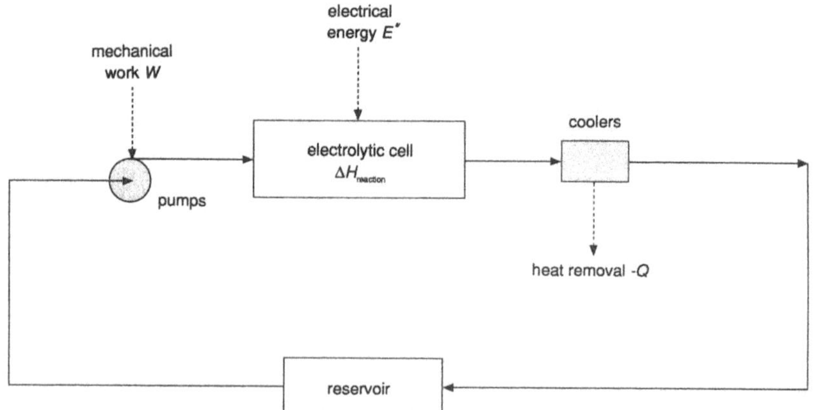


**FIGURE 6.6.** Flow sheet for batch reactor.

TABLE 6.10. Variable Costs

| | |
|---|---|
| *Capital:* | |
| Reactor | Installed £8000 $S^{0.9}$<br>$S$ = cathode area, $m^2$ |
| Power supply | Installed £200 $P^{0.8}$<br>$P$ = rating, kW |
| Cooler | Installed £600 $A^{0.8}$<br>$A$ = heat transfer area, $m^2$ |
| Pumps | Installed £2000 $R^{0.6}$<br>Plastic-lined centrifugal<br>$R$ = rating, kW |
| *Running:* | |
| Power to reactor and pumps | £0.03 per kWh |
| Cooling water | £0.02 per $m^3$ |

assumption that the reactor itself will transfer heat from the anolyte to the cooler catholyte stream.

We must identify costs that vary as we change the optimizing variable (current density). These costs are listed in Table 6.10 (see also Section 6.5.1).

From Table 6.10 we see that the variable capital items are the reactor, the power supply, the cooler, and the pumps, but not the storage tanks. The running costs affected by $i$, the current density, are the cost of the power consumed by the reactor and pumps, and of the cooling water supplied to the cooler.

The optimization procedure is plain. We select a value of $i$, indicated by the first box in Fig. 6.7, and calculate the corresponding value of area $S$. $\bar{E}$ and $E_{max}$ are the average and maximum values of the combined potential drop at the anode and cathode and are indicated in the second box. The journey from one box to the other is shown in Fig. 6.7 as a line, but is in reality an entire program (Fig. 6.8).

We insert some fixed process and physical data, after which we select a current density and initialize the variables which will change as the calculation proceeds. We go through an integration loop (current density → reactor model → $\Delta S$) repeatedly until the required number of loops has been completed. The increments of $S$ are accumulated, and the final value is printed out or fed directly into the main chart (Fig. 6.7).

While calculating $S$ we can pick up values for $\bar{E}$ and $E_{max}$, the average being used in the calculation of the energy balance and the maximum in the sizing of the power supply. Having determined $S$, $\bar{E}$ and $E_{max}$, we determine

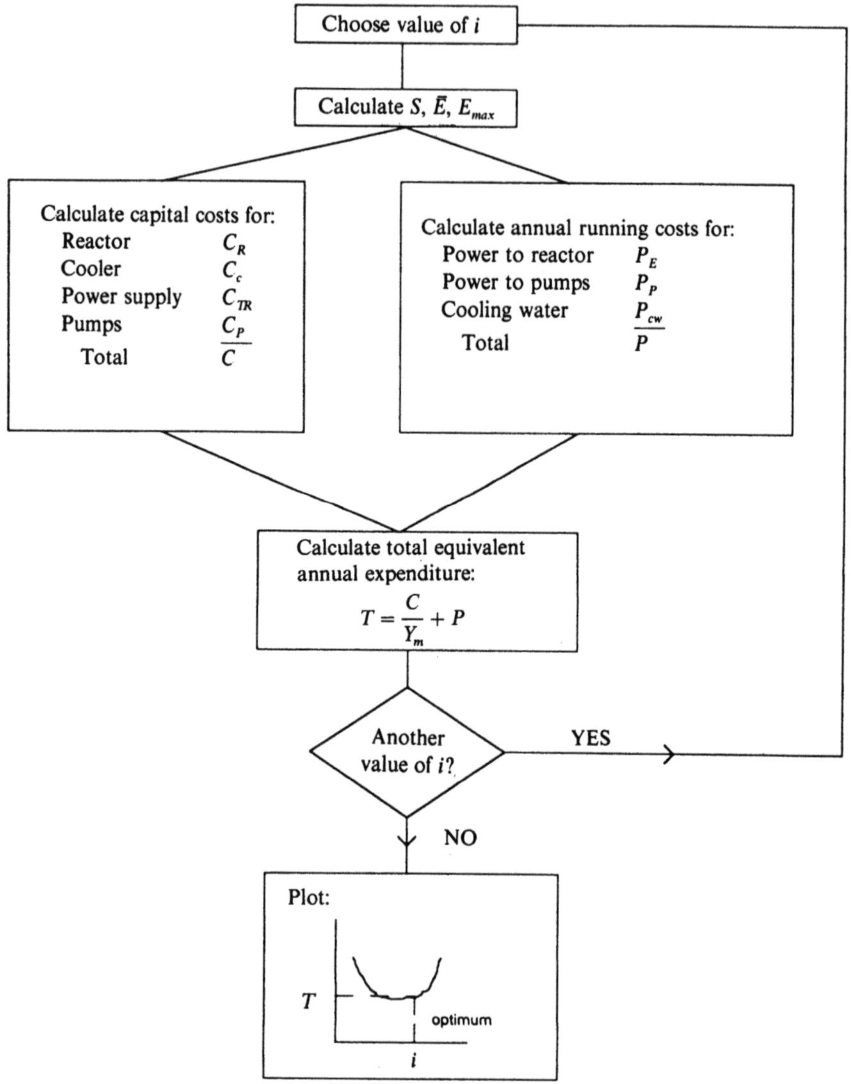


FIGURE 6.7. Computer flow diagram for optimization of current density in a batch reactor.

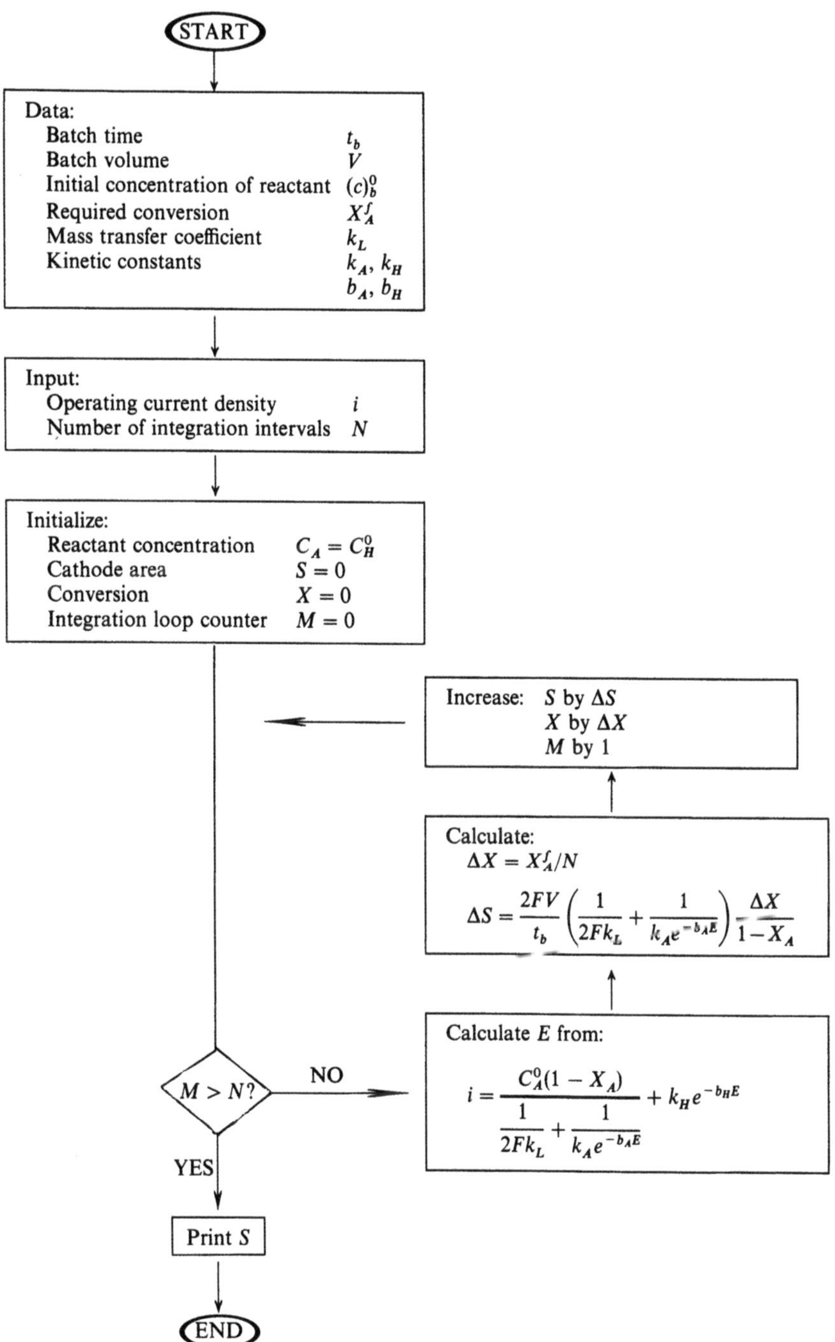


**FIGURE 6.8.** Computer program for optimization of current density in a batch rector.

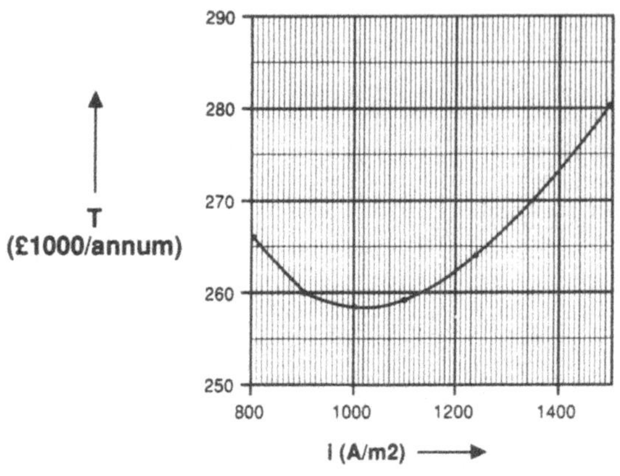


Optimum: Current density 1000 A/m$^2$

Cathode Area 124m$^2$

**FIGURE 6.9.** Variation of annual expenditure with current density.

the capital costs and running costs, which are summed and combined into a total annual expenditure, $T$ using $Y_m$, the maximum payback period. The procedure is repeated as a loop until values of $T$ can be plotted against $i$ to give a minimum, which corresponds to the optimum value of $i$ and the associated costs $T$.

We obtain the curve in Fig. 6.9, which as it happens gives an optimum value of $i$ of 1000 A/m$^2$. The corresponding electrode area is 124 m$^2$.

In Example 6.3.1 the optimum current density was determined while maintaining fixed conditions in the feed and product stream. These conditions might be seen as "boundary conditions" imposed on the reactor. They are also imposed on unit operations just upstream and downstream of the reactor. They cannot be changed without altering the duty imposed on plant external to the reactor. For instance, if the conversion in the reactor is changed the duty and associated costs of downstream separation units are affected. This "knock-on" effect is called interaction; it is important when optimizing an interactive variable to take into consideration all items of plant which are affected. This will be illustrated in our next example, but first we will consider interaction in more detail.

### 6.3.2. Interactions Between an Electrochemical Reactor and Associated Unit Processes

If interactions are strong they will have pronounced effects on process control and optimization, so it is important that interactive behavior be recognized in process design and development. Let us return to Fig. 6.5. Clearly, changing the flow rate of feed and recycle through the reactor will alter conditions in the distillation unit: the two interact. Using this simple flow sheet we will use a numerical example to show that the optimum operating conditions for a reactor cannot be determined in isolation from its associated unit operation. The example will also illustrate the building up of a process model. First, though, let us generalize some concepts we met in Section 6.2.1.

For process optimization, costing models must be used, on the basis of minimizing the cost of product. An "optimization" in terms of energy conservation alone, for example, would not be appropriate. If a plant comprises $N$ items of equipment to which individual capital and production costs $C$ and $P$ can be assigned, such a model will produce a number of equations:

$$C_1 = f_1(a_1, a_2, \ldots, a_i) \tag{6.6}$$

$$C_2 = f_2(a_1, a_2, \ldots, a_i) \tag{6.7}$$

$$\vdots$$

$$C_N = f_N(a_1, a_2, \ldots, a_i) \tag{6.8}$$

and

$$P_1 = g_1(a_1, a_2, \ldots, a_i) \tag{6.9}$$

$$P_2 = g_2(a_1, a_2, \ldots, a_i) \tag{6.10}$$

$$\vdots$$

$$P_N = g_N(a_1, a_2, \ldots, a_i) \tag{6.11}$$

where $a_1, a_2, \ldots, a_i$ are process variables.

Capital and production costs associated with the plant are:

$$C = \sum_{n=1}^{N} C_n \tag{6.12}$$

and

$$P = \sum_{n=1}^{N} P_n \tag{6.13}$$

The optimum value of the process variable $a_j$ is:

$$\frac{\partial C}{\partial a_j} = -Y_m \frac{\partial P}{\partial a_j} \tag{6.14}$$

where $Y_m$ is the allowable payback time (Section 6.2). This criterion of optimization can be rewritten as:

$$\frac{\partial C_1}{\partial a_j} + \frac{\partial C_2}{\partial a_j} + \cdots + \frac{\partial C_n}{\partial a_j} + \cdots + \frac{\partial C_N}{\partial a_j} = -Y_m \left[ \frac{\partial P_1}{\partial a_j} + \frac{\partial P_2}{\partial a_j} + \cdots + \frac{\partial P_n}{\partial a_j} + \cdots + \frac{\partial P_N}{\partial a_j} \right] \tag{6.15}$$

If the process variable $a_j$ affects the performance of only one item of equipment, item $n$, say, then Eq. (6.15) reduces to:

$$\frac{\partial C_n}{\partial a_j} = -Y_m \frac{\partial P_n}{\partial a_j} \tag{6.16}$$

from which it follows that the optimum value of $a_j$ can be found from the cost of plant item $n$ taken alone. Such a process variable $a_j$ is called noninteractive. But if $a_j$ affects the cost of more than one item of equipment, the cost of these items must be taken together when seeking an optimum.

### 6.3.3. Example of a Process Model to Demonstrate Interaction

We shall use the flow sheet of Fig. 6.5 to develop a first-order costing model.[6] Table 6.11 lists the numerical data on which the model is based. Again, the costing data quoted in this book do not necessarily represent up-to-date values that can be used by the reader for his own calculations. (See, however, Section 6.5.1 for estimated 1990 figures.)

#### 6.3.3.1. The Electrochemical Reactions

The model is based on the cathodic reduction of a material $A$ to product $B$,

TABLE 6.11. Data Employed in the Costing Model

| | |
|---|---|
| Rate of production of $B$ (i.e., $B_5$) | 1.00 kmol/h |
| Cost of feedstock, $A$ | £200/kmol |
| Cost of cooling water | £0.02/m$^3$ |
| Latent heat of all components | $4 \times 10^4$/mol |
| Allowable rise in the temperature of cooling water | 10 K |
| Cost of steam | £0.004/kg |
| Cost of installed cell | £10000 $S^{0.9}$ ($S$ = electrode area in m$^2$) |
| Cost of installed power supply | £190 $P^{0.8}$ ($P$ = power rating in kW) |
| Superficial vapor velocity in the distillation column | 1.0 m/s |
| Specific volume of vapor in the distillation column | 0.04 m$^3$/kmol |
| Relative volatilities with respect to water, $\alpha_{AH}$ | 1.5 |
| $\alpha_{BH}$ | 2.25 |
| Cost of installed distillation column (including reboiler and condenser) | £1850 $Nd$ ($d$ = column diameter in $m$, $N$ = number of plates) |
| Payback time, $Y_m$ | 3 years |

in aqueous solution:

$$A + 2H^+ + 2e \rightarrow B \tag{6.17}$$

Competing with Eq. (6.17) will be hydrogen evolution:

$$2H^+ + 2e \rightarrow H_2 \tag{6.18}$$

At the anode the only product is oxygen:

$$4OH^- \rightarrow O_2 + 2H_2O + 4e \tag{6.19}$$

Considerations along the lines of Section 5.2 lead to the choice of a filter press cell.

### 6.3.3.2. Mass Balance

Building the process model starts with a mass balance (Fig. 6.10). The distillation unit recovers and recycles unconverted feedstock and electrolyte

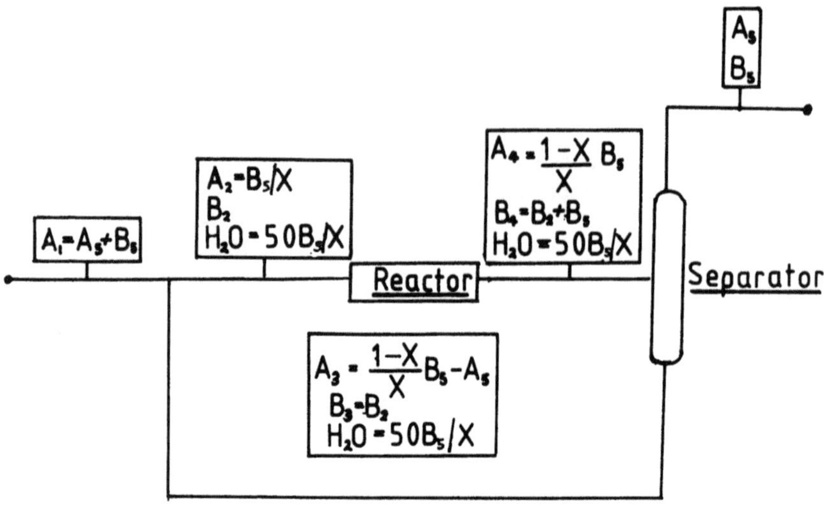


**FIGURE 6.10.** Mass balance (letters denote flow rates).

and also delivers product. Note the following:

1. For simplicity's sake the concentration of $A$ at the inlet to the reactor has been fixed at 1 kmol/m$^3$, giving a molar ratio of water : $A$ of about 50.
2. Following a convention usually employed in reactor design, the fraction of $A$ converted in the reactor is called $X$.
3. Consumption of water and the by-products hydrogen and oxygen has been omitted, partly for simplicity and partly because only items of major significance should be included in a first-order model.
4. The mass balance round the distillation column implies that $B$ is more volatile than $A$, both being more volatile than water.

The purpose of the mass balance in a process model is to express each stream in terms of the relevant process variables $B_2$, $X$, and the current density $i$. A fourth parameter $B_5$, is required but is not a process variable, since it is the rate of production of product and will be fixed as a process specification.

Having established the mass balance relationships the first stage of the process model is complete and can be shown in a box diagram (Fig. 6.11), which charts the influence of the process variables through each item of plant. The mass balance in Fig. 6.11 has been divided in three parts: the overall mass balance, the reactor mass balance, and the distillation mass

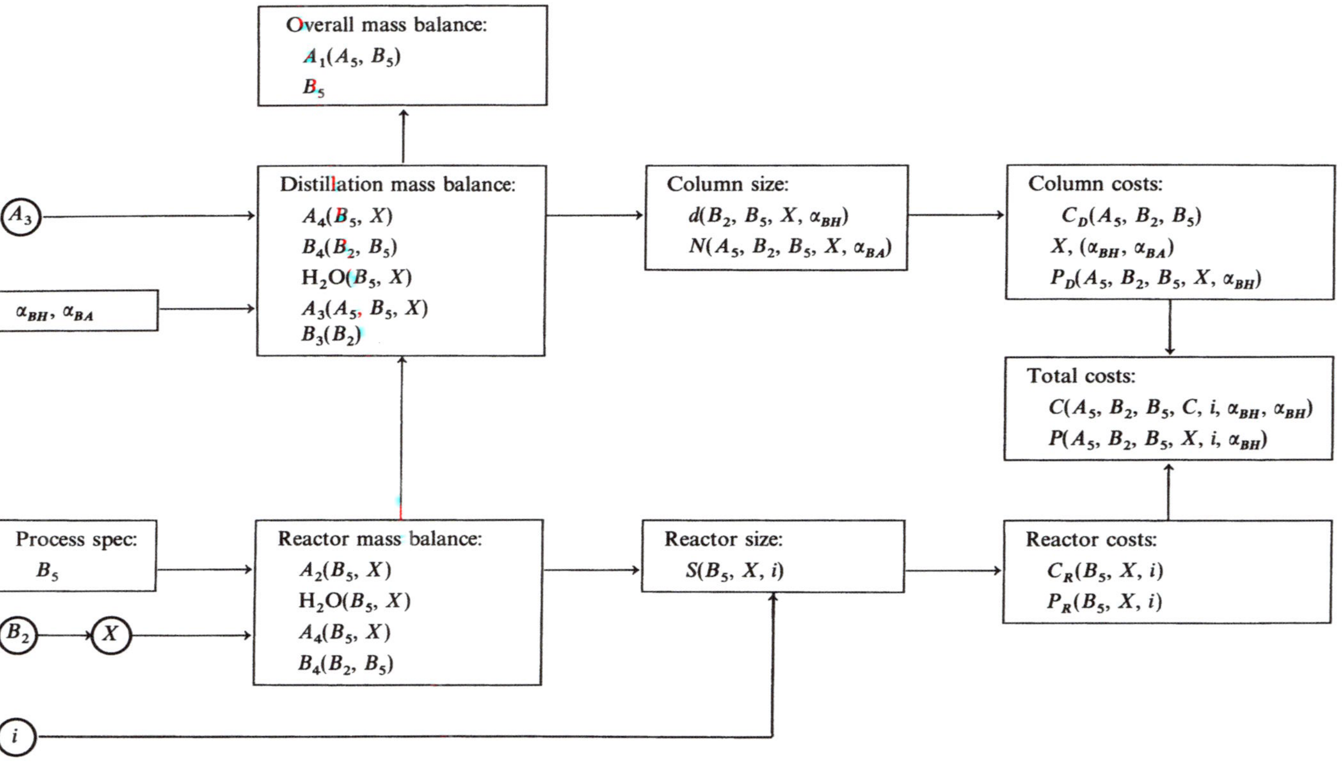


**FIGURE 6.11.** Relationship between process streams, process economics, and process variables (process variables are identified by circles).

balance; each box shows the functional dependence between the process streams and the process variables. The next step is to model each item of equipment so that capital and running costs can be estimated.

#### 6.3.3.3. The Reactor

(i) Reaction model. The partial current densities for the reduction of $A$ and hydrogen evolution are given by Eqs. (3.147) and (3.148):

$$i_A = \frac{c_A}{\dfrac{1}{2\mathfrak{F}k_L} + \dfrac{1}{k_A \exp(-b_A E')}}$$

$$i_{\mathrm{H}} = k_{\mathrm{H}} \exp(-b_{\mathrm{H}} E')$$

$i$, the total current density, is:

$$i = i_A + i_{\mathrm{H}} \tag{6.20}$$

Inserting typical numerical values for the kinetic constants (which in an actual process development would be determined by the experimental methods of Section 3.2.2), Eqs. (6.20) and (3.147) become:

$$i = \frac{c_A}{\dfrac{1}{5000} + \dfrac{1}{10e^{-19E'}}} + 0.1e^{-19E'} \tag{6.21}$$

$$i_A = \frac{c_A}{\dfrac{1}{5000} + \dfrac{1}{10e^{-19E'}}} \tag{6.22}$$

In an electrochemical reactor $c_A$ is a variable that can be expressed in terms of conversion $X$:

$$c_A = (1 - X)c_A^0 \tag{6.23}$$

where $c_A^0$ is the initial concentration of reactant. Equations (6.21) and (6.22) can be written as:

$$i = \frac{(1 - X)c_A^0}{\dfrac{1}{5000} + \dfrac{1}{10e^{-19E'}}} + 0.1e^{-19E'} \tag{6.24}$$

$$i_A = \frac{(1 - X)c_A^0}{\dfrac{1}{5000} + \dfrac{1}{10e^{-19E'}}} \tag{6.25}$$

(ii) Reactor model. As in Chapter 4, the reaction model can be used to develop a model for the reactor. The first aim is to specify the reactor in terms of the electrode area it must provide to produce at the specified rate. The plate and frame cell will be regarded as an ideal plug-flow reactor. Considering the charge balance in Fig. 6.12 where the rectangle is intended to represent the entire cathode surface and not just one frame, you can write:

$$2\mathfrak{F}Q\,dc_A = -i_A\,dS \tag{6.26}$$

where $Q$ is the electrolyte flow rate (in $m^3/s$) and $S$ is the electrode area.

Substituting for $i_A$ and solving for $S$:

$$S = -2\mathfrak{F}Q \int_{c_A^0}^{c_A} \left[\frac{1}{5000} + \frac{1}{10e^{-19E'}}\right] \frac{dc_A}{c_A} \tag{6.27}$$

or in terms of conversion $X$:

$$S = -2\mathfrak{F}Q \int_0^X \left[\frac{1}{5000} + \frac{1}{10e^{-19E'}}\right] \frac{dX}{1 - X} \tag{6.28}$$

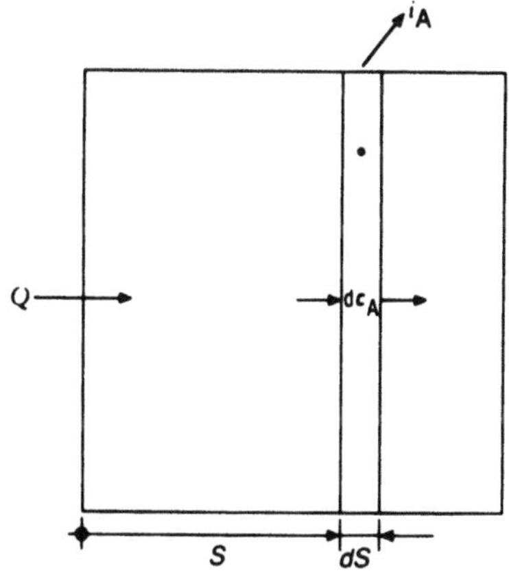


**FIGURE 6.12.** Charge balance between electrolyte and electrode.

Solving these equations is not straightforward. Plate and frame cells are often operated in bipolar mode at constant cell current, resulting in a nominally constant current density $i$. It follows from Eqs. (6.21) and (6.24) that $E'$ is a function of $c_A$ or $X$ and that Eqs. (6.21) and (6.27) or Eqs. (6.24) and (6.28) must be solved in parallel. Either pair of equations can be regarded as constituting the reactor model, but their solution, in general, requires use of a computer. From Eqs. (6.24) and (6.28) it is evident that $S$ is functionally dependent on $Q$, $X$, $c_A^0$, and $i$. Since $Q$ is a function of $B_5$ and $X$, and $c_A^0$ is fixed, it follows that:

$$S = f(B_5, X, i) \tag{6.29}$$

as recorded in Fig. 6.11. The situation is more complex if current distribution in the cell is significantly uneven.

(iii) Reactor costs. The reactor model is now used to develop a model for reactor costs. The costing procedure is simple enough for a first-order model. Capital investment is the sum of the installed costs of the reactor and its associated transformer/rectifier. The major production cost is taken as the consumption of electricity. The justification for this simplification can be found in the functional dependence of $C_R$ and $P_R$ developed in Table 6.12 and recorded in the box diagram of Fig. 6.11.

Remembering that $B_5$ and $c_A^0$ are defined as fixed parameters, it is evident from Table 6.12 that $C_R$ and $P_R$ are dependent only on the current density and conversion. The effect of conversion, for a current density of 2000 A/m$^2$, is shown in Fig. 6.13. As $X$ tends to zero and unity, $C_R$ and $P_R$ tend to infinity and exhibit fairly flat minima at intermediate values. $C_R$ is plotted against $P_R$ in the curve in Fig. 6.14. When costing data are presented in this form it is tempting to draw a "payback" tangent to the curve to determine an optimum, but this would be premature since the variable $X$ might influence the costs of the distillation unit, a point yet to be considered. Let us note, however, that "optimization" of the reactor in isolation from the other subsystems would lead us to a "payback" period of 3 years, indicating an optimum capital investment of £210,000 with production costs of £36,000 per annum. In Fig. 6.13, this combination of figures corresponds to an apparent "optimum" conversion of 0.1–0.3. Intuitively, this value looks rather low.

#### 6.3.3.4. The Distillation Column

(i) Column model. To size a distillation unit a model must be able to calculate the reflux ratio and the required number of plates, because these largely determine the capital investment and production costs.

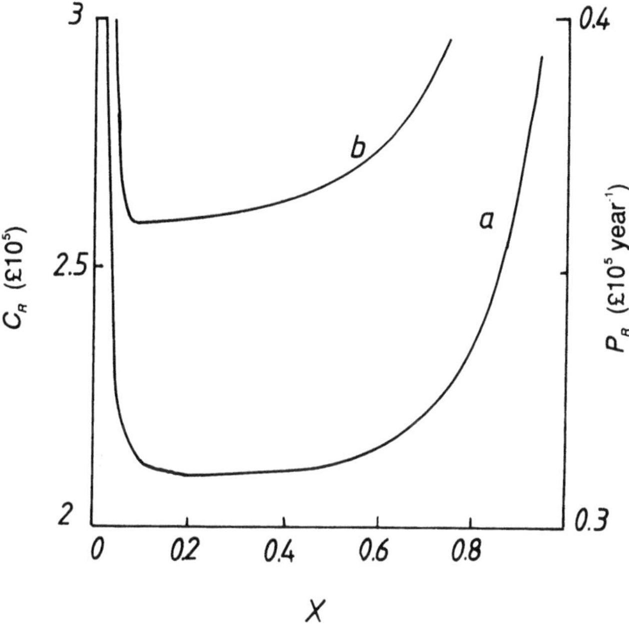


**FIGURE 6.13.** Effect of conversion on (a) the capital cost, $C_R$ and (b) the production costs $P_R$ of the electrochemical reactor. (current density = 2000 A/m$^2$).

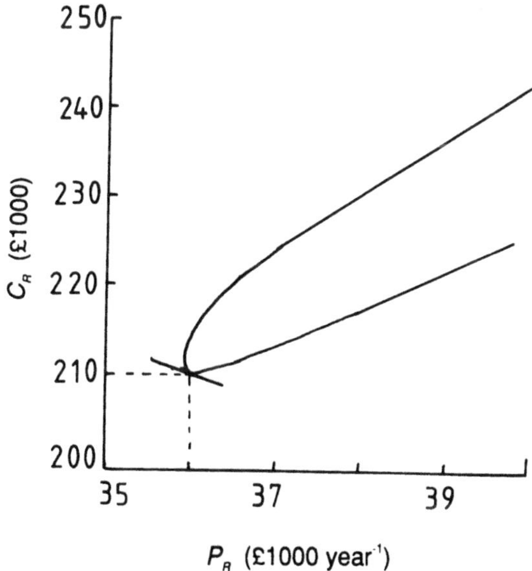


**FIGURE 6.14.** Variation of capital costs $C_R$ with production costs $P_R$ of the electrochemical reactor. (current density = 2000 A/m$^2$).

TABLE 6.12. Reactor Costs

| Item | Symbol | Value | Functional dependence |
|---|---|---|---|
| Availability | $T$ | 8500 h | |
| Electrolyte resistance | $R$ | $1.5 \times 10^{-2}\ \Omega/m^2$ of cross section | |
| Mean concentration of A | $c_A$ | $c_A^0[1-(X/2)]$ | $c_A(c_A^0, X)$ |
| Average cathode potential | $E'_C$ | Fixed by $i$ and the polarization curve for $c_A$ | $E'(X, i, c_A^0)$ |
| Anode potential | $E'_A$ | $(1/19)\ \ln\ [i/8]$ | $E'_A(i)$ |
| Voltage drop anode to cathode | $V^C$ | $R_i + E'_A - E'_C$ | $V^C(X, i, c_A^0)$ |
| Power consumed by cell | $w_c$ | $V^C Si$ | $w_c(X, i, B_5, c_A^0)$ |
| Power consumed in transmission | $w_t$ | $0.06 w_c$ | $w_t(X, i, B_5, c_A^0)$ |
| Total power consumed | $w$ | $w_c + w_t$ | $w(X, i, B_5, c_A^0)$ |
| Rating of transformer/rectifier | $P$ | $w_c + w_t$ | $w(X, i, B_5, c_A^0)$ |
| Capital investment in transformer/rectifier | $C_1$ | $190P^{0.8}$ | $C_1(X, i, B_5, c_A^0)$ |
| Capital investment in cell | $C_2$ | $10000S^{0.9}$ | $C_2(X, i, B_5)$ |
| Total capital investment in reactor | $C_R$ | $C_1 + C_2$ | $C_R(X, i, B_5, c_A^0)$ |
| Cost of power | | 0.02 £/kWh | |
| Production cost for the reactor | $P_R$ | $0.02Tw$ | $P_R(X, i, B_5, c_A^0)$ |

The conditions around a simple distillation column are shown in Fig. 6.15. Products $D$ and $W$ leave the top and bottom of the column respectively and feed $F$ is introduced at some intermediate point. A liquid stream $L$ is returned to the top of the column and a vapor stream $V$ to the bottom. The stream $L$ is the reflux and $L/D$ is the reflux ratio $R$. The number of plates or stages in the column is $N$. Heating and cooling services are provided by a reboiler and condenser.

If $F$ and $D$ are fixed, the only variable open to choice is $R$: its optimum value is determined, as usual, by comparing capital investment with production costs. Since $R$ is a noninteractive variable, the comparison is confined to costs associated with the column.

Capital investment is the installed costs of the column, its internals, the condenser, and the reboiler. Production costs are the cost of cooling water, steam, and losses of material due to imperfect separation. $R$ can vary from a minimum $R_{min}$ to infinity. Capital investment is infinite and production

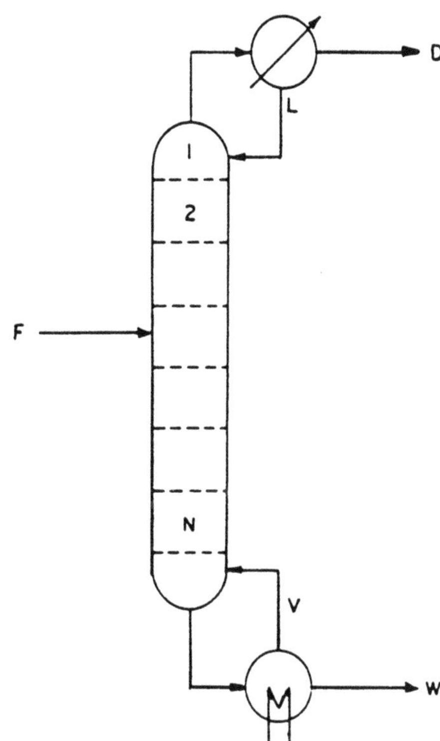


**FIGURE 6.15.** Process conditions around a simple distillation column. $F$ = feed; $D$ = top product; $W$ = bottom prduct; $L$ = liquid returned to top of column; $V$ = vapor returned to bottom of column; $N$ = number of plates.

costs are finite at $R_{min}$, but both tend to go to infinity as $R$ goes to infinity. Plotting capital investment against production costs allows an optimum to be located by a tangent of slope equal to minus the payback period. The model is developed using the rule of thumb[7] that the optimum reflux ratio $R_{opt}$ is roughly twice $R_{min}$. With this approximation the problem reduces to the determination of $R_{min}$, for which the rigorous Fenske–Underwood method for multicomponent systems can be employed.

For conditions where the relative volatilities remain constant, $R_{min}$ can be calculated[8] from the following two equations by determining $\Theta$ from Eq. (6.30) and substituting this value into Eq. (6.31):

$$\frac{\alpha_A x_{FA}}{\alpha_A - \Theta} + \frac{\alpha_B x_{FB}}{\alpha_B - \Theta} + \frac{\alpha_C x_{FC}}{\alpha_C - \Theta} + \cdots = 1 - q \quad (6.30)$$

and

$$\frac{\alpha_A x_{DA}}{\alpha_A - \Theta} + \frac{\alpha_B x_{DB}}{\alpha_B - \Theta} + \frac{\alpha_C x_{DC}}{\alpha_C - \Theta} + \cdots = R_{min} + 1 \quad (6.31)$$

where $x_{FA}$, $x_{FB}$, $x_{FC}$, $x_{DA}$, $x_{DB}$, $x_{DC}$ are mol fractions, suffixes $A$, $B$, and $C$ refer to components $A$, $B$, and $C$ and suffixes $F$ and $D$ to the feed and top product; $q$ is the ratio of the heat required to vaporize 1 mol of the feed to its molar latent heat; $\alpha_A$, $\alpha_B$, $\alpha_C$ are the volatilities with respect to the least volatile component; and $\Theta$ is the root of Eq. (6.30) that lies between the values of $\alpha_A$ and $\alpha_B$.

Taking as a first approximation that $q = 1$ (the feed is at its boiling point), Eqs. (6.30) and (6.31) become:

$$\frac{\alpha_{AH} x_{FA}}{\alpha_{AH} - \Theta} + \frac{\alpha_{BH} x_{FB}}{\alpha_{BH} - \Theta} + \frac{x_{FH}}{1 - \Theta} = 0 \tag{6.32}$$

and

$$R_{min} + 1 = \frac{\alpha_{AH} x_{DA}}{\alpha_{AH} - \Theta} + \frac{\alpha_{BH} x_{DB}}{\alpha_{BH} - \Theta} + \frac{\alpha_{HH} x_{DH}}{\alpha_{HH} - \Theta} \tag{6.33}$$

From the mass balance in Fig. 6.10 we obtain values for the parameters in Eq. (6.32) and (6.33):

$$x_{FA} = \frac{(1 - X)B_5}{51B_5 + XB_2} \tag{6.34}$$

$$x_{FB} = \frac{(B_2 + B_5)X}{51B_5 + XB_2} \tag{6.35}$$

$$x_{FH} = \frac{50B_5}{51B_5 + XB_2} \tag{6.36}$$

$$x_{DA} = \frac{A_5}{A_5 + B_5} \approx 0 \tag{6.37}$$

$$x_{DB} = \frac{B_5}{A_5 + B_5} \approx 1 \tag{6.38}$$

$$x_{DH} = 0 \tag{6.39}$$

Substituting Eqs. (6.34) to (6.36) into Eq. (6.32):

$$\frac{\alpha_{AH}(1 - X)B_5}{(\alpha_{AH} - \Theta)D} + \frac{\alpha_{BH}(B_2 + B_5)X}{(\alpha_{BH} - \Theta)D} + \frac{50B_5}{(1 - \Theta)D} = 0 \tag{6.40}$$

where $D = 51B_5 + XB_2$.

Substituting Eqs. (6.37) to (6.39) into Eq. (6.33):

$$R_{min} + 1 = \frac{\alpha_{AH} A_5}{(\alpha_{AH} - \Theta)(A_5 + B_5)} + \frac{\alpha_{BH} B_5}{(\alpha_{BH} - \Theta)(A_5 + B_5)} \tag{6.41}$$

Equation (6.40) can be solved by computer for $\Theta$ and the value substituted in Eq. (6.41) to get a value for $R_{min}$.

It is good to check computer results when possible. This can be done by substituting the approximate values of Eqs. (6.37) and (6.38) into Eq. (6.33), which reduces to:

$$R_{min} + 1 = \frac{\alpha_{BH}}{\alpha_{BH} - \Theta} \tag{6.42}$$

$\Theta$ is again determined from Eq. (6.32), where since $\alpha_{AH} < \Theta < \alpha_{BH}$ the first term is negative. Remembering that $x_{FA} \ll x_{FH}$ we can ignore the first term in comparison with the last and Eq. (6.32) reduces to:

$$\frac{\alpha_{BH} x_{FB}}{\alpha_{BH} - \Theta} + \frac{x_{FH}}{1 - \Theta} = 0 \tag{6.43}$$

giving:

$$\Theta = \frac{\alpha_{BH}(x_{FB} + x_{FH})}{\alpha_{BH} x_{FB} + x_{FH}} \tag{6.44}$$

Substituting Eqs. (6.35) and (6.36) into Eq. (6.44):

$$\Theta = \frac{\alpha_{BH}(B_2 + B_5 + 50B_5/X)}{\alpha_{BH}(B_2 + B_5) + 50B_5/X} \tag{6.45}$$

Finally, using this value of $\Theta$ in Eq. (6.42) gives:

$$R_{min} + 1 = \frac{\alpha_{BH}(B_2 + B_5) + 50B_5/X}{(\alpha_{BH} - 1)(B_2 + B_5)} \tag{6.46}$$

Hence:

$$R_{opt} \approx 2R_{min} = f(B_2, B_5, X, \alpha_{BH}) \tag{6.47}$$

The functional dependence is recorded in the box diagram in Fig. 6.11.

Having evaluated $R_{opt}$, the corresponding number of plates $N$ can be determined. The calculation is truncated by employing another useful rule of thumb[7] that when $R = 2R_{min}$ the number of plates required in the column will be $N = 2N_{min}$, where $N_{min}$ is the number of plates required at total reflux. Determination of $N$ begins with Fenske's relationship:

$$N_{min} + 1 = \ln\left[\frac{x_{DB}}{x_{DA}}\frac{x_{WA}}{x_{WB}}\right] \ln\,[\alpha_{BA}] \tag{6.48}$$

where $x$ represents a mol fraction and subscripts refer to the presence of components $A$ and $B$ in the top and bottom products $D$ and $W$. Also, $\alpha_{BA}$ is the volatility of $B$ with respect to $A$.

Substituting from the mass balance in Eq. (6.48) gives:

$$N_{min} + 1 = \ln\left\{\left[\frac{B_5}{A_5}\right]\frac{[B_5(1-X)/X] - A_5}{B_2}\right\}\Big/\ln\,[\alpha_{BA}] \tag{6.49}$$

Anticipating that the content of $A$ in the top product will be low, i.e., $B_5(1 - X) \gg A_5$, results in:

$$N_{min} + 1 = \ln\left[\frac{(1-X)B_5^2}{XA_5B_2}\right]\Big/\ln\,[\alpha_{BA}] \tag{6.50}$$

giving:

$$N = 2N_{min} = f(A_5, B_2, X, \alpha_{BA}) \tag{6.51}$$

The functional relationship in Eq. (6.51) is again transferred to the box diagram of Fig. 6.11.

(ii) Column costs. The expressions obtained for $R_{opt}$ and $N$ are a model for sizing the column, and provide the basis for a costing model as developed in Table 6.13.

### 6.3.3.5. Interactive Behavior

The block diagram of Fig. 6.11, now finished, is a convenient means of summarizing a process model because it eliminates mathematical detail and quickly identifies the process variables and their interactive or noninteractive influence. The variables are $X$, $i$, $B_2$, and $A_5$, of which $X$ and $B_2$ affect column and reactor costs and are therefore interactive, whereas $i$ affects only the reactor costs and $A_5$ only the column costs; these are noninteractive.

TABLE 6.13. Cost of the Distillation Unit

| Item | Symbol | Value | Functional dependence |
|---|---|---|---|
| Reflux ratio | $R$ | | $f(B_2, B_5, X, \alpha_{BH})$ |
| Heat load on the condenser | $H_{con}$ | $\propto(R+1)(A_5+B_5)$ $\propto(R+1)B_5$ | $f(B_2, B_5, X, \alpha_{BH})$ |
| Heat load on the reboiler | $H_{reb}$ | $\propto(R+1)B_5$ | $f(B_2, B_5, X, \alpha_{BH})$ |
| Column diameter | $d$ | $d^2 \propto$ vapor flow rate $d^2 \propto (R+1)B_5$ | $f(B_2, B_5, X, \alpha_{BH})$ |
| Number of plates | $N$ | | $f(A_5, B_2, B_5, X, \alpha_{BA})$ |
| Cost of cooling water | $P_{cw}$ | $\propto H_{con}$ | $f(B_2, B_5, X, \alpha_{BH})$ |
| Cost of steam | $P_s$ | $\propto H_{reb}$ | $f(B_2, B_5, X, \alpha_{BH})$ |
| Cost of lost $A$ | $P_A$ | $\propto A_5$ | $f(A_5)$ |
| Total production costs | $P_D$ | $P_{cw}+P_s+P_A$ | $f(A_5, B_2, B_5, X, \alpha_{BH})$ |
| Cost of installed column | $C_3$ | $1850\,Nd$ | $f(A_5, B_2, B_5, X, \alpha_{BH}\alpha_{BA})$ |
| Total capital cost | $C_D$ | $C_3$ | $f(A_5, B_2, B_5, X, \alpha_{BH}\alpha_{BA})$ |

Complete optimization would embrace all the variables, but with the effect of conversion alone we can illustrate that optimization procedures which ignore interactive behavior can lead to wrong conclusions. Figure 6.16 shows the variation of capital and production costs with conversion $X$, again for a current density $i$ of 2000 A/m$^2$, a ratio $B_2/A_2$ of 0.1, and optimum values for $A_5$.

To get $A_5$, we begin with the capital cost of the distillation column $C_D$ (see Table 6.13):

$$C_D = 1850k_1N[(R+1)B_5]^{1/2} \tag{6.52}$$

$P_D$, the corresponding production cost, is:

$$P_D = (k_2+k_3)(R+1)B_5 + k_4A_5 \tag{6.53}$$

Differentiating:

$$\frac{dC_D}{dA_5} = 1850k_1[(R+1)B_5]^{1/2}\frac{dN}{dA_5} \tag{6.54}$$

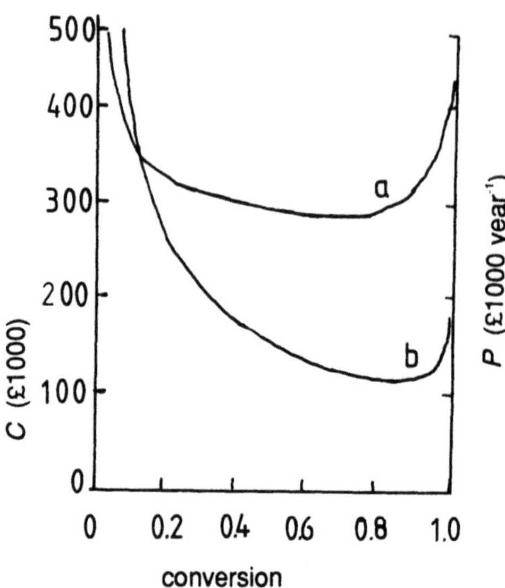


**FIGURE 6.16.** Effect of conversion on process (a) capital costs $C$ and (b) production costs $P$. The process costs are those of the reactor and the distillation column. ($i = 2000\ \text{A/m}^2$)

$$\frac{dP_D}{dA_5} = k_4 \tag{6.55}$$

$$\frac{dC_D}{dP_D} = \frac{1850k_1[(R+1)B_5]^{1/2}}{k_4}\frac{dN}{dA_5} = \frac{-2}{A_5 \ln[\alpha_{BA}]}\frac{1850k_1[(R+1)B_5]^{1/2}}{k_4} \tag{6.56}$$

Since the slope of the "payback tangent" $dC_D/dP_D$ is $-3$, we write:

$$3 = \frac{2}{A_5\ \ln\ [\alpha_{BA}]}\frac{1850k_1[(R+1)B_5]^{1/2}}{k_4} \tag{6.57}$$

and finally:

$$A_5 = \frac{2}{3\ \ln\ [\alpha_{BA}]}\frac{1850k_1[(R+1)B_5]^{1/2}}{k_4} \tag{6.58}$$

If capital cost is plotted against production cost (Fig. 6.17), a "payback" tangent to the curve, again of slope $-3$, identifies the optimum capital investment as £296,000 with annual production costs of £116,000. In Fig. 6.16 these figures point to an optimum conversion of 0.85. Taking the reactor in isolation an optimum conversion of roughly 0.3 was indicated, which corresponds to a capital investment of £310,000 and an annual production cost of £203,000. Thus by ignoring interactive behavior we

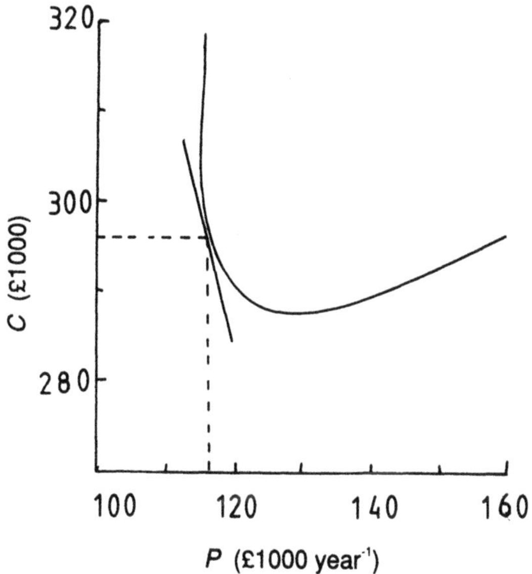


**FIGURE 6.17.** Variation of process capital costs $C$ with production costs $P$. ($i = 2000\ A/m^2$)

would have prescribed a plant incurring nearly twice the production costs of the true optimum.

This sort of design error would not occur now that process modeling in design organizations has become routine. This is not true in research and development, although its role is equally important, for the following reasons[9]:

1. Because it identifies the most probable commercial process conditions, a process model will help concentrate research effort in areas where data are most needed. Here, experimental emphasis should be placed on conversions in the region of 0.85. It will be recalled that conclusions drawn from a model based on the reactor alone rather than from a global process model would have placed this emphasis erroneously on a conversion of 0.3.
2. A process model will quantify the sensitivity of process economics to uncertainty in physical data and will therefore help set priorities in supporting research programs. Here, the model will show that process economics are particularly sensitive to the values assigned to the relative volatilities, which would therefore be given a high priority in the research program directed toward measurement of physical data.

### 6.3.4. Example of Optimization and a Choice between Alternatives

In Section 6.2 we considered a qualitative choice between alternatives. We end this chapter with a quantitative example of a choice between (1) raising the current density to cover the loss of a cell during maintenance and (2) installing a standby cell. The example starts with optimizing the current density for a filter press type cell as in Example 6.3.1, but this time the cell is operated continuously. The required conversion is 50%.

Data for optimization are given in Table 6.14.

TABLE 6.14. Data Employed in Example of Section 6.3.4

*Reactions:*

"Tafel-type" reduction $A + 2H^+ + 2e \rightarrow B$ } cathode reactions
Side reaction $2H^+ + 2e \rightarrow H_2$ }

Anode reaction $H_2O - 2e \rightarrow 2H^+ + 0.5O_2$

*Production rate:*

80,000 kmol/annum of product $B$

*Reactor:*

Continuous reactor operating with a conversion $X_p$ of 0.5, a recycle of 50%, and an availability 360 days per annum.

*Pertinent reactor and reaction models:*

$$S = (q + Q)\int_{[q/(Q+q)]X_A^f}^{X_A^f}\left[\frac{1}{k_L} + \frac{1}{k\exp[-bE']}\right]\frac{dX_A}{(1 - X_A)}$$

$$i = \frac{c_A^0(1 - X_A)}{\dfrac{1}{n\mathfrak{F}k_L} + \dfrac{1}{n\mathfrak{F}k\exp[-bE']}} + 2\mathfrak{F}k_H\exp[-b_H E']$$

*Reaction data (50°C)*

$b = b_H = 19\ V^{-1}$
$2\mathfrak{F}k'_A = 10\ A/m^2$ per unit concentration
$2\mathfrak{F}k'_H = 0.4\ A/m^2$
$2\mathfrak{F}k_L = 5000\ A/m^2$
$A + H_2O \rightarrow B + 0.5O_2$, $\Delta H_R = 2.0 \times 10^5$ kJ/kmol $A$
$H_2O \rightarrow H_2 + 0.5O_2$, $\Delta H_R = 3.0 \times 10^5$ kJ/kmol $H_2$

*Reactor data:*

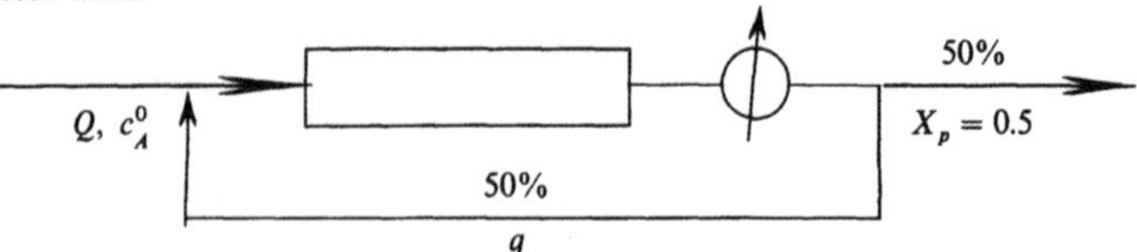


Reactant concentration in feed, $c_A^0 = 1$ kmol/m$^3$
Feed flow rate consistent with production rate, $Q = 5.144 \times 10^{-3}$ m$^3$/s

TABLE 6.14. (*Continued*)

*Recycle*, $q = 5.144 \times 10^{-3}$ $m^3/s$

Reactor type: "filter press" type with 1 $m^2$ of active cathode and 1 $m^2$ of active anode surface per module

"Ohmic resistance": $2 \times 10^{-3}$ ohm per module

Anode electrode potential assumed equal to the electrode potential $E'$ at the cathode (calculations later showed that $E'$ could be adequately represented by $E' = 0.0675i + 0.207$ V with $i$ in $kA/m^2$.)

| | |
|---|---|
| *Heat exchanger data:* | |
| Overall heat transfer coefficient | $h = 1.7$ $kW/m^2K$ |
| Cooling water supply temperature | 25°C |
| Allowable rise in cooling water temperature | 10°C |
| Mean $\Delta T$ in the heat exchanger(s) | 30°C |
| *Costing data*[a]: | |
| Installed reactor | £$12{,}000S^{0.9}$<br>$S$ = cathode area in $m^2$ |
| Installed power supply | £$200P^{0.8}$<br>$P$ = rating in kW |
| Installed heat exchanger(s) | £$500A^{0.8}$<br>$A$ = heat transfer area in $m^2$ |
| Cooling water | £0.02 per $m^3$ |
| Electrical power | £0.03 per kWh |
| Nominal plant life | 10 years |
| Required DCFRR, $i$ | 25% |
| Payback time $Y_m$: | $\dfrac{(1+i)^n - 1}{i(1+i)^n} = 3.57$ years |

[a] Possible replacement costs of membranes and electrodes are not considered. In practice, these might be important.

Capital investment and production costs, using the techniques of Example 6.3.1 for a range of current densities, are plotted in Fig. 6.18. The optimum investment is identified by a tangent slope $-3.57$, which is the maximum payback time corresponding to the specified nominal plant life of 10 years with a minimum rate of return of 0.25. The optimum magnitude of capital investment is £2,550,000 with annual production costs of £580,000.

Figure 6.19 shows capital investment $C$ together with cathode area $S$, plotted against current density $i$. Optimum investment of £2,550,000 corresponds to an optimum current density of 1550 $A/m^2$ with a cathode area of 372 $m^2$.

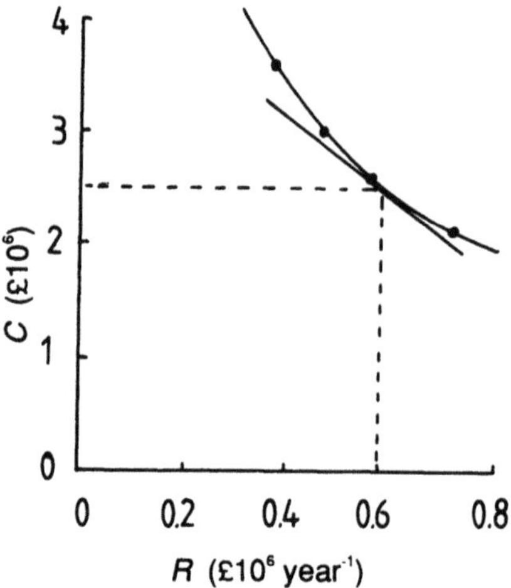


**FIGURE 6.18.** Variation of process capital costs $C$ with production costs $P$.

The preferred current density may not be feasible in practice because the cells are made up of discrete modules. Here, we intended to make up cells from modules offering an active cathode area of $1\,m^2$. It might be decided to provide the necessary electrode area using 20 cells, each containing 19 modules. This would provide a cathode area of $380\,m^2$ compared to the optimum $372\,m^2$, with a current density of $1521\,A/m^2$ instead of $1550\,A/m^2$. The capital and production costs would change to £2,728,600 and £571,400.

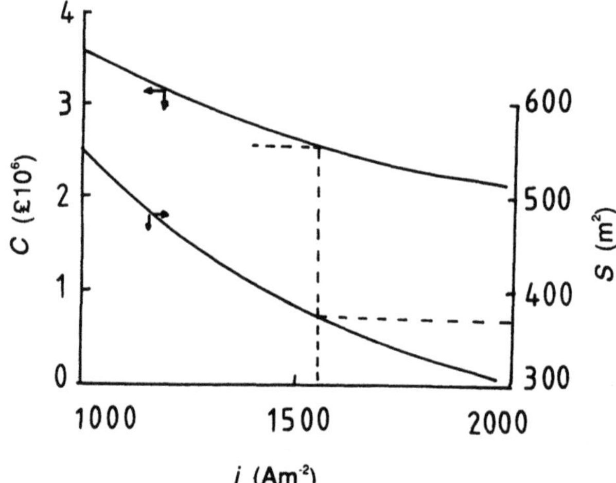


**FIGURE 6.19.** Variation of capital investment $C$ and cathode surface area $S$, with current density. (optimum: current density $1550\,A/m^2$, cathode area $372\,m^2$).

A problem of alternatives: Suppose that one of the cells will be out of commission due to maintenance for 10% of running time. Should we install a standby cell or raise the current density for the remaining 19 cells? The choice is determined by relative cost. To raise the current density of operation, reserve capacity may be required in the power supply and the cooling unit. The cost of the alternatives is examined in Table 6.15: additional investment for a standby unit is not justified in view of the high payback time of alternative 1.

## 6.4. CONCLUDING REMARKS

The reader has learned the basic principles underlying the optimization of a chemical plant design or the justification of an industrial strategy. Now the reader can confidently use software optimization packages.

The reader can see that a number of "optimizations" met in the literature are unsound and to be avoided.

Finally, we hope the reader will face the interesting challenges posed by the constantly developing field of electrochemical process engineering with confidence and will enjoy using this book as much as we have enjoyed writing it.

## 6.5. AFTERWORD

### 6.5.1. Costs Specific to Electrolytic Processes

Although capital costs of most items in an electrolytic plant are fairly standard, e.g., for tanks, pumps, and separation devices, there are two exceptions: the electrolytic reactor itself and the DC power source. Supply of electrical power to the reactor to bring about chemical reaction is also specific to electrolytic processes, and introduces an element into the energy balance which is not otherwise encountered.

#### 6.5.1.1. Capital Costs

(i) The reactor. The installed cost of a plate-and-frame cell $C_2$, still the most versatile form of reactor, can be scaled up roughly according to:

$$C_2 = a'N^{0.9} = aS^{0.9} \tag{6.59}$$

$N$ being the number of modules and $S$ the working electrode surface area in $m^2$. The factors $a'$ and $a$ depend on electrode materials and the materials

TABLE 6.15. Choice between Standby Cell and Increased Operating C.D.

*Alternative 1: standby cell:*

| | |
|---|---|
| Operating electrode area | 380 $m^2$ |
| Installed electrode area | 399 $m^2$ (21 cells) |
| Operating current density | 1521 $A/m^2$ |

| Capital costs (£) | | Production costs (£) | |
|---|---|---|---|
| Cells | 2,630,600 | Electrical power | 548,500 |
| Power supply | 91,500 | Cooling water | 22,900 |
| Cooler(s) | 6,500 | | |
| Total | 2,728,600 | Total | 571,400 |

*Alternative 2: increased operating current density:*

| | |
|---|---|
| Installed electrode area | 380 $m^2$ (20 cells) |
| Capital cost of cells | £2,517,600 |

When operating with 19 cells:

| | |
|---|---|
| Operating current density | 1615 $A/m^2$ |
| Capital cost of power supply | £96,100 |
| Capital cost of cooler(s) | £7,000 |
| Cost of electrical power | £583,200 |
| Cost of cooling water | £24,900 |

When operating with 20 cells the production costs are as in alternative 1.

Weighted production costs over one year's operation:

| | |
|---|---|
| Electrical power $(0.9 \times 548{,}500) + (0.1 \times 583{,}200)$ | £552,000/annum |
| Cooling water $(0.9 \times 22{,}900) + (0.1 \times 24{,}900)$ | £23,100/annum |

*Comparison between alternatives 1 and 2:*

| | Capital (£) | Production (£/annum) |
|---|---|---|
| Alternative 1 | 2,728,600 | 571,400 |
| Alternative 2 | 2,620,700 | 575,100 |
| | $\Delta C = 107{,}900$ | $\Delta P = -3{,}700$ |

Payback time $Y$ $$-\frac{\Delta C}{\Delta P} = 29 \text{ years}$$

Since this figure is greater than the required payback time of 3.57 years, the increased capital expenditure required by alternative 1 would not be justified; alternative 2 is preferred.

of construction, but an average figure for $a'$ would lie between £12,000 and £18,000 (as of 1990).

If more cost data were available it might be possible, and would be better, to split the cost into two parts, one for frames and press, and the other for electrodes. After estimating a reliable purchase (as opposed to installed) cost, the normal procedure of applying an installation factor could be followed.

The high value of the exponent (0.9) reflects that these cells are not so much scaled up as multiplied up.

(ii) The transformer/rectifier. For a relatively small plant, up to say 20 or 30 $m^2$ of electrode area, the capital cost of the installed transformer/rectifier scales up in kW according to the power rating $P$ raised to a power of 0.7 to 0.8. The value of the constant (1990 figures) to be used with the exponent 0.8, for an oil-cooled unit, installed with transmission lines, is 1000 to 1500. The lower figure is quoted here:

$$C_1 = £1000P^{0.8} \tag{6.60}$$

In Examples 6.3.3 and 6.3.4 the constants were 190 and 200 respectively. Even for 1975 figures these are on the low side but in the context of Chapter 6 it doesn't matter.

Power is the product of volts and amps supplied to the cell (or cells), with a 5% addition for transmission losses. The current $I$ can be calculated as the product of the area $S$ and the current density $i$, and $V^C$ as the sum of the anodic and cathodic potentials, $E'_A$ and $E'_C$, the resistance in the separator (if any) $V_D$, and the potential drop in the electrolyte, which is dependent on $i$ and the resistance of the electrolyte $R$. This dependence on functional process parameters is encapsulated in the functional statement of Eq. (6.61):

$$P = 1.05V^C I = f(S, i, E'_A, E'_C, V_D, R) \tag{6.61}$$

### 6.5.2. Duty of Heat Exchangers

The reader has met examples of reactor energy balances in Chapter 2 and this chapter. It may be useful to consider this a little further with reference to optimization. Inclusion of a substantial electrical term makes the energy balance, and therefore the calculation of the cooling or heating requirements, somewhat different from that of other chemical plants.

### 6.5.2.1. Continuous Operation

The energy balance in Eq. (2.88)—

$$H_2 - H_1 = W + E'' - Q - r_1\Delta H_{R1} - r_2\Delta H_{R2} - \cdots$$

—relates the change in enthalpy flow between the inlet and outlet streams $H_2 - H_1$ with the duty on the coolers $Q$, the power supplied via the pumps $W$, the electrical energy supplied to the cell $E''$, the rate of reaction $r$, and the enthalpy of reaction per kmol $\Delta H_R$.

The terms of the energy balance have the functional dependencies shown in Eq. (6.62) to (6.65):

$$W = f(S) \tag{6.62}$$

$$E'' = V^C I = f(S,\ i,\ E'_A,\ E'_C,\ V_D,\ R) \tag{6.63}$$

$$r_1 : r_2 : \cdots = f(i) \tag{6.64}$$

Hence:

$$Q = f(S,\ i,\ E'_A,\ E'_C,\ V_D,\ R) \tag{6.65}$$

Pumping requirements are roughly proportional to the number of modules, i.e., they are a function of electrode area. As we have shown in connection with the transformer/rectifier, $W$ is a function of $S$, $i$, $E'_A$, $E'_C$, $V_D$, and $R$. The relative rates at which the primary and secondary reactions proceed varies with $i$. The process variables which affect $W$, $E''$, and $r$ will, by virtue of the energy balance, also affect $Q$, which leads us to Eq. (6.65). Since the capital cost of the coolers $C_C$ and the annual cost of cooling water $P_{cw}$ are directly related to $Q$, they will also be functionally dependent upon these variables.

All items which contribute significantly to the costs of the process are ultimately decided by the resistance of the electrolyte $R$, the kinetics ($E'_A$ and $E'_C$), the choice of current density $i$, and the consequential electrode area $S$. If the feed compositions is fixed, so is $R$, and it follows that:

$$E'_A = f(i) \qquad \text{and} \qquad E'_C = f(i)$$

If annual production rate is defined, then $S = f(i)$ and from the foregoing dependences it is clear that cost will depend on $i$, which becomes the obvious choice for $Z$, the optimizing variable.

### 6.5.2.2. Batch Operation

As we saw in Chapter 2, batch operation is essentially unsteady-state. The energy balance is drawn up at some instant $t$; the average cooling rate over the entire batch $\bar{Q}$ is given by integration with respect to time:

$$\bar{Q} = \frac{1}{t_b} \int_0^{t_b} Q\, dt \tag{6.66}$$

where $t_b$ is total batch time.

As in continuous operation, we see that $Q$ and $\bar{Q}$, and therefore $C_C$ and $P_{cw}$ (the capital cost of coolers and the cost of cooling water), are related through the energy balance to $W$, $E''$, and $r$. In consequence they are functionally dependent on $S$, $i$, $E'_A$, $E'_C$, and $R$. If the size, composition, and conversion of the batch are fixed, then $i$ is again, the obvious choice for optimizing variable.

In contrast to the continuous reactor, there is an extra degree of freedom in that it is possible to vary $i$ as the batch proceeds, which can be advantageous if a change in the kinetics at low reactant concentration requires counteraction. This possibility adds a new and interesting dimension to the process of optimization—and, inevitably, a complicating one.

## REFERENCES


1. "A New Guide to Capital Cost Estimating," 1977, Instit. Chem. Eng. & Assoc. Cost Eng.
2. Bauman, H. C., 1964, *Fundamentals of Cost Engineering in the Chemical Industry*, Reinhold Publishing, Chapman & Hall.
3. Peters, M. S. and Timmerhaus, K. D., 1980, *Plant Design and Economics for Chemical Engineers*, 3d ed., McGraw-Hill, New York.
4. Backhurst, J. R. and Harker, J. H., 1983, *Process Plant Design*, Heinemann Educational Books.
5. Coulson, J. M. and Richardson, J. F., 1990, *Chemical Engineering*, 4th ed., vol. 6, Pergamon Press.
6. Plimley, R. E., 1980, "Interactions between an electrochemical reactor and associated unit processes," *J. Chem. Tech. Biotechnol.*, **30:** 233–243.
7. Sawistowski, H. and Smith, W., 1963, *Mass Transfer Process Calculations*, Interscience Publishers, New York.
8. Coulson and Richardson, 1978, 3d ed. vol. 2, p. 471.
9. Rose, L. M., 1974, *The Application of Mathematical Modelling to Process Development and Design*, Applied Science Publications Ltd., London.

# Appendix

# Notation of Variables

An asterisk (*) indicates that dimensions are dependent on order of reaction. A dash indicates that the variable is dimensionless.

| Abbr. | Term | Units |
|---|---|---|
| $A$ | Cross-sectional area | $m^2$ |
| $A$ | Heat transfer area of cooler | $m^2$ |
| $A$ | Frequency of transformation of activated complex into products $= k_B T/h$ | 1/s |
| $A_-$ | Frequency of transformation of activated complex during a reduction | 1/s |
| $A_+$ | Frequency of transformation of activated complex during an oxidation | 1/s |
| $A'$ | Frequency factor in the Arrhenius equation | * |
| $A_n$ | Concentration of species $A_n$ | $kmol/m^3$ |
| $a$ | Stoichiometric constant for reactant $A$ | no. molecules |
| **a** | Lumped kinetic constant | * |
| $a$ or $a'$ | Specific area $= S/V$ | $m^2/m^3$ |
| $a$ | Constant in Eq. (6.59) | $£/m^{1.8}$ |
| $a'$ | Constant in Eq. (6.59) | £ |
| $a_e$ | Effective specific electrode area | $m^2/m^3$ |
| $a_j$ | Process variable | |
| $a_j$ | Relative activity of species $j$ | — |

| Abbr. | Term | Units |
|---|---|---|
| $a_+$ | Activity of the cation | — |
| $a_-$ | Activity of the anion | — |
| $a_\pm$ | Mean ionic activity $= (a_+a_-)^{1/2}$ | — |
| $B_5$ | Rate of production of $B$ | kmol/s |
| $b$ | Adsorption coefficient $= k_a/k_d$ | $m^3$/mol |
| $b$ | Stoichiometric constant for reactant $B$ | no. molecules |
| **b** | Lumped kinetic constant | * |
| $b_a$ | Linear polarization coefficient for the anode reaction $= dE'/di$ | V $m^2$/A |
| $b_c$ | Linear polarization coefficient for the cathode reaction $= dE'/di$ | V$m^2$/A |
| $b_j$ | Slope of polarization curve for current involving species $j$ | $mV^{-1}$ or $V^{-1}$ |
| $C$ | Capital investment | £ |
| $C_C$ | Capital cost of coolers | £ |
| $C_D$ | Total capital cost of distillation column $= C_3$ | £ |
| $C_E$ | Capacity associated with an electrochemical process | A s/V |
| $C_1$ | Cost of transformer/rectifier | £ |
| $C_2$ | Cost of electrolytic cell | £ |
| $C_3$ | Cost of installed distillation column | £ |
| $C_P$ | Cost of pump | £ |
| $C_R$ | Capital cost of reactor $= C_1 + C_2$ | £ |
| $C_f$ | Friction factor | — |
| $C_p$ | Molar heat capacity | J/mol K |
| $C_{p(j)}$ | Molar heat capacity of species $j$ | J/mol K |
| C.E. | Current efficiency or current yield $= i_j/i$ | — |
| $c$ | Concentration | kmol/$m^3$ |
| **c** | Lumped kinetic constant | * |
| $C^*$ | Concentrations of an activated complex | kmol $m^{-3}$ |
| $c_c$ | Concentration of reactant in the continuous phase | kmol/$m^3$ |
| $c_d$ | Concentration of reactant in the dispersed phase | kmol/$m^3$ |
| $c_j$ | Bulk concentration of species $j$ | kmol/$m^3$ |
| $c^0$ | Initial concentration | kmol/$m^3$ |
| $c_j^0$ | Initial bulk concentration of species $j$ | kmol/$m^3$ |
| $c_j^f$ | Final bulk concentration of species $j$ | kmol/$m^3$ |
| $c_j^i$ or $c_{j1}$ | Inlet concentration of species $j$ | kmol/$m^3$ |
| $c_{j1}^0$ | Initial inlet concentration of species $j$ | kmol/$m^3$ |
| $c_{j1}^f$ | Final inlet concentration of species $j$ | kmol/$m^3$ |

| Abbr. | Term | Units |
|---|---|---|
| $c_{j2}$ | Outlet concentration of species $j$ | $kmol/m^3$ |
| $c_{j2}^0$ | Initial outlet concentration of species $j$ | $kmol/m^3$ |
| $c_{j2}^f$ | Final outlet concentration of species $j$ | $kmol/m^3$ |
| $c_{js}$ | Concentration of species $j$ near the electrode surface | $kmol/m^3$ |
| $c_{jx}$ | Concentration of species $j$ at level $x$ | $kmol^{-3}$ |
| $c'_{js}$ | Surface concentration of species $j$ | $kmol/m^3$ |
| $D$ | Molecular diffusivity | $m^2/s$ |
| $D$ | Distillation rate of top product | mol/s |
| $Da_j$ | Damköhler number $= k_{jf}/n\mathfrak{F}k_L$ | — |
| $d$ | Distance or half distance between electrodes | m |
| $d$ | Diameter of distillation column | m |
| **d** | Lumped kinetic constant | * |
| $d_G$ | Thickness of graphite | m |
| $d_I$ | Thickness of insulation | m |
| $d_e$ | Equivalent diameter of channel $= 2d$ or $2Wd/(W + d)$ or $2(r_o - r_i)$ | m |
| $d_p$ | Particle diameter | m |
| $E$ | Activation energy | J/kmol |
| $E_1$ | Activation energy for the forward reaction | J/kmol |
| $E_2$ | Activation energy for the backward reaction | J/kmol |
| $E$ | Equilibrium, rest potential or open-circuit voltage | V |
| $\bar{E}$ | Average value for the combined potential drop at the anode and cathode | V |
| $E_A$ | Anodic equilibrium or rest potential | V |
| $E_C$ | Cathodic equilibrium or rest potential | V |
| $E_C$ | Energy consumption $= 1/Y_E$ | kWh/kmol or Wh/kg |
| $E_D$ | Decomposition voltage | V |
| $E'$ | Electrode potential $= E_m - E_s$ | V |
| $E'_A$ | Anodic electrode potential | V |
| $E'_C$ | Cathodic electrode potential | V |
| $E^o$ | Standard electrode potential or open-circuit voltage | V |
| $E_A^o$ | Standard anode electrode potential | V |
| $E_C^o$ | Standard cathode electrode potential | V |
| $E^*$ | Defined in Eq. (5.3) $= \pi L/2d$ | — |
| $E_f$ | Effectiveness factor for the uniformity of current distribution in parallel plate cells | — |

| Abbr. | Term | Units |
|---|---|---|
| $E_m$ | Metal potential | V |
| $E_{max}$ | Maximum value for the combined potential drop at the anode and cathode | V |
| $E_s$ | Solution potential | V |
| $E_{s,\infty}$ | Potential of the bulk solution | V |
| $E_{s1}$ | Potential at tip of Luggin capillary | V |
| $E_{tn}$ | Thermoneutral voltage | V |
| $E''$ | Electrical energy supplied to the cell | W |
| e | Denotes an electron | — |
| $e$ | Voidage | — |
| $e_{mf}$ | Voidage at minimum fluidization | — |
| $F$ | Faraday | C/g-equiv. |
| $\mathfrak{F}$ | $1000F$ | C/k-equiv. |
| $F$ | Flow of cooling water | kg/s |
| $F$ | Feed rate to distillation column | mol/s |
| $F(f_g)$ | Function of $f_g$ in Eq. (5.8) | — |
| $f$ | $nF/RT$ or $n\mathfrak{F}/RT$ | As/J |
| $f_g$ | Gas voidage | — |
| $f_{gx}$ | Gas fraction as a function of distance $x$ | — |
| $G$ | Gibbs free energy | J/mol |
| $G^0$ | Standard Gibbs free energy | J/mol |
| $G_j^0$ | Standard Gibbs free energy of formation of species $j$ | J/mol |
| $\Delta G_A^0$ | Standard Gibbs free energy charge for anode reaction | J/mol |
| $\Delta G_C^0$ | Standard Gibbs free energy charge for cathode reaction | J/mol |
| $G^*$ | Free energy of activation | J/mol |
| $G_-^*$ | Free energy of activation for the cathodic reaction | J/mol |
| $G_+^*$ | Free energy of activation for the anodic reaction | J/mol |
| $G_{ads}$ | Free energy of adsorption | J/mol |
| $G_{sp}$ | Specific part of the free energy of adsorption | J/mol |
| $G_{nsp}$ | Nonspecific part of the free energy of adsorption | J/mol V |
| $g$ | Acceleration due to gravity | m/s$^2$ |
| $H$ | Enthalpy | J/mol |
| $H$ | Heat of reaction | J/mol |
| $H^0$ | Standard heat of formation | J/mol |

| Abbr. | Term | Units |
|---|---|---|
| $H_1^o$ | Standard heat of formation; subscript 1 denoting standard state at temperature $T_1$ | J/mol |
| $H_j^o$ | Standard heat of formation of species $j$ | J/mol |
| $\Delta H_R$ | Enthalpy of reaction | J/mol |
| $\Delta H_A^0$ | Standard enthalpy for anode reaction | J/mol |
| $\Delta H_C^0$ | Standard enthalpy for cathode reaction | J/mol |
| $H_{con}$ | Heat load on condenser | W |
| $\Delta H_{reaction}$ | Rate of change of enthalpy due to reaction $= r\Delta H_R$ | W |
| $H_{reb}$ | Heat load on reboiler | W |
| $h$ | Planck's constant | Js |
| $h$ | Overall heat transfer coefficient | J/h $m^2$ K or W/$m^2$ °C |
| $I$ | Current | A |
| $I$ | Income | £ |
| $I_c$ | Net cathodic current $= I_- - I_+$ | A |
| $I_-$ | Cathodic current | A |
| $I_+$ | Anodic current | A |
| $i$ | Minimum required discounted cash flow rate of return | — |
| $i$ | Current density | A/$m^2$ |
| $i(x)$ | Local current density | A/$m^2$ |
| $i'$ | Mean current density in Eq. (2.81) | A/$m^2$ |
| $i_c$ | Net cathodic current density $= i_- - i_+$ | A/$m^2$ |
| $i_d$ | Dimensionless current density $= i_m/i$ | — |
| $i_j$ | Partial current density for a reaction involving species $j$ | A/$m^2$ |
| $i_{lim}$ | Limiting current density | A/$m^2$ |
| $i_m$ | Current density in the metal phase | A/$m^2$ |
| $i_n$ | Current density for forward reaction | A/$m^2$ |
| $i_{-n}$ | Current density for backward reaction | A/$m^2$ |
| $i_s$ | Current density in the electrolyte phase | A/$m^2$ |
| $i_x$ | Current density associated with gas evolution | A/$m^2$ |
| $i_{av}$ | Average current density | A/$m^2$ |
| $i_{jx}$ | Local partial current density of species $j$ | A/$m^2$ |
| $i_-$ | Cathodic current density | A/$m^2$ |
| $i_+$ | Anodic current density | A/$m^2$ |
| $i_o$ | Exchange current density | A/$m^2$ |
| $J$ | Parameter in Eq. (2.77) | — |

| Abbr. | Term | Units |
|---|---|---|
| $j$ | Number of moles of species $j$ | — |
| $j_D$ | $j$ factor $= StSc^{2/3}$ | — |
| $j_b$ | Species $j$ in the bulk of the electrolyte | — |
| $j_s$ | Species $j$ near the electrode | — |
| $K$ | Equilibrium constant $= k/k'$ | * |
| $K$ | Gas effect parameter $= (RT\kappa V^C L)/(nFPd^2 V_g)$ | — |
| $K^*$ | Equilibrium constant for activated complex | * |
| $K[m]$ | Complete elliptical integral of the first kind | — |
| $K_2$ | Adsorption coefficient $= k_{-2}/k_2$ | * |
| $k$ | Rate constant for the forward reaction | * |
| $k'$ | Rate constant for the backward reaction | * |
| $k_-$ | Rate constant for the cathodic reaction | * |
| $k_+$ | Rate constant for the anodic reaction | * |
| $k_B$ | Boltzmann constant | J/K |
| $k_G$ | Thermal conductivity of graphite | J/h m K |
| $k_I$ | Thermal conductivity of insulation | J/h m K |
| $k_L$ | Mass transfer coefficient | m/s |
| $k_{L,x}$ | Local mass transfer coefficient | m/s |
| $k'_L$ | Coefficient for mass transfer due to the disruption of boundary layer by the dispersed phase | m/s |
| $k_a$ | Rate constant for adsorption | $m^3$/kmol s |
| $k_d$ | Rate constant for desorption | $s^{-1}$ |
| $k_e$ | Coefficient for mass transfer enhancement due to liquid extraction | m/s |
| $k_j$ | Rate or velocity constant | * |
| $k_{jf}$ | Charge transfer rate $= k_j \exp[-b_j E']$ of species $j = nFk'_j \exp[-b_j E']$ | Am/kmol |
| $k_n$ | Rate constant for forward reaction | * |
| $k_{-n}$ | Rate constant for backward reaction | * |
| $k'_j$ | Rate or velocity constant | * |
| $L$ | Liquid reflux returned to the top of the distillation column | kmol/s |
| $L$ | Electrode length or half length, reactor length or the bed width of a three-dimensional electrode | m |
| $L$ or $L'$ | Characteristic length of the cell | m |
| $L_{max}$ | Maximum electrode bed width in the direction of current flow | m |

| Abbr. | Term | Units |
|---|---|---|
| $l$ | Length of current path | m |
| $l$ | Distance between the tip of the Luggin capillary and the electrode surface | mm |
| $l_a$ | Thickness of the anode | m |
| $l_c$ | Thickness of the cathode | m |
| ln | Natural logarithm = $\log_e$ | — |
| $M$ | Molecular weight | — |
| $M$ | Surface site | — |
| M | Molarity | gmol/l |
| $M$ | Integral loop counter | — |
| m | Molality | kmol/kg |
| $m$ | Hydraulic radius of diaphragm = $e/a$ | m |
| $m$ | Amount of deposition | gmol |
| $m$ | Eq. (5.19) = $(\kappa_e L^2)/(dl\sigma)$ | — |
| $m_j$ | Molality of species $j$ | kmol/kg |
| $m_j$ | Amount of species $j$ formed or converted | gmol, kmol or k-equiv./m$^2$ |
| $m_{ACT}$ | Actual amount of product | kmol or gmol |
| $m_{MAX}$ | Maximum amount of product relating to 100% conversion | kmol or gmol |
| $m_{MX}$ | Molality of salt $MX$ | kmol/kg |
| $N$ | Flux to and reaction rate at an electrode surface | kmol/s |
| $N$ | Number of plates in a distillation column | — |
| $N$ | Number of integral loops | — |
| $N$ | Number of modules | — |
| $N_D$ | Diffusional flux | kmol/m$^2$s |
| $N_j$ | Flux of species $j$ | kmol/m$^2$s |
| $N_{min}$ | Minimum number of plates required at total reflux | — |
| $n$ | Number of electrons | — |
| $n$ | Depreciation period | years |
| $n_j$ | Number of molecules or moles of species $j$ | |
| $P$ | Total pressure | N/m$^2$ |
| $\Delta P$ | Pressure loss | N/m$^2$ |
| $P$ | Polarization parameter = $Wa$ | — |
| $P$ | Production or running costs | £ |
| $P$ | Rating of power supply | kW |
| $Pr$ | Profit | £ |

| Abbr. | Term | Units |
|---|---|---|
| $P_A$ | Cost of lost species $A$ | £ |
| $P_D$ | Total running costs of distillation column | £ |
| $P_R$ | Running costs of reactor | £ |
| $P_{cw}$ | Cost of cooling water | £ |
| $P_s$ | Cost of steam | £ |
| $p$ | Hydraulic permeability | $m^2$ |
| $p_j$ | Partial pressure of species $j$ above its solution | $N/m^2$ |
| $p_j^0$ | Vapor pressure of pure $j$ | $N/m^2$ |
| $Q$ | Volumetric fluid flow rate | $m^3/s$ |
| $Q$ | Heat removed from the system | J/mol or W |
| $Q'$ | Heat absorbed by the system | J/mol or W |
| $\bar{Q}$ | Average cooling rate | J/mol s |
| $Q_F$ | Heat flux through the floor of an aluminum cell | J/h $m^2$ |
| $Q_c$ | Heat flux by conduction | J/h $m^2$ |
| $Q_j$ | Amount of species $j$ transported to the electrode surface | kmol |
| $Q_n$ | Heat flux by natural convection | J/h $m^2$ |
| $Q_r$ | Heat flux by radiation | J/h $m^2$ |
| $q$ | Recycle rate | $m^3/s$ |
| $q$ | Ratio of heat required to vaporize 1 mol of feed to its molar latent heat | — |
| $R$ | Universal gas constant | J/mol K |
| $R$ | Rating of pump | kW |
| $R$ | Reflux ratio $= L/D$ | — |
| $R$ | Resistance | ohm |
| $R_C$ | Resistive load in the lead to the electrode | ohm |
| $R_c$ | Production of product | kmol |
| $R_D$ | Production of by-product | kmol |
| $R_E$ | Resistance associated with an electrode process | ohm |
| $R_G$ | Resistance of the electrolyte between the electrode and the tip of the Luggin capillary | ohm |
| $R_L$ | Resistive load in the lead to the electrode | ohm |
| $R_c$ | Production of product | kmol |
| $R_{min}$ | Minimum reflux ratio | — |
| $R_{opt}$ | Optimum reflux ratio | — |
| $R_a$ | Polarization resistance | ohm $m^2$ |
| $R_e$ | Resistance of electrolyte per $m^2$ of cross-sectional area | ohm $m^2$ |
| $R_j$ | Reaction rate of species $j = Sr_j$ | kmol/s |

| Abbr. | Term | Units |
|---|---|---|
| $R_j$ | Production of species $j$ | kmol |
| $r$ | Radial distance | m |
| $r$ | $=r_i/r_o$ | — |
| $r$ | Reaction rate, conversion rate | kmol/m²s |
| $r_A$ | Rate of conversion of $A$ | kmol/m²s |
| $r_a$ | Rate of adsorption | kmol/m²s |
| $r_d$ | Rate of desorption | kmol/m²s |
| $r_i$ | Radius of inner cylinder | m |
| $r_j$ | Conversion rate of species $j$ | kmol/m²s |
| $r_-$ | Chemical reduction rate | * |
| $r_+$ | Chemical oxidation rate | * |
| $r_n$ | Forward rate of reaction | * |
| $r_{-n}$ | Backward rate of reaction | * |
| $r_o$ | Radius of outer cylinder | m |
| $Re$ | Reynolds number $= ud_e/v$ | — |
| $Re_a$ | Axial Reynolds number $= [u_a 2(r_o - r_i)/v]$ | — |
| $Re_p$ | Particle Reynolds number $= ud_p/v$ | — |
| $Re_w$ | Radial Reynolds number $= 2u_i r_i/v = 2r_i^2 w/v$ | — |
| $Re'_w$ | Radial Reynolds number $= r^2 w/v$ | — |
| $Re_x$ | Local Reynolds number $= ux/v$ | — |
| $S$ | Electrode area | m² |
| $S$ | Half distance from the cell wall | m |
| $S$ | Entropy | J/mol K |
| $S^0$ | Standard entropy | J/mol K |
| $S_1^0$ | Standard entropy; subscript 1 denoting standard state at temperature $T_1$ | J/mol K |
| $S_{AM}$ | Surface concentration of occupied sites | kmol/m² |
| $S_M$ | Surface concentration of empty sites | kmol/m² |
| $S_{MAX}$ | Surface concentration of total number of available sites | kmol/m² |
| $Sc$ | Schmidt number $= v/D$ | — |
| $Sh$ | Average Sherwood number $= k_L d_e/D$ | — |
| $Sh_w$ | Radial Sherwood number $= k_L r_i/D$ | — |
| $Sh_x$ | Local Sherwood number $= k_{L,x} d_e/D$ | — |
| $Sh'_x$ | Local Sherwood number $= k_{L,x} x/D$ | — |
| $St$ | Stanton number $= ShRe^{-1}Sc^{-1} = k_L/u$ | — |
| $St_w$ | Radial Stanton number $= Sh_w Re_w^{-1} Sc^{-1} = k_L/u_i$ | — |
| $2s$ | Distance separating the insulating walls of a cell | m |
| $T$ | Temperature | K |

| Abbr. | Term | Units |
|---|---|---|
| $T$ | Total costs | £ |
| $T$ | Availability of plant | hours |
| $T_A$ | Temperature of air around aluminum cell | K |
| $T_G$ | Temperature of boundary between graphite cathode and solid alumina | K |
| $T_M$ | Temperature of alumina melt | K |
| $T_W$ | Temperature of steel mantle of an aluminum cell | K |
| $\Delta T_{lm}$ | Log mean temperature difference | °C |
| $\Delta T_m$ | Mean temperature difference | °C |
| $Ta$ | Taylor number defined by Eq. (2.23) | — |
| $Ta_m$ | Taylor number in Eq. (2.25) and (2.26) | — |
| $t$ | Time | s |
| $t_b$ | Batch time | s |
| $t^+$ | Transport number | — |
| $U$ | Internal energy | J/mol |
| $U_L$ | Fluid flow rate | l/s |
| $u$ | Fluid velocity | m/s |
| $u_a$ | Axial fluid velocity | m/s |
| $u_c$ | Calculated value of electrolyte velocity | m/s |
| $u_e$ | Experimental value of electrolyte velocity | m/s |
| $u_i$ | Peripheral velocity | m/s |
| $u_x$ | Fluid velocity in the direction $x$ at point $y$ | m/s |
| $u_{av}$ | Average fluid velocity | m/s |
| $u_{mf}$ | Minimum fluidization velocity | m/s |
| $V$ | Volume | $m^3$ |
| $V$ | Fluid velocity | m/s |
| $V$ | Vapor stream returned to the bottom of the distillation column | kmol/s |
| $V$ | Measured potential | V |
| $V^C$ | Cell voltage | V |
| $V^C_{min}$ | Minimum cell voltage | V |
| $V_D$ | Voltage drop in diaphragm | V |
| $V_G$ | Voltage drop in the electrolyte between the electrode and the tip of the Luggin capillary | V |
| $V_{LC}$ | Voltage drop in the lead to the electrode | V |
| $V_{RA}$ | Voltage drop in anolyte | V |
| $V_{RC}$ | Voltage drop in catholyte | V |
| $V_c$ | Volume of reactor | $m^3$ |

| Abbr. | Term | Units |
|---|---|---|
| $V_g$ | Rise velocity of bubbles | m/s |
| $V_r$ | Volume of reservoir | $m^3$ |
| $V_{ohm}$ | Voltage drop through electrolyte | V |
| $W$ | Distillation rate of bottom product | kmol/s |
| $W$ | Work done on the system | J/mol or W |
| $W$ | Electrode width | m |
| $W'$ | Work done by the system | J/mol or W |
| $W_c$ | Energy derived from pumping the catholyte | W |
| $Wa$ | Wagner number $= \frac{\kappa}{L'} \frac{dE'}{di} = \frac{R_a}{R_e}$ | — |
| $w$ | Angular velocity | rad/s |
| $w$ | Total power consumed $= w_c + w_t = P$ | kW |
| $w_c$ | Power consumed by cell | kW |
| $w_t$ | Power consumed by transmission | kW |
| $X$ | Fractional chemical conversion | — |
| $X_j$ | Fractional chemical conversion of species $j$ | — |
| $X_j^f$ | Final fractional conversion of species $j$ | — |
| $X_j^0$ | Initial fractional conversion of species $j$ | — |
| $x$ | Coordinate | m |
| $x$ | Cell height | m |
| $x$ | Stoichiometric constant for product $X$ | no. molecules |
| $x_{dj}$ | mol fraction of component $j$ in top product | — |
| $x_{Fj}$ | mol fraction of component $j$ in feed | — |
| $x_H$ | Thickness of compact part of the electrical double layer | m |
| $x_j$ | Mol fraction of species $j$ | — |
| $Y$ | Defined by Eq. (3.199) | $A/m^2$ |
| $Y$ | Payback time | years |
| $Y_C$ | Chemical yield $= m_{ACT}/m_{MAX}$ | — |
| $Y_E$ | Energy yield $= 0.036$ C.E.$/nV^c$ | kmol/kWh |
| $Y_{ST}$ | Spacetime yield $=$ C.E. $iaM/n\mathfrak{F}$ | $kg/m^3s$ |
| $Y_m$ | Maximum acceptable payback time | years |
| $y$ | Coordinate | m |
| **y** | Stoichiometric constant for product $Y$ | no. molecules |
| $y'$ | Dimensionless coordinate $= x/L$ | — |
| $y_0$ | Bed length in the direction of electrolyte flow | m |
| $Z$ | Convenient variable for optimization | |
| $z$ | Coordinate | m |
| $\alpha$ | Transfer coefficient | — |

| Abbr. | Term | Units |
|---|---|---|
| $\alpha_j$ | Volatility of component $j$ with respect to the least volatile component | — |
| $\alpha_{jA}$ | Relative volatility of component $j$ with respect to $A$ | — |
| $\beta$ | $=\alpha nF/RT$ or $(l-\alpha)nF/RT$ | C/J |
| $\gamma$ | Aspect ratio $= d/W$ | — |
| $\gamma_j$ | Activity coefficient of species $j$ | — |
| $\gamma_\pm$ | Mean ionic activity coefficient | kg/kmol |
| $\delta$ | Boundary layer thickness | m |
| $\delta'$ | Defined by Eq. (5.33) | — |
| $\delta_N$ | Thickness of the diffusion layer | m |
| $\delta_{Pr}$ | Thickness of the Prandtl boundary layer | m |
| $\varepsilon'$ | Defined by Eq. (5.34) | — |
| $\eta$ | Overpotential $= E' - E$ | V |
| $\eta_A$ | Anodic overpotential | V |
| $\eta_C$ | Cathodic overpotential | V |
| $\Delta\eta_{max}$ | Maximum allowable value of overpotential | V |
| $\bar{\eta}$ | Effectiveness factor for the uniformity of current distribution in three-dimensional electrodes | — |
| $\theta$ | Fractional coverage of surface | — |
| $\theta$ | Integration constant defined by Eq. (5.38) | — |
| $\Theta$ | Root of Eq. (6.30) | — |
| $\kappa$ | Specific electrolyte conductivity or specific conductance | mho/m |
| $\kappa_d$ | Effective conductivity of diaphragm | mho/m |
| $\kappa_e$ | Effective conductivity of electrolyte | mho/m |
| $\lambda$ | Number of moles of aluminum reacting per hour in the cell | mol/h |
| $\mu$ | Viscosity | $Ns/m^2$ |
| $\mu_j$ | Chemical potential of species $j$ | J/mol |
| $\mu_j^{sol}$ | Chemical potential of species $j$ in the solution | J/mol |
| $\mu_j^{vap}$ | Chemical potential of species $j$ in the vapor | J/mol |
| $\mu_j^o$ | Standard chemical potential of species $j$ | J/mol |
| $\nu$ | Kinematic viscosity $= \mu/\rho$ | $m^2/s$ |
| $\nu_j$ | Number of ions in molecule $j$ | — |
| $\rho$ | Specific resistance | ohm m |
| $\rho$ | Density | $kg/m^3$ |
| $\rho_s$ | Density of solid particle | $kg/m^3$ |

| Abbr. | Term | Units |
|---|---|---|
| $\sigma$ | Specific metal phase conductivity | mho/m |
| $\sigma_a$ | Specific conductivity of the anode | mho/m |
| $\sigma_c$ | Specific conductivity of the cathode | mho/m |
| $\tau$ | Residence time in reactor = $L/u$ or $x/u$ | s |
| $\tau_r$ | Residence time in reservoir = $V_r/Q$ | s |
| $\Phi$ | Parameter defined by Eq. (2.22) or by Eq. (5.18) | — |
| $\phi$ | Parameter defined by Eq. (5.18) | — |
| $\psi$ | Integration constant defined by Eq. (5.39) | — |

# Index

www.ingramcontent.com/pod-product-compliance
Ingram Content Group UK Ltd.
Pitfield, Milton Keynes, MK11 3LW, UK
UKHW020432200726
13857UKWH00002B/384

* 9 7 8 1 4 8 9 9 0 2 2 5 2 *